Internet of Things –
The Call of the Edge
Everything Intelligent Everywhere

RIVER PUBLISHERS SERIES IN COMMUNICATIONS

Series Editors:

ABBAS JAMALIPOUR
The University of Sydney
Australia

MARINA RUGGIERI
University of Rome Tor Vergata
Italy

JUNSHAN ZHANG
Arizona State University
USA

Indexing: All books published in this series are submitted to the Web of Science Book Citation Index (BkCI), to SCOPUS, to CrossRef and to Google Scholar for evaluation and indexing.

The "River Publishers Series in Communications" is a series of comprehensive academic and professional books which focus on communication and network systems. Topics range from the theory and use of systems involving all terminals, computers, and information processors to wired and wireless networks and network layouts, protocols, architectures, and implementations. Also covered are developments stemming from new market demands in systems, products, and technologies such as personal communications services, multimedia systems, enterprise networks, and optical communications.

The series includes research monographs, edited volumes, handbooks and textbooks, providing professionals, researchers, educators, and advanced students in the field with an invaluable insight into the latest research and developments.

For a list of other books in this series, visit www.riverpublishers.com

Internet of Things –
The Call of the Edge
Everything Intelligent Everywhere

Editors

Ovidiu Vermesan

SINTEF, Norway

Joël Bacquet

EU, Belgium

River Publishers

Published, sold and distributed by:
River Publishers
Alsbjergvej 10
9260 Gistrup
Denmark

www.riverpublishers.com

ISBN: 978-87-7022-196-2 (Hardback)
        978-87-7022-195-5 (Ebook)

©2020 River Publishers

Dedication

"Simplicity is the ultimate sophistication."

– Leonardo Da Vinci

"Logic will take you from A to B. Imagination will take you everywhere."

"Creativity is intelligence having fun."

– Albert Einstein

Acknowledgement

The editors would like to thank the European Commission for their support in the planning and preparation of this book. The recommendations and opinions expressed in the book are those of the editors and contributors, and do not necessarily represent those of the European Commission.

Ovidiu Vermesan

Joël Bacquet

Contents

Preface

Internet of Things Intelligence and Senses

The next wave of the Internet of Things (IoT) technological development grows with radical advances in artificial intelligence (AI), edge computing processing, new sensing capabilities and autonomous functions accelerating progress towards the ability to self-develop, self-maintain and self-optimise.

IoT applications' collective intelligence will eventually enhance the autonomous capabilities of the IoT while creating new possibilities for humans to connect seamlessly, enabling new ways to collaborate among machines and between humans and machines.

Strengthening human–machine interaction, collaboration and cooperation using hyper autonomous IoT technologies and applications provides new opportunities for economic development and the digitisation of industries in the new digital age, extending the wave of continuous innovation and disruption of IoT business models.

The emergence of hyper autonomous IoT with enhanced sensing, distributed intelligence, edge processing and connectivity, combined with human augmentation, has the potential to power both the transformation and optimisation of industrial sectors and change the innovation landscape.

IoT sensing, actuating and computing processing at the edge provide IoT systems the capabilities to deliver disconnected or distributed functions to the embedded physical environment and provide the digital representation and modelling, simulation and augmented functions through 'digital twins' in digital, virtual and cyber environments.

Combining AI and distributed ledger technologies (DLTs), the IoT and edge computing continuum create a distributed intelligent architecture consisting of a wide range of sensing/actuating intelligent things and services linked in a dynamic mesh connected by a set of distributed and federated edge/cloud services. The significant advances of hyper autonomous IoT include collaborative robotic things integrated into real-time industrial and intelligent business processes across different sectors that create a multi experience combining product design, visualisation, field service, operations, modelling, virtualisation, simulations into extended reality and virtual environments.

AI-driven IoT autonomous capabilities increase the decision capabilities of IoT edge devices, creating new business scenarios that incorporate the use of intelligent things into traditional manual and semiautomated tasks.

Protecting AI-based hyper autonomous IoT from malicious cyberattacks requires leveraging AI to enhance security defence, predicting the use of AI by attackers and creating techniques for quarantining the IoT devices as well as preventing the attack security solutions based on Machine Learning (ML) at the training and prediction stages.

Editors Biography

Dr. Ovidiu Vermesan holds a PhD degree in microelectronics and a Master of International Business (MIB) degree. He is Chief Scientist at SINTEF Digital, Oslo, Norway. His research interests are in the area of mixed-signal embedded electronics and cognitive communication systems. Dr. Vermesan received SINTEF's 2003 award for research excellence for his work on the implementation of a biometric sensor system. He is currently working on projects addressing nanoelectronics, integrated sensor/actuator systems, communication, cyber–physical systems and the IoT, with applications in green mobility, energy, autonomous systems and smart cities. He has authored or co-authored over 85 technical articles and conference papers. He is actively involved in the activities of the Electronic Components and Systems for European Leadership (ECSEL) Joint Technology Initiative (JTI). He has coordinated and managed various national, EU and other international projects related to integrated electronics. Dr. Vermesan actively participates in national, H2020 EU and other international initiatives by coordinating and managing various projects. He is the coordinator of the IoT European Research Cluster (IERC) and a member of the board of the Alliance for Internet of Things Innovation (AIOTI).

Joël Bacquet is a senior official of DG CONNECT of the European Commission, taking care of the research and innovation policy for the Internet of Things. Before working in this field, he was programme officer in "Future Internet Experimental Platforms", head of the sector "Virtual Physiological Human" in the ICT for health domain. From 1999 to 2003, he was head of the sector "networked organisations" in the eBusiness unit. He started working with the European Commission in 1993, in the Software Engineering Unit of the ESPRIT Programme. He started his carrier as visiting scientist for Quantel a LASER company in San José, California in 1981. From 1983 to 1987, he was with Thomson CSF (Thales) as software development engineer for a Radar System. From 1987 to 1991, he worked with the European Space Agency as software engineer on the European Space shuttle and international Space platform programmes (ISS). From 1991 to 1993 he was with Eurocontrol where he was Quality manager of an Air Traffic Control system. He is an engineer in computer science from Institut Supérieur d'Electronique du Nord (ISEN) and he has a MBA from Webster University, Missouri.

List of Figures

List of Tables

1

The EU IoT Policy and Regulatory Strategy – the Way Forward

Nikolaos Isaris

European Commission, Belgium

1.1 Introduction

Already in the Digitising European Industry (DEI) strategy [1], the Commission set up the ambitious goal of making Europe the world leader in the Internet of Things (IoT).

In 2017, we launched five IoT Large-Scale Pilot (LSP) projects [2] in the areas of smart cities, health, wearables, cars and food and agriculture, with a total amount of EU funding of 100 million EUR. In 2018, a set of eight IoT security and privacy research projects was launched with a budget of 37 million EUR. In 2019, six additional LSP projects (two on agriculture, one on energy and three on digital health and care) have been launched. These pilot deployment projects are addressing both the technology aspects and the regulatory and societal issues around IoT, demonstrating that IoT technology and digitisation have the potential of solving societal challenges as well as stimulating the creation of open European and global standards.

However, these research and innovation activities need to be taken in parallel with the relevant policy and regulatory steps, which can contribute to the delivery of our goals. The current work that the Commission is undertaking on the legal framework for liability for emerging digital technologies, should help to stimulate investment and to enhance users' trust in them, especially for IoT and Artificial Intelligence (AI). In addition, the recently adopted ICT Cybersecurity Certification Framework will enable the development of IoT innovation, while providing security at the expected level of assurance.

However, the next Commission will need to be able to face the bigger challenge of European businesses' contribution to the global digital supply chain. Despite the fact that the region remains one of the world's largest markets for digital products and services, Europe is increasingly dependent on foreign technologies in key parts of its economy. The risk is that the next digital transformation wave will be entirely shaped by third countries.

The current chapter aims to explain the importance of the right policy and regulatory strategy to overcome the challenges in the next decade.

1.2 Safety and Liability for Emerging Digital Technologies

Emerging digital technologies, such as IoT and AI, will create new opportunities for our economy and society. The increased autonomy of the products and services incorporating emerging digital technologies will result in beneficial effects, in particular in terms of increased productivity, positive societal outcomes, prevention of human error and potentially improved safety. For example, a home equipped with sensors, robots and connected devices enables elderly people to live in their homes safely and independently. This smart environment can monitor their health status and prevent them from becoming frail, as well as keeping them remotely in touch with their doctor. In another application area, precision farming, these technologies promise to reduce pesticide use and increase yields.

However, certain characteristics of emerging digital technologies bring new challenges concerning safety and liability. The complexity of these technologies is reflected on both the plurality of actors involved in the value chain and the multiplicity of components, parts, systems or services, which together form a joint ecosystem. The vast amounts of data involved and the reliance on algorithms make it more difficult to understand the potential causes of damage. In addition, the increased autonomy makes it more difficult to predict the behaviour of the product compared to the functions that were attributed initially by the producer. Finally, connectivity can also expose the products to cyber-threats.

Therefore, the challenge is to assess whether the EU and national legal frameworks on safety and liability are able to cope with the challenges brought by these technologies. Ensuring safety in a connected world is primordial. Moreover, safer products mean less need for liability actions. There is a need for clear, predictable and interoperable legal frameworks able to address the new technological challenges, in order to guarantee trust for all users, while encouraging continuous innovation and providing a level playing field for the industry.

The Commission Staff Working Document on Liability for emerging digital technologies [3] kicked off a first assessment of liability issues. As a follow-up, the Expert Group on Liability and New Technologies [4] was created, in order to provide the Commission with expertise on the applicability of the Product Liability Directive 85/374/EEC and with assistance in developing guiding principles for possible adaptations of applicable laws related to new technologies.

The Communication on Artificial Intelligence for Europe [5] announced that in 2019 the Commission would publish two deliverables:

- A guidance document on the interpretation of the Product Liability Directive in light of technological developments; and
- A report on the broader implications for, potential gaps in and orientations for, the liability and safety frameworks for AI, IoT and robotics.

The aim is to ensure legal clarity for consumers and producers in case of defective products, to facilitate the uptake of emerging digital technologies and to ensure users' trust and protection.

1.3 Cybersecurity Certification for IoT Products, Services and Processes

IoT implementation comes with a number of challenges, the most important of which are: security, privacy, data protection, increasing trust and consumer acceptance in IoT. Some of the challenges are due to the increased scale and scope of IoT with billions of devices potentially connected to the Internet. This number may pose a commensurate number of security risks.

In addition, IoT brings along new types of concerns on top of the safety aspects that exist in any consumer product or service. Moreover, the life cycle of some of the connected devices varies considerably and this can be a source of additional complexity when clarifying questions of security, upgradeability, liability and others. Finally, as the recent cyber-attacks have shown, consumer IoT vulnerabilities can be the source of damage in critical infrastructures and industrial IoT.

That is why the Cybersecurity Act [6], which entered into force on 27 June 2019, established the European cybersecurity certification framework, putting forward the instruments that will allow the development of IoT innovation. Cybersecurity certification will become less expensive, more effective and commercially attractive.

The main thrust of the legislation is establishing clear rules and a governance framework that would allow to set up European schemes for the cybersecurity certification of ICT products, services and processes. The framework allows also the potential development of labels accompanying a specific scheme. This would allow, for example, the creation of a Trusted IoT label for a particular type of IoT products, services or processes.

As next steps according to the Cybersecurity Act, the Commission needs to set up the governance framework and set out the policy priorities for the future. The 2017 Communication that accompanied the Cybersecurity Act indicated three priority areas for certification schemes: (1) IoT, (2) critical or high-risk applications, and (3) security products, networks, systems and services (e.g. VPNs, firewalls). Other technologies, including 5G, have also made the priority list since then.

Therefore, the 'Union rolling work programme for European Cyber-security Certification' is a forward-looking document which shall identify strategic priorities for future schemes. In particular, the rolling work programme will include a list of ICT products, services and processes or categories thereof that may benefit from being included in the scope of a European Cybersecurity Certification Scheme.

In the context of the ongoing work on liability in emerging digital technologies, certification could be also used to demonstrate that the required standard of care has been met by manufactures of ICT products or providers of ICT services.

1.4 Globally Competitive European Digital Supply Chains for the Next Generation Internet

Emerging digital technologies and services such as AI, 5G, IoT or edge computing are transforming many economic sectors significantly (e.g. health/care, manufacturing, agriculture, smart cities, energy, mobility). They are creating new markets with enormous potential. However, Europe currently still depends on foreign technologies for key parts of the digital supply chain. In essence, while connectivity infrastructure is mostly European, the necessary hardware and software is often made elsewhere. This technological dependence could translate into dependence for the next wave of data solutions: there is a risk that Europe will become a simple consumer of products and services made elsewhere.

This concerns mainly investments and rapid deployment of connectivity and data infrastructure, as well as accompanying policy measures required to create the right conditions for scalable and viable business models as well as critical mass of investments.

Europe is particularly exposed to substantial global competition and market concentration by foreign players in the field of cloud infrastructure and telecom networks supply, which may adversely affect its position for the Next Generation Internet, in particular concerning 5G, IoT and cloud services. A 'do nothing' option would only serve to increase the global imbalance, will foster the dominance of a few global non-European players and let Europe drift further behind, with negative consequences in terms of growth and undesired global dependencies for critical components, technologies and infrastructures.

If Europe really wants to become a global leader in IoT in the near future, it is crucial to look at connecting IoT ecosystems across different sectors, to continue supporting piloting and testbeds at scale and to check whether the regulatory and policy framework across different policy aspects are fit for purpose.

This is why the way forward is to focus on how data is gathered, managed and shared across the European Union and internationally, rather than looking into a full harmonisation of IoT services. Globally there are major players battling to control the supply and delivery of future digital products and services which will be ever more critical for the proper functioning of our society and economy. Europe is, therefore, at a crossroad in terms of ensuring freedom of choice and promoting values such as user control, ethics, privacy and security embodied in the future digital solutions. This requires new partnerships cutting across traditional value chains, which were formed during the first phases of digital transformation with the emergence of the Internet.

The Commission has proposed as part of the new Horizon Europe programme to explore the idea of a partnership on Smart Networks and Services, bridging the major connectivity and service infrastructures required for the Next Generation Internet, including 5G/6G, Internet of Things and distributed Cloud Computing.

Such a partnership would be driven by industrial agendas and coordinated with the EU Member States. This kind of endeavour requires that all relevant stakeholders react to this idea and formulate together the best way of partnering to ensure that Europe can continue to play a prominent role in taking the IoT and digital transformation forward into the next decade.

1.5 Conclusion

We must now build on top of the achievements of the Digitising European Industry (DEI) strategy, and the policy and regulatory strategy mentioned above. The European Union should strive for, and promote internationally, a third way of doing digital policy that is human-centric and founded on respect for fundamental rights, distinct from both a *laissez-faire* approach, privately ruled digital economy and society, and a top-down controlled model. This third way will enhance trust, promote an inclusive digital society and become a competitive advantage for European companies acting worldwide. It will allow Europe to create economic value in accordance with its core values.

References

[1] Digitising European Industry, online at: https://ec.europa.eu/digital-singl e-market/en/policies/digitising-european-industry

[2] IoT European Large-Scale Pilots, online at: https://european-iot-pilots. eu/

[3] Commission Staff Working Document on Liability for emerging digital technologies SWD/2018/137, online at: https://ec.europa.eu/digital-singl e-market/en/news/european-commission-staff-working-document-liabi lity-emerging-digital-technologies

[4] Expert Group on Liability and New Technologies. Register of Commission Expert groups online at: http://ec.europa.eu/transparenc y/regexpert/index.cfm?do=groupDetail.groupDetail&groupID=3592

[5] Communication on Artificial Intelligence for Europe COM(2018)237, online at: https://ec.europa.eu/digital-single-market/en/news/commu nication-artificial-intelligence-europe

[6] Regulation (EU) 2019/881 of the European Parliament and of the Council of 17 April 2019 on ENISA (the European Union Agency for Cybersecu-rity) and on information and communications technology cybersecurity certification and repealing Regulation (EU) No. 526/2013 (Cybersecurity Act), PE/86/2018/REV/1, OJ L 151, 7.6.2019, p. 15–69, online at: https: //eur-lex.europa.eu/legal-content/EN/TXT/?qid=1561545219894&uri=C ELEX:32019R0881

2

Focus Area on Digitization Further Deployment of Digital Technologies in Industrial Sectors

Joël Bacquet, Rolf Riemenschneider and Peter Wintlev-Jensen

European Commission, Belgium

2.1 Introduction

In 2016, the European Commission launched a strategic investment in the Internet of things (IoT) as part of the Digitising European Industry (DEI) policy [1]. The overall DEI objective is to put in place the necessary mechanisms to ensure that every industry in Europe, in whichever sector, wherever situated, and no matter of which size can fully benefit from digital innovations.

The investment in IoT materialised into the set-up of the IoT Large-Scale Pilots Programme in support of societal challenges like health, agri-food, mobility and smart city. In this programme five IoT Large-Scale Pilots (LSPs) projects [2] were funded and started in 2017 in the areas of smart cities, smart living environments for ageing well, wearables for smart ecosystems, autonomous vehicles in a connected environment and smart farming and food security, with a total amount of EU funding of 100 million EUR.

Based on the achievements of the first wave of pilots, in 2019, 10 large-scale pilots projects have been launched with a total amount of EU funding of 150 million EUR. These new pilot deployment projects are addressing both the technology aspects and the regulatory and societal issues around Digital technologies, demonstrating that digital technologies and digital innovation have the potential of solving societal challenges as well as stimulating the creation of open European and global standards. This chapter provides an overview of these new pilots in three sectors: agriculture, energy and healthcare.

2.2 Deployment in Agriculture Sector

The topic on "Agricultural digital integration platforms" targeted agricultural digital platforms for knowledge creation and innovation in agricultural sector with 3 pillars: (a) building digital platforms solving interoperability issues; (b) sharing data and generating knowledge; (c) developing decision support systems that could provide advices to farmers. From this topic, two projects, ATLAS and DEMETER were funded.

The goal of ATLAS – Agricultural Interoperability and Analysis System is to achieve a new level of interoperability of agricultural machines, sensors and data services. ATLAS enables farmers to have full control over their data: farmers decide which data is shared with whom in which place. ATLAS will build an open, distributed and extensible data platform based on a microservice architecture which offers a high level of scalability from a single farm to a global community.

The technology developed in ATLAS will be tested and evaluated within pilot studies on a multitude of real agricultural operations across Europe along four relevant use cases: precision agriculture tasks, sensor-driven irrigation management, data-based soil management and behavioural analysis

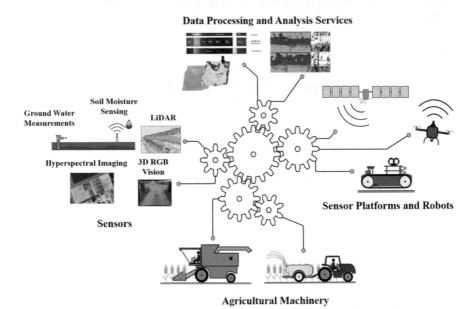

Figure 2.1   Interoperability of sensors, machines and data services through the ATLAS platform.

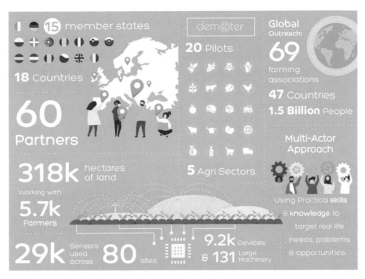

Figure 2.2   Scale of the DEMETER project.

of livestock. ATLAS will involve all actors along the food chain, simplifying and improving the processes from farm to fork. Through the support of innovative start-ups, SMEs and farmers, ATLAS will enable new business models for and with the farmers and establish sustainable business ecosystems based on innovative data-driven services. Started in October 2019, ATLAS will run for 3 years, involving 30 partners from 8 European countries, including research institutions and universities, the agricultural manufacturing industry, SMEs as well as commercial agricultural operations and agricultural cooperatives.

The second new LSP in this sector is DEMETER. This project will show how field and plant sensors, weather stations, monitoring and control devices and much more will help support sustainable and safe farming and food production systems. Through its multi-actor approach, the project also aims to improve farmer wellbeing and generally support farmers in precision decision making. DEMETER will demonstrate the real-life potential of advanced interoperability in the IoT technologies across the value chain in multiple agri-food operational environments, involving different production sectors, production systems and farm sizes. DEMETER will also display how an integrated approach to business, behaviour and technology can support farmers and the sector. It will provide further opportunities, including new business models on the farm and in the wider agri-food economy, while also helping to safeguard Europe's precious natural resources.

DEMETER will ease the uptake of future (not yet developed) services, data sources, technologies by farmers, thereby allowing the farmers and relevant other stakeholders to increase the range of choices for the most appropriate combination of tools from different suppliers in order to support their expected innovation, all the while limiting vendor lock-in. An open call for interested farmers, technology solution providers and other interested parties will be launched in 2020 and will have a €1 million budget.

2.3 Deployment in Energy Sector

Changes in climate, digital technologies and geopolitics are already having a profound effect on the lives of Europeans. In striving for digitising and transforming European industries under the new focus area, we must focus on making energy markets work better for consumers, business and society, and must support industry to adapt to globalisation and better achieve climate goals.

The internet of things is an enabler for a digital energy world that reinvents power, generation, transmission, distribution, and consumer services. IoT makes use of synergies that are generated by the convergence of Consumer, Business and Industrial Internet.

Tomorrow's energy grids consist of heterogeneous interconnected systems, of an increasing number of small-scale and of dispersed energy generation and consumption devices, generating huge amounts of data. Digital platforms may overcome fragmentation between different sectors, and lead to intelligent services and a new level playing field. The distribution network as such runs into Local Flex dilemma – more renewable energy requires more flexibility at system level and at the local edge; more flexibility is needed through an agile charging infrastructure for electric vehicles, both will result into more complexity for system design. Distribution System Operators (DSOs) are transformed more into system operators rather than just distributors of energy.

Launch of 4 new pilot projects under the Focus Area domain in the energy vertical under the topics: "Interoperable and smart homes and grids" and "Big data solutions for energy". With the pilot **InterConnect**, we aim at open platform concepts that demonstrate a transversal approach connecting different sectors and different systems to achieve energy efficiency and manage flexibility through volatile energy sources and increasing demands from electric vehicles. Digital technologies like IoT and big data analytics will be central to the new pilot projects in order to break down silos between different

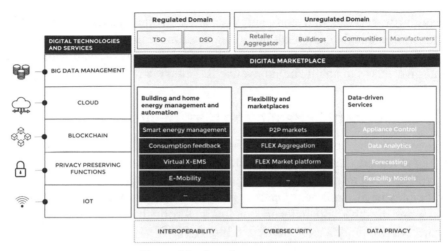

Figure 2.3 InterConnect concepts.

sectors like living, mobility, smart city and utilities to achieve cross-sector optimization in terms of resource-efficiency and sustainability: It is time for Change: "The new Energy Platforms of the Future won't be built on silos of the past."

With 3 more pilots called **BD4OPEM, SYNERGY, PLATOON**, data analytics is central to support the development of a wide range of energy services, at least to increase the efficiency and reliability of the operation of the electricity network, to optimize the management of assets connected to the grid, and to de-risk investments in energy efficiency (e.g. by reliably predicting and monitoring energy savings).

The aim of the large-scale pilot **InterConnect** is to exploit IoT reference architectures models that allow for combining services for home or building comfort and energy management. InterConnect's vision is to produce a digital marketplace, using an interoperability toolbox and SAREF complaint *IoT reference architecture* as main backbone, through which energy services, compliant devices, platform enablers (i.e., blockchain, smart contracts) and applications can be downloaded onto IoT and smart grid digital platforms (see Figure 2.3). Main focus will be on the design of *a Distributed System Operators standardized interface* to interact with market platforms and ensure connectivity with DSO legacy systems.

InterConnect will rely on *Co-creation involving citizens* which is the basis for the design of applications that turn energy management technologies into easy-to-use services and non-energy services, while ensuring comfortable, efficient, sustainable and healthier living environments. The envisaged architecture will allow for *third party contributions by means of open calls* which will give access to small and innovative ICT players (SMEs and start-ups).

2.4 Deployment in Health and Care Sector

The topic on "Smart and healthy living at home" targeted a platform for smart living at home to integrate a mix of advanced ICT ranging from biophotonics to robotics, from artificial intelligence to big data and from IoT to smart wearables in an intelligent manner to promote early risk detection of health risks and healthy and independent living. From this topic, the following projects were funded as visualised in Figure 2.3.

PHARAON

PHARAON's overall objective is to make a reality smart and active living for Europe's ageing population by creating a set of integrated and highly customizable interoperable open platforms with advanced services, devices, and tools including IoT, artificial intelligence, robotics, cloud computing, smart wearables, big data, and intelligent analytics. Platform interoperability

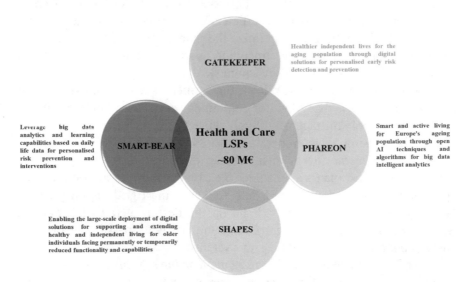

Figure 2.4   Health and Care LSPs.

will be implemented within PHARAON ecosystems and platforms, as well as other standardised platforms within health and other domains (energy, transport and smart cities). Pharaon will consider relevant standards and will contribute to them with the help of the two standardisation bodies of the consortium. Data privacy, cybersecurity, interoperability and openness will be key design principles to pursue through the requirements generated by PHARAON experts.

PHARAON will be built upon mature existing state-of-the-art open platforms and technologies/tools provided by the partners, which will be customised and will implement cloud technologies, AI techniques and traditional algorithms for big data intelligent analytics. A user-centric approach will be followed. PHARAON will evolve based on the user feedback and the results from a MAFEIP framework that will be implemented for impact assessment. Both inputs will be used to find innovative solutions through two "open calls": (1) single solutions, and (2) solutions to be demonstrated in small-scale pilots. Pharaon's integrated platforms will be validated in two stages: pre-validation and LSPs, in six different pilot sites: Murcia and Andalusia (Spain), Portugal, The Netherlands, Slovenia and Italy. A team of partners in each pilot will ensure its right development.

A set of development tools will be created and made publicly available to simplify the customisation and integration. These tools and the results of dissemination will spread the generated knowledge to promote the development of new solutions similar to PHARAON.

GATEKEEPER

The main objective of the GATEKEEPER Project is to create a ecosystem, that connects healthcare providers, businesses, entrepreneurs, elderly citizens and the communities they live in, in order to originate an open, trust-based arena for matching ideas, technologies, user needs and processes, aimed at ensuring healthier independent lives for the ageing populations.

By 2022, GATEKEEPER will be embodied in an open source, European, standard-based, interoperable and secure framework available to all developers, for creating combined digital solutions for personalised early detection and interventions that (i) harness the next generation of healthcare and wellness innovations; (ii) cover the whole care continuum for elderly citizens, including primary, secondary and tertiary prevention, chronic diseases and co-morbidities; (iii) straightforwardly fit "by design" with European regulations, on data protection, consumer protection and patient protection (iv) are subjected to trustable certification processes; (iv) support value generation through the deployment of advanced business models based on the VBHC paradigm.

GATEKEEPER will demonstrate its value by scaling up, during a 42-months work plan, towards the deployment of solutions that will involve ca 40.000 elderly citizens, supply and demand side (authorities, institutions, companies, associations, academies) in 8 regional communities, from 7 EU Member States.

SMART BEAR

It is a fact that the European population growth is slowing down, while the population ageing accelerates. Rapid increases in the elderly population are predicted for the coming decades due to the ageing of post-war baby births. Within Europe's ageing population, Hearing Loss, Cardiovascular Diseases, Cognitive Impairments, Mental Health Issues and Balance Disorders, as well as Frailty, are prevalent conditions, with tremendous social and financial impact. Preventing, slowing the development of or dealing effectively with effects of the above impairments can have a significant impact on the quality of life and lead to significant savings in the cost of healthcare services. Digital tools hold the promise for many health benefits that can enhance the independent living and well-being of the elderly.

Motivated by the above, the aim of the SMART BEAR platform is to integrate heterogeneous sensors, assistive medical and mobile devices to enable the continuous data collection from the everyday life of the elderly, which will be analysed to obtain the evidence needed in order to offer personalised interventions promoting their healthy and independent living. The platform will also be connected to hospital and other health care service systems to obtain data of the end users (e.g., medical history) that will need to be considered in making decisions for interventions.

SMART BEAR will leverage big data analytics and learning capabilities, allowing for large scale analysis of the above-mentioned collected data, to generate the evidence required for making decisions about personalised interventions. Privacy-preserving and secure by design data handling capabilities, covering data at rest, in processing, and in transit, will cover comprehensively all the components and connections utilized by the SMART BEAR platform. The SMART BEAR solution will be validated through five large-scale pilots involving 5.000 elderly living at home in Greece, Italy, France, Spain, and Romania.

SHAPES

SHAPES aims to create the first European open Ecosystem enabling the large-scale deployment of a broad range of digital solutions for supporting and extending healthy and independent living for older individuals who are facing permanently or temporarily reduced functionality and capabilities. SHAPES builds an interoperable Platform integrating smart digital solutions to collect and analyse older individuals' health, environmental and lifestyle information, identify their needs and provide personalised solutions that uphold the individuals' data protection and trust. Standardisation, interoperability and scalability of SHAPES Platform sustain increased efficiency gains in health and care delivery across Europe, bringing improved quality of life to older individuals, their families, caregivers and care service providers. SHAPES Large-scale Piloting campaign engages over 2000 elderly persons in 15 pilot sites across 10 EU Member States, including 6 EIP on AHA Reference Sites, and involves hundreds of key stakeholders to bring forth solutions to improve the health, wellbeing, independence and autonomy of older individuals, while enhancing the long-term sustainability of health and care systems in Europe. SHAPES's multidisciplinary approach to large-scale piloting is reflected across 7 themes that, together, provide a clear understanding of the reality of European health and care systems and enable the validation of cost-efficient, interoperable and reliable innovations capable of effectively supporting healthy and independent living of older individuals within and outside the home.

Building an ecosystem attractive to European industry and policy-makers, SHAPES develops value-based business models to open and scaleup the market for AHA-focused digital solutions and provides key recommendations for the far-reaching deployment of innovative digital health and care solutions and services supporting and extending healthy and independent living of older population in Europe.

2.5 Conclusion

The rise of global platforms and 'Digital champions' drives the need to change the rules of the game. In large-scale piloting, pilots are set up that make use of the digital platforms, develop prototype applications on top of the platforms, and validate the platforms in both reduced, controlled environments and in real-life use cases. Supporting pilots and platforms we build on Europe's strong industrial companies and this track record of collaboration to foster cooperation across industry's vertical and horizontal boundaries, across manufacturing, farming, energy, homes & buildings and health.

With intangible investment eclipsing the tangible kind, the rise of platforms and the ability to scale and to do so quickly seem to matter more for a significant part of the innovation ecosystem. In supporting ecosystem building, the take-up of open APIs and standards as integral part of digital platforms, is fostered by expanding the ecosystem of players involved and through opportunities for entrepreneurs by promoting new market openings allowing also smaller and newer players to capture value. With cascading grants, the selected pilot projects are open for medium-size companies as well as small entrepreneurial firms and start-ups in a straight and non-complicated manner.

Significant contributions should be made to suitable standardisation bodies or pre-normative activities, as outlined in the Communication on Priorities of ICT Standardisation for the Digital Single Market. With this major investments under the Focus Area in Horizon 2020 [3], pilots represent a unique and rich source of use cases – by which we expect increased prospects for future digital industrial platforms, validation of technological choices, sustainability and reproducibility, through architecture models, standards, and interoperability, as well as of verification of non-functional characteristics such as security and privacy.

References

[1] Digitising European Industry, online at: https://ec.europa.eu/digital-singl e-market/en/policies/digitising-european-industry
[2] IoT European Large-Scale Pilots, online at: https://european-iot-pilots.eu/
[3] Platforms and Pilots, online at https://ec.europa.eu/digital-single-market /en/industrial-platforms-and-large-scale-pilots
[4] ETSI releases 3 new SAREF specifications, online at https://ec.europa.eu /digital-single-market/en/news/etsi-releases-three-new-saref-ontology-specifications-smart-cities-industry-40-and-smart

3

New Waves of IoT Technologies Research – Transcending Intelligence and Senses at the Edge to Create Multi Experience Environments

Ovidiu Vermesan[1], Marcello Coppola[2], Mario Diaz Nava[2],
Alessandro Capra[3], George Kornaros[4], Roy Bahr[1],
Emmanuel C. Darmois[5], Martin Serrano[6], Patrick Guillemin[7],
Konstantinos Loupos[8], Lazaros Karagiannidis[9]
and Sean McGrath[10]

[1]SINTEF, Norway
[2]STMicroelectronics, France
[3]STMicroelectronics, Italy
[4]Hellenic Mediterranean University, Greece
[5]CommLedge, France
[6]Insight Centre for Data Analytics, NUI Galway, Ireland
[7]ETSI, France
[8]Inlecom Innovation, Greece
[9]Institute of Communication and Computer Systems, Greece
[10]University of Limerick, Ireland

Abstract

The next wave of Internet of Things (IoT) and Industrial Internet of Things (IIoT) brings new technological developments that incorporate radical advances in Artificial Intelligence (AI), edge computing processing, new sensing capabilities, more security protection and autonomous functions accelerating progress towards the ability for IoT systems to self-develop, self-maintain and self-optimise. The emergence of hyper autonomous IoT applications with enhanced sensing, distributed intelligence, edge processing and connectivity, combined with human augmentation, has the potential to power the transformation and optimisation of industrial sectors and to change the innovation landscape. This chapter is reviewing the most recent advances in the next wave of the IoT by looking not only at the technology enabling the IoT but also at the platforms and smart data aspects that will bring intelligence, sustainability, dependability, autonomy, and will support human-centric solutions.

3.1 Next Wave of Internet of Things Technologies and Applications

IoT is defined [172] as "a dynamic global network infrastructure with self-configuring capabilities based on standard and interoperable communication protocols where physical and virtual things have identities, physical attributes, and virtual personalities using intelligent interfaces for seamless integration into the information network". In the IoT, things are expected "to become active participants in business, information and social processes where they interact and communicate among themselves and with the environment by exchanging data and information sensed about the environment, while reacting autonomously to the real/physical world events and influencing it by triggering actions and creating services with or without direct human intervention". Interfaces in the form of services facilitate interactions with these intelligent things over the Internet, query and change their state and any information associated with them, considering security and privacy issues.

The IoT is enabled by heterogeneous technologies used to sense, collect, act, process, infer, transmit, notify, manage, and store data. IoT technologies and applications are evolving from a network of objects towards intelligent social capabilities meant to address the interactions between humans and autonomous/automated systems.

The cognitive transformation of IoT applications allows the use of optimised solutions for individual applications and the integration of immersive technologies, i.e. virtual reality (VR), augmented reality (AR), and Digital Twin (DT). Such concepts transform the way individuals and robots interact with one another and with IoT platform systems.

The powerful combination of AI, IoT, Distributed Ledger Technologies (DLTs) can form the foundation of improved and – potentially – entirely new products and services. It also brings new challenges that require addressing, in a holistic manner distributed IoT architectures, decentralised security mechanisms and the evolution of IoT/IIoT devices towards intelligence-by-default and their deeper integration into platforms that aggregate the valuable data involved in a variety of processes. Additionally, recent research outcomes in data processing and edge high-performance computing (eHPC) further support basic foundations for AI.

Artificial Intelligence of Things (AIoT) is seamlessly controlling and optimising the systems and their environment, analysing data from devices to applications. It is integrated into IoT/IIoT digital platforms that support artificial/augmented intelligence, and control and optimise the automation processes and thus support the next wave of innovation. This includes the adoption of federated learning concepts for the creation of low-latency intelligent IoT applications and new AI business models, by extracting common knowledge from participating IoT devices, while enabling reduced bandwidth use, localised personalisation of models, and granular data security and compliant privacy policy.

In the last ten years, new breakthrough technologies, such as IoT, AI and 5G have impacted our daily life at home, at work, in the public spaces, etc. They have allowed the emergence of more and more rich and complex applications developed and deployed in different market areas such as smart home and buildings, smart industry, smart mobility, smart city, smart healthcare, smart farming, and wearables. This evolution has been supported by progress done by the semiconductor industry in the design and fabrication of more complex integrated circuits, allowing the development of cost-effective smart devices, systems, and applications.

The new capabilities of 5G will lead to more complex value chains where more actors provide connectivity and bundle it with vertical-specific services. These actors, eager to benefit from new revenue streams, can include telecommunication companies looking for to diversification, equipment providers betting on small cell networks, over-the-top (OTT) players looking at unlicensed networks or purely vertical players integrating connectivity into their new services.

The IoT systems allow data gathering from the environment through different sensor types (according to the application) and data analysis to facilitate decision making by the end-user. Although the control-command concept used by IoT system is not revolutionary, IoT has allowed the gathering of a huge amount of data (big data), through wireless and Internet networks, and their storage in databases/data lakes/data spaces in the cloud for further analysis and the creation of valuable information.

At the end of 2019, there were 7.6 billion active IoT devices, a figure which is expected to grow to 24.1 billion in 2030, a compound annual growth rate (CAGR) of 11%, according to the research published by Transforma Insights [4].

Short range connectivity solutions, (e.g. Wi-Fi, Bluetooth and ZigBee), will dominate connections, accounting for 72% in 2030, largely unchanged compared to the 74% it accounts for today. Public networks, which are dominated by cellular networks, will grow from 1.2 billion connections to 4.7 billion in 2030, growing market share from 16% to 20%. Private networks account for the balance of connections, 10% in 2019 and 8% in 2030. Services, including connectivity, will account for 66% of spend, with the remainder accounted for by hardware, in the form of dedicated IoT devices, modules and gateways [4].

In IoT deployments, the collection and processing of a large amount of information at the edge of the network where the IoT end-devices are deployed and where information is captured and collected has been a major evolution. New IoT systems use smart solutions with embedded intelligence performing real-time analysis of information and connectivity at the edge. These new IoT systems are departing from centralised cloud-computing solutions towards distributed intelligent edge computing systems. Traditional centralised cloud computing solutions are generally fit for non-real-time applications that require high data rates, huge amounts of storage and processing power, that do not require low latency, and can be used for heavy data analytics and AI processing jobs. On the other hand, distributed edge solutions introduce computations at the edge of the network where information is generated supporting real-time services with very low latency (in the order of milliseconds) that can be used for simple to medium ultra-fast analytics. The collection, storage and processing of data at the edge of the network in a distributed way also contributes to the increased privacy of the user data since raw personal information no longer needs to be stored in backbone centralised servers and each user can retain the full control of the data. Even more, for some applications, data can be processed online without need of storage capabilities. However, the convergence between cloud computing, edge, and IoT requires smart and efficient management of resources, services, and data, whose elements can move across different heterogeneous infrastructures.

The densification of the mobile and associated services network strongly challenges the connection with the core network. Next-generation IoT networks need greater cloud systems flexibility and implement cloud utilisation mechanisms to maximise efficiency in terms of latency, security, energy efficiency and accessibility. Cloud technologies should combine software-defined networks and network function virtualisation to enable network flexibility in order to integrate new applications and adequately configure network resources (e.g. by sharing computing resources, splitting data traffic, enforcing security rules, implementing QoS parameters, ensuring mobility). Global connection growth is mainly driven by IoRT devices on both the consumer side (e.g. smart homes) and the enterprise/business-to-business (B2B) side (e.g. connected machinery). These devices require an extension of the spectrum in the 10–100 GHz range and unlicensed bands and technologies, such as WiGig or 802.11ax, which are mature enough for massive deployment and can be used for cell backhaul, point-to-point or point-to-multipoint communication.

Edge processing requires more specialised hardware knowledge (e.g. security and privacy protection, data analysis through Machine Learning (ML) techniques and more precisely implementation of Deep Learning (DL) algorithms) to ensure that the IoT devices operate efficiently and capture data in ways supporting more complex and sophisticated processing at the edge (such as distributed ledgers, increased trust, and identity management, data analysis, etc.). These advances in IoT technologies allow the integration of IoT capability directly into modules or hardware so that, once the IoT device is deployed, it can automatically connect to any network available and start provisioning data – to an on-premise solution, an edge computing intelligent infrastructure or the hyperscale cloud services – that can be used by a rich set of applications.

In this context, the wireless connectivity for IoT deployments will continue to diversify with different communication protocols being used according to the needs and performance required by different IoT applications and services. At the same time, these communication technologies have opened the door to potential malicious cyber-attacks and threats, and the risk to stole personally identifiable information (PII). End-to-end security and privacy mechanisms have been developed over time to avoid these issues. However, standard solutions are needed. Furthermore, end-devices processing capabilities have been fostered by the maturation of AI, and the flexibility offered by ML and DL techniques regarding the place/level where the data analysis must be performed. Performing data analysis closer to the data sources has important benefits such as reduced overall system power consumption, reduced communication bandwidth and throughput, increased data security and privacy (personal data protection), and reduced processing latency. However, state of the art techniques and requirements impose for more data processing and "smartness" at the edge, including at sensor level, end-device (sensor nodes) level, server level with gateway capabilities, and other network components.

From the IoT system architecture perspective to obtain the mentioned benefits, this implies to move the data analysis (for many applications) from Cloud to Fog and Edge architectures. It also creates the need for greater interoperability between the different systems at cloud, edge and fog levels and a key role for the associated standards.

These changes are made possible especially if the devices involved in the edge domain (e.g. edge servers, end-devices, and sensors) are energy efficient and can support AI techniques at low power consumption to ensure high autonomy/longer battery life, system availability and reliability. This will become possible by developing a new generation of more performant Integrated Circuits (ICs) with the best trade-off between increased processing power vs ultra-low power consumption.

Table 3.1 gives an overview of the IoT evolution phases and indicates, the focus areas of edge devices evolution. Finally, it should be also considered the impact that 5G will have in the more intelligent edge devices for future applications requiring high throughputs such as vehicle to everything (V2X), augmented reality (AR), virtual reality (VR), Industrial 4.0, autonomous systems and e-health.

Table 3.1   IoT evolution and associated technologies

Main Features	Phase 1 (2016–2018)	Phase 2 (2019–2020)	Phase 3 (2021–2025)
Reference Architecture Domains: Cloud, Edge	Cloud Computing	Cloud and Fog Computing	Cloud, Fog, Edge Computing /Private Networks
Data analysis	On the Cloud	On the Cloud, starting in the Edge servers, Hybrid	Cloud, Edge (Server, End-Device and Sensor), Hybrid
Artificial Intelligence	No	Cloud, Edge Servers	Cloud, Edge (Server, End-Device and Sensor)
End-to-End security	No	Yes	Yes
End-to-End privacy	Not considered	Not yet supported every where	Yes
Connectivity between domains	Wired, 3G/4G	Wired, 3G/4G	Wired, 3G/4G, 5G

(Continued)

Table 3.1 Continued

Main Features	Phase 1 (2016–2018)	Phase 2 (2019–2020)	Phase 3 (2021–2025)
Connectivity edge domain	Wi-Fi, BT, Z-Wave, ZigBee, LoRa, Sigfox,	Wi-Fi, BT, Z-Wave, ZigBee, LoRa, Sigfox,NB-IoT, LTE-M	Wi-Fi, BT, Z-Wave, ZigBee, LoRa, Sigfox.NB-IoT, LTE-M, 5G
Edge devices	Gateway with routing capabilities	Edge Servers with Gateway and routing capabilities	Distributed Edge Servers with Gateway and routing capabilities
End devices	Smart Sensors	Smart Sensors Cyber-Physical Systems (Drones, Robots)	Smart Sensors, Smart Cameras, Cyber-Physical Systems (Drones, Robots, Autonomous cars)
Architecture	Sensor + Computing processing + Connectivity	Sensor + Computing processing + Connectivity + Security	Sensor + Computing processing + Connectivity + Security + Privacy + AI
Security	Not supported	Yes	Yes
Privacy	Not supported	Not supported	Yes
AI capabilities for data analysis and decision making	No supported	No supported	Yes
Data throughput	Low (< 200 Kbytes)	Low (< 500 Kbytes)	> 500 Kbytes
Computing processing	Low	Medium	High
Sensor	Simple	Multiple sensors	Multiple-sensors, smart cameras
Actuator	Not supported	Not supported	Yes
Autonomy	Low	Medium	High Autonomy
Reliability	No	Improved	Mandatory
IoT Digital Twin (DT)	Status IoT DTs, simulation DTs,	Operational IoT DTs with events and simulations (domain specific)	Autonomous IoT DTs, collaborative twins, at the edge, cross-domain

The next waves of IoT developments are presented in Figure 3.2. IoT/IIoT research and innovation is building on advancements in semiconductors, photonics, sensing technologies AI techniques and embedded systems for the "things". The creation of next-generation information and communication platforms will be strongly driven by wireless, cellular (e.g. 5G and beyond) technologies, which will generate massive amounts of information as input for hyper computing processing capabilities and enabling new applications in services such as robotics, autonomous vehicles, AI and intelligent systems of systems.

3.2 Energy Efficient and Green IoT 3D Architectural Approach

Climate change and environmental degradation are an existential threat to Europe and, more globally, to the world. The European Green Deal roadmap for economic sustainability entails a new growth strategy to transform Europe "into a modern, resource-efficient and competitive economy where there are no net emissions of greenhouse gases by 2050, economic growth is decoupled from resource use" [15], and no persons and no place are left behind. This will be done by "turning climate and environmental challenges into opportunities across all policy areas and making the transition just and inclusive for all" [15]. The Green Deal is an integral part of the EC's strategy to implement the UN's 2030 Agenda and its sustainable development goals [16]. To implement this strategy, a circular economy action plan [17] has detailed measures to make sure that sustainable products are the norm in the EU, which means that products on the EU market are designed to last longer, are easier to reuse, repair, and recycle, and incorporate recycled material as much as possible, put a major focus on electronics, ICT and batteries, and drastically reduce waste by transforming it into high-quality secondary resources [17].

Digital technologies are a significant enabler for attaining the sustainability goals of the Green Deal in many different sectors. The EC is launching initiatives and taking measures to ensure that digital technologies such as AI, 5G, cloud and edge computing and the IoT can accelerate and maximise the impact of policies that deal with climate change and protect the environment. In digitalisation, IoT in particular offers new opportunities for remote monitoring of air and water pollution, or for monitoring and optimising how energy and natural resources are used [15].

ICT is consuming increasing amounts of energy and this is also true for IoT/IIoT systems and applications. Saving energy and fuel-costs by using energy-efficient demand-side technologies can lead to increased consumption of those services, an outcome known as the "rebound effect". For ICT, this means that if the demand for the services increases more than 11% due to rebound, all of the environmental or natural resource impacts considered will rise even if overall energy consumption declines [13].

Disruptive technological change can enable sustainable development with global benefits for closing the emissions gap but can also exacerbate unsustainable patterns of resource use. This is clearly evidenced by the promises and risks of the digital revolution and the ongoing advances in ICT, ML and AI, connectivity, IoT, additive manufacturing (e.g., 3D printing), virtual and augmented reality, blockchain, robotics and synthetic biology [12].

The next waves of IoT technologies and system architectures are designed to improve energy efficiency and circular economy performance of the IoT/IIoT systems and use the IoT applications to support all industrial sectors to increase the energy efficiency, reduce their CO_2 footprint and provide new innovative value chains, value networks and business models to achieve economic sustainability.

New innovative models are needed to accelerate circularity and the dematerialisation of the economy to make Europe less dependent on primary materials [17]. These models, based on better involvement of customers, information sharing, mass customisation, sharing and collaborative economy, will be powered by digital technologies, such as IoT, big data, blockchain and AI.

IoT must consider the increase of energy efficiency with an optimal use of resources and infrastructure supported by AI techniques with the use of real-time analytics related to energy and resources, the analysis of event and information streams across the architectural layers, the optimisation of the location of processing, and the transfer of intelligence where the application needs it (e.g. edge, IoT gateway, device, etc.). Energy-efficient, and green IoT requires a holistic end-to-end strategy across the information value chain through the IoT architectural layers. This requires the design of energy-efficient and green IoT components at each layer level and at the IoT applications level by combining AI technologies and optimising IoT capabilities across the architectural layers to unify the complete IoT and analytics life cycle, streaming the information, filtering, scoring, exchanging storing what's relevant, analysing and using the results to continuously improve and optimise the IoT system.

As it is defined in [21] the green IoT must address the study and practice of designing, using, manufacturing, and disposing of IoT devices, systems, subsystems, communication network systems, HW/SW platforms efficiently and effectively with minimal or no impact on the environment. In this context, a holistic approach is needed to follow the IoT green design (e.g. advanced and adequate semiconductor technologies, efficient design, energy efficient SW/HW platforms) for providing environmentally sound components at all IoT architectural layers and functions, energy efficient and low CO_2 footprint at IoT infrastructure and technical solutions levels (e.g. edge, cloud, data centre, AI-based learning/training, etc.), green manufacturing (e.g. manufacture IoT electronic components, HW/SW platforms, and IoT systems with minimal or no impact on the environment.), green use (e.g. reduce/minimise the energy consumption of IoT systems and support the monitoring and optimisation of the energy consumption for other systems in various industrial sector applications and use them in an environmentally sound manner) and green disposal (refurbish and reuse the IoT systems and properly recycle these systems and support through IoT applications by monitoring the implementation of solutions for recycling, and circular economy in different industrial sectors).

Green and energy efficient IoT strategies involve the use of IoT technologies and the design of applications in such a way that the IoT systems are optimised across all the IoT architectural layers. These strategies must include the embedded carbon generated by producing and moving the materials for the IoT electronics, producing the devices, installing the IoT systems, etc. In addition, IoT technologies and applications enable and leverage applications in different industrial sectors that support cutting CO_2 emissions increasing energy efficiency, monitoring supply chains for providing circular economy solutions, reducing resources consumption, preserving natural resources, minimising the technology impact on the environment and human health and reducing the costs.

In order to reach the previous goals, the use of the IoT 3D reference architecture provides the tool to optimise the IoT systems developments for energy efficiency and green design by identifying the specific (energy and green design) techniques that need to be applied to the design and implementation of various HW/SW components across the different IoT architectural layers as part of the systems implementation and IoT platforms used.

The architecture implementation depends on the application domains and their requirements, specific QoS factors, systems characteristics, and the cross-cutting functions involved. The architecture and application designers can identify the major energy intensive components, nodes, protocols, middleware elements, SW algorithms, HW blocks/sub-systems in each layer and optimise them at each layer, across several layers and at the application level.

For this, CAD tools and design flows are required to optimise each individual component and the overall system. Furthermore, the analysis of the energy efficiency, the green methods and techniques to be applied, can be done by defining, at each layer, the main functions and components and identifying adequate optimisation parameters to be used for energy efficiency and green IoT design.

The approach utilises the 3D IoT layered architecture [62] based on eight IoT domain layers as illustrated in Figure 3.1. The architecture is composed of two other views that include the system properties and the cross-cutting functions.

The green attributes and energy efficiency are correlated with various tasks (e.g. sensing/actuating, communicating, processing, analysing, storing, transferring, learning, etc.) performed in the different architecture layers by various HW/SW components and algorithms integrated into IoT platforms running at the edge or on cloud infrastructure.

The implementation of green and energy efficient techniques and methods (e.g. optimisation, trade-off analyses among cross-cutting functions/system properties vs. energy, green IoT metrics, performance, measurement, test-beds, energy harvesting, wireless power transfer, etc.) depends on the

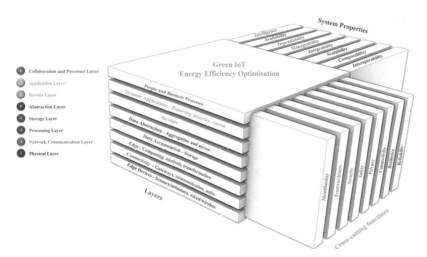

Figure 3.1 3D IoT layered architecture. Adapted from [62].

functions performed by different HW/SW components integrated in the IoT layers, which include energy management, wake-up scheduling mechanisms, selective sensing, HW-SW partitioning, energy efficient algorithms, communication techniques and distribution of tasks, minimisation of data path length, data buffer delivery, wireless communication, and processing trade-offs [18–20]. To optimise energy efficiency, CAD tools and design flows are required at each layer.

3.2.1 Physical Layer

The physical layer includes the sensors/actuators, processing (SW/HW), and connectivity for all IoT end-devices. The design of the HW/SW components, functional blocks, sub-systems define the relevant functions that should be implemented through a set of ICs on a board, a system-on-chip (SoC), a system-in-package (SiP) or an embedded system intelligent component. The intelligence and complexity of a physical object can vary from simple RFID tags with sensing capabilities to Cyber physical systems (CPS) such as intelligent robots, drones and autonomous vehicles. The requirements for these physical layer components are correlated with the elements from the other two dimensions of the IoT 3D architecture such as system properties (intelligence, availability, dependability, manageability, integrability, scalability, composability, interoperability) and cross-cutting functions (identifiability, trustworthiness, security, safety, privacy, connectivity, resilience, reliability). The IoT application requirements will include the optimisation of energy efficiency and green properties of the HW/SW components in the physical layer aligned with the degree of system properties embedded (e.g. intelligence, scalability, interoperability, etc.) and the stringent implementation of the cross-cutting functions (e.g. end-to-end security, safety, resilience, etc.). The energy efficiency and green properties optimisation, with the right tools at this level in the early design phase of the IoT systems developments offers flexibility in designing energy efficient and green IoT devices aligned with the requirements of the IoT use cases and applications. Furthermore, technology considerations are also key aspects to ensure high performances at low power (autonomy is critical for many IoT applications) and provide high integrated solutions at cost effective. They are important factors to accelerate the introduction of IoT devices on the market.

The tasks performed by the devices in this layer focus on collecting the information from the physical environment, self-organised sensing, processing, load balancing and preparing data to be exchanged with other layers.

The components of this layer include sensors, microcontrollers (MCUs), wireless modules, actuators, energy management, energy sources (batteries, solar panels, etc.), SW embedded components, operating systems, etc. which all together constitute the edge processing nodes or well known as end-points or end-devices. For intelligent edge devices, performing data analysis, more intelligence is added to the IoT edge devices and AI-based HW/SW modules or HW accelerations associated to MCUs are included in the design. For the intelligent devices, the overall CO_2 footprint including the green characteristic and energy efficiency of the AI inference at the edge must be considered over the lifetime of the device compare with the same function and the learning/training performed in the cloud or data centres. For the external infrastructure elements, the type of energy (e.g. renewables, fossil, etc.) used for powering the facilities must be included in the design optimisation. Moreover, the energy management techniques, the selection of materials, energy sources (e.g. batteries, solar panels, socket, etc.), energy harvesting, SW optimisation for data sensing, monitoring, filtering, prediction and compression with processing, sleep/wake-up techniques, energy efficient task scheduling algorithms, selection of quality of information (QoI), allocation of workload distribution at edge, wireless communication optimisation (send/receive), power down mechanisms, dynamic wireless network behaviour (e.g. IoT devices-move-in and IoT devices-move-out), cooperation/information exchange between the edge nodes while processing are elements that need to be consider to optimise the green and energy profile of the IoT devices. It should also be considered that the benefits brought by performing data analysis in the physical layer such as reduce bandwidth use have a direct impact in the energy consumption saving in the overall communication path, reduce the size of the data memory storage considering that the data were already processed by the physical device, finally reduce processing power and energy saving in the cloud/data centres. The energy and resources saving could be very important in an edge computing approach versus a cloud computing one.

3.2.2 Network Communication Layer

The network and communication layer plays a central role in all IoT applications and deployments that support the exchange of information between different layers and create value due to the interaction in real-time between IoT devices. The transmission of information between IoT devices, the network, across networks, and between the networks and high-level information processing infrastructure (e.g. cloud, data centres, etc.) is one of the main features of IoT applications. The evolution from centralised IoT architectures to decentralised and distributed increase the importance of the continuous "6As" connectivity (anything, anyone, anywhere, anytime, any path/network, any service).

This layer provides communication between the physical layer (IoT devices) and the components of other layers including the interactions between different types of networks (e.g. wireless, wireless sensor networks, cellular, optical, satellite, etc.) that connect the IoT devices to gateways, edge, cloud, and data centres infrastructure. The selection of connectivity infrastructure constituted of hubs switches, gateways, routers, radio access network architecture, edge processing for software-defined and cognitive-based radios and network functions virtualisation, SW algorithms, remote transmitting, cloud servers and data centres architectures, define the CO_2 footprint, green, and energy profile of the network and communication layer IoT components. This layer plays the role of a dynamic channel for transferring/exchanging information from the physical layer to processing, storage layers and beyond base on more than 50 wireless and cellular communication protocols (e.g. Wireless Personal Area Network (WPAN), Wireless Local Area Network (WLAN), Wireless Neighbourhood Area Network (WNAN), Wireless Wide Area Networks (WMAN). Cellular WWAN, Low Power Wide Area Networks (LPWAN), etc.).

The optimisation of the network and communication layer components must consider the functionalities required for transmission of information, neighbourhood communication, retransmission and avoidance of packet collision, handshaking mechanisms, etc. For the new cellular technologies such as 5G and beyond the network densification via small cells need new models for optimising the green effect and energy efficiency for IoT applications as the 5G network infrastructure will increase the energy consumption. The 5G networks will have higher energy efficiency (less energy per transmitted bit) but will consume more power than 4G networks. The energy consumption is a result of the growth in mobile data consumption, deployment of additional, network elements (e.g. small cells, MIMO antennas) and the use of mobile edge computing, cloud computing and data centres processing.

A trade-off must be considered among multi-hop and cooperative multi-hop routing in wireless sensor networks communication to gateways and base stations, path optimisation to reduce energy consumption, network traffic control, dynamic downloading of packets using access points, scheduling of the communication tasks to reduce energy consumption in network communication, etc. The optimisation techniques to be applied at the IoT network and communication layer support are achieving a much higher energy efficiency of the IoT network with limited bandwidth provisioning and low transmit power, by utilizing advanced capabilities of IoT (e.g. in-network storage and caching, offload the IoT data to release the traffic scale in the cellular networks, provide low-latency IoT services in an energy efficient manner, etc.), designing lightweight context-aware security schemes (e.g. encryption, authorisation, etc.) to reduce the energy consumption of secure IoT networks.

Last but not least, the IoT system architecture choice between cloud vs fog vs edge computing will be fundamental in the energy savings and determinant in the complexity of the Network Communication Layer.

3.2.3 Processing Layer

This layer addresses edge computing, information element analysis and transformation, analytics, mining, machine learning, in a pervasive manner considering that autonomic services must be provided through ubiquitous machines in both "autonomic" and "smart" way. The processing layer provides the ability to process and act upon events created by the edge devices and store the data into a database in the storage layer. The processing layer can be closely interlinked with data analytics platform(s) based on edge-/cloud-scalable platform that supports information processing technologies. The green IoT and energy efficiency optimisation depend on the type of information proccesion, location, real-time requirements, AI components, platforms, and system architecture (e.g. centralised, decentralised, distributed). Complex event processing can be supported for edge, cloud, data centre solutions to initiate near real-time activities and actions based on data from the edge devices and from the rest of the system. The requirements for the processing layer are connected to the need for highly scalable, column-based data storage for storing events, map-reduce for long-running batch-oriented processing of data and complex event processing for fast in-memory processing and near real-time reaction and autonomic actions based on the data and activity of devices and the interconnected systems.

Edge computing performs data analysis at sensors, end-devices, the gateways, micro servers, edge/embedded high-performance computing levels and the enterprise applications leverage edge devices data in end-to-end value streams involving edge devices and people within digitised processes.

Developments of a new range of IoT applications is enhancing the role of this layer that is being used to analyse, process and in many cases store the information at the edge level. This layer provides support to the lower layers, allocates resources for efficient storage in virtual and physical machines and converts the data into the required form. The components of this layer include HW/SW processing elements/units (e.g., operating systems, embedded software, MCUs, CPUs, GPUs, FPGAs, other AI-based accelerators, etc.), semantic-based and service-based middleware, databases virtual machines, resource allocators, information convertors and translators.

Processing the information at the edge where data is being generated, and providing insights from the information kept at the network and processing level has the benefits of reducing both latency and the demand on network bandwidth, preventing downtime, reducing CO_2 emissions, and improving the overall system efficiency. The integration of ML and AI methods support the optimisation of edge computing-based green and energy efficient functions providing solutions for moving optimally the processing from cloud to the edge and decarbonising the whole value chain of IoT information.

3.2.4 Storage Layer

The storage layer addresses the efficient storage and organisation of data and the continuous update with new information, as it is made available through the capturing and processing of IoT channels. The functions implemented include archiving the raw and processed data for offline long-term storage and the storage of information that is not needed for the IoT system's real-time operations. Centralised storage considers the deployment of storage structures that adapt to the various data types and the frequency of data capture. Decentralised and distributed storage is implemented with the functions in the processing layer. Relational database management systems that involves the organisation of data into table schemas with predefined interrelationships and metadata for efficient retrieval for later use and processing are used for different IoT systems.

For many autonomous IoT applications, the storage is decentralised, and data is kept at the edge or at the device that generates it and is not sent up across the other top layers. This model is used in conjunctions with the mobile edge computing and fog computing implementations and offer energy, computing, and connectivity resource optimisation advantages. Scalable storage platforms can be used to process very large data sets (including the ones for AI operations) across many computing nodes that operate in parallel, which provides a cost-effective storage solution for large data volumes with no format requirements and used by several IoT solutions and connected to different existing IoT platforms.

3.2.5 Abstraction Layer

The abstraction layer provides the interfaces, the event and action management through simple rules engine to allow mapping of low-level sensor events to high level events and actions, while assuring the basic analytics for data normalisation, reformatting, cleansing and simple statistics. It provides the visualisation of processed data in the form of intelligent tasks. The layer incorporates the components and SW/HW elements to implement these functions and allow the IoT systems to scale using multiple storage systems to accommodate IoT device information and data from traditional Enterprise Resource Planning (ERP), Customer Relationship Management (CRM), Supplier Relationship Management (SRM), Supply Chain Management (SCM), Product Lifecycle Management (PLM) and other systems that interact with the IoT platforms and applications. The data abstraction functions are rendering data and its storage in ways that enable developing simpler, performance-enhanced applications. Green and energy efficient HW/SW components at data abstraction layer must ensure the data processing optimisation and reconcile multiple data formats from different sources, assuring consistent semantics of data across sources, confirming that data is complete to the higher-level application, consolidating data, providing access to multiple data stores through data virtualisation, normalising or denormalising and indexing data to provide fast application access, protecting data with appropriate authentication and authorisation.

3.2.6 Service Layer

The service layer integrates the middleware on top of networks and IoT device streams and provides data management and data analytics that are key functions for IoT systems where large amounts of sensor generated data and events must be logged, stored and processed to generate new insights or events, AI-based processing on which business decisions can be made.

The service layer functions are provided at the edge or cloud level. Several cloud service providers are extending the offerings to implement IoT solutions by using the existing "infrastructure as a service (IaaS)" ecosystems to provide new IoT services. In many cases these systems are not optimised for energy efficiency and green usage and an evaluation must be performed to analyse the CO_2 footprint and appropriateness to be used for specific IoT applications.

Various IoT platforms are providing full stack solution for ingesting data from IoT devices and linking them to edge-/cloud-based storage and processing services. The platform as a service (PaaS) offering integrate the functions of the IoT service layer and are not always optimised for low energy consumption and CO_2 footprint.

3.2.7 Application Layer

The application layer is offering the software platforms that are suited to deliver the key components for implementing various IoT applications that are connecting users, business partners, devices, machines, and enterprise systems with each other to provide information interpretation to different applications. The SW/HW components at this layer interact with the service layer, while the software applications are based on vertical markets, the nature of device data, and business needs. At this layer, many applications are addressed such as mission-critical business applications, specialised industry solutions, mobile applications, analytic applications that interpret data for business decisions, etc. The optimisation for energy efficiency and green IoT requires the use of federation and orchestrations techniques that creates dynamic and distributed energy control frameworks for IoT applications that need large capacity, higher delivery efficiency, reduced energy consumption and low costs. The implementation of energy efficient search engines, the cooling systems and use of energy harvesting techniques and of renewables must be considered when the HW/SW components of the IoT application layer are evaluated. New IoT applications including AR/VR, digital twins, virtual simulations, real-time searching engines and discovery services will bring new challenges to optimising the energy efficiency.

3.2.8 Collaboration and Processes Layer

The collaboration and processes layer includes the enterprise systems and large data platforms and the exchange of information among these platforms based on high-level collaborative processes. This layer addresses the processes that involves assets, people and organisations that use IoT applications and associated information for their specific needs or for a range of different purposes, to provide the right information, at the right time, to perform the right tasks. The energy efficiency and green IoT functions have to consider end-to-end optimisation for each layer and as the data is moved across the layers to secure the optimal communication, exchange of information and energy consumption per functions at each level and between layers.

The overall optimisation required to address the energy efficiency and green IoT capabilities must be addressed at the architecture level by considering the aggregation, over the technology stack, of the functions that are required to fulfil a given IoT task. This includes estimating the energy used for learning/training of different algorithms implemented in various IoT architectural layers by employing large data sets from different databases.

3.3 IoT Strategic Research and Innovation at the Horizon

The IERC brings together EU-funded projects with the aim of defining a common vision for IoT technology and addressing European research challenges. The rationale is to leverage the large potential for IoT-based capabilities and promote the use of the results of existing projects to encourage the convergence of ongoing work; ultimately, the endpoints are to tackle the most important deployment issues, transfer research and knowledge to products and services, and apply these to real IoT applications.

The objectives of IERC are to provide information on research and innovation trends, to present state-of-the-art IoT technologies and societal trends, to apply the developments to IoT-funded projects and to market applications and EU policies. The main goal is to test and develop innovative and interoperable IoT solutions in areas of industrial and public interest. The IERC objectives are addressed as an IoT continuum of research, innovation, development, deployment, and adoption.

The IERC launches every year the Strategic Research and Innovation Agenda (SRIA), which is the outcome of discussion involving the projects and stakeholders involved in IERC activities.

Enabled by the activities of the IERC, IoT is bridging physical, digital, virtual, and human spheres through networks, connected processes, and data, and turning them into knowledge and action, so that everything is connected in a large, distributed network. New technological trends bring intelligence and cognition to IoT technologies, protocols, standards, architectures, data acquisition, processing approaches, and analysis, all with a societal, industrial, business, and/or human purpose in mind. The IoT technological

Figure 3.2 Next wave IoT technology developments.

trends are presented in the context of integration; hyperconnectivity; digital transformation; and actionable data, information, and knowledge.

The IERC SRIA addresses these IoT technologies and covers in a logical manner the vision, technological trends, applications, technological enablers, research agendas, timelines, and priorities, and finally summarises in two tables future technological developments and research needs.

The IoT technologies and applications will bring fundamental changes in individuals' and society's views of how technology and business work in the world. The IERC supports the expansion of common European IoT, Industrial IoT (IIoT) and data/knowledge ecosystems, strengthen the impact of research and innovation in developing, supporting and implementing the policies address global challenges, including climate change and the Sustainable Development Goals. The next wave of human-centred IoT/IIoT green technologies, applications and the development of competitive European ecosystems are aligned with the priorities such as the European Green Deal, an economy that works for people, and a stronger Europe fit for the Digital Age. This has an important impact on the research activities that need to be accelerated without compromising the thoroughness, rigorous testing and time required for commercialisation, business impact and economic growth.

Knowledge-driven technologies may bring future risks, that can potentially include erosion of individual privacy, misinformation, misuse of data, increase of cybersecurity threats, and increase the digital divide.

Developing research and innovation agendas that blends the priorities to foster technology excellence while encouraging the respect to rights and values of citizens become of paramount relevance. This balancing act involves costs and legitimate private, national, public interests and the rights and interests of producing and using data.

Research and development are tightly coupled. Thus, the IoT research topics should support the further development of IoT ecosystems, partnerships, stakeholders networking and implementation of secured and trusted IoT solutions based on reference implementations for smart devices into self-adaptive, robust, safe, interconnected smart network and service platforms, proof-of-concept, demonstrations driven by realistic use cases in different sectors. A hyperconnected society is converging with a consumer-industrial-business Internet that is based on hyperconnected IoT environments. The latter requires new IoT systems architectures that are integrated with network architecture (a knowledge-centric network for IoT), system design and horizontal interoperable platforms that manage more intelligent things that are digital, automated and connected, functioning in real-time, having remote access and being controlled based on Internet-enabled tools.

The next-generation IoT/IIoT developments, including human-centred approaches, are interlinked with the evolution of enabling technologies (AI, connectivity, security, etc.) that require strengthening trustworthiness with electronic identities, service and data/knowledge portability across applications and IoT platforms. This ensures an evolution towards distributed IoT architectures with better efficiency, scalability, end-to-end security, privacy, and resilience. The virtualisation of functions and rule-based policies will allow for free, fair flow of data and sharing of data and knowledge, while protecting the confidentiality, integrity, and privacy of data. Vertical industry stakeholders will become more and more integrated into the connectivity-network value chain. Moreover, unified, heterogeneous, and distributed applications, combining information and operation technologies (IT and OT), will expose the network to more diverse and specific demands.

Intelligent/cognitive connectivity networks provide multiple functionalities, including physical connectivity that supports the transfer of information and adaptive features that adapt to user needs (context and content). These networks can efficiently exploit network-generated data and functionality in real-time and can be dynamically instantiated close to where data are generated and needed. The dynamically instantiated functions are based on intelligent algorithms that enable the network to adapt and evolve to meet changing requirements and scenarios and to provide context- and content-suitable services to users. The intelligence embedded in the network allows the functions of IoT platforms to be embedded within the network infrastructure and data, and the knowledge generated by the intelligent connectivity network and by the users/things can be used by the network itself. This knowledge can be taken advantage of in applications outside of the network.

The connectivity networks for next-generation IoT/IIoT are transforming into intelligent platform infrastructures that will provide multiple functionalities and will be ubiquitous, pervasive, and more integrated, further embedding telephone/cellular, Internet/data and knowledge networks.

Advanced technologies are required for the Next Generation Internet (NGI) to provide the energy-efficient, intelligent, scalable, high-capacity and high-connectivity performance required for the intelligent and dynamically adaptable infrastructure to provide digital services – experiences that can be developed and deployed by humans and things. In this context, the connectivity networks provide energy efficiency and high performance as well as the edge-network intelligence infrastructure using AI, ML, NNs and other techniques for decentralised and automated network management, data analytics and shared contexts and knowledge.

New solutions are needed for designing products to support multiple IoT standards or ecosystems and further research is required to investigate new standards and related APIs.

Summarising, although huge efforts have been made within the IERC community for the design and development of IoT technologies, the continuously changing IoT landscape and the introduction of new requirements and technologies create new challenges or raise the need to revisit existing well-acknowledged solutions. Thus, below is a list of the main open research challenges for the future of IoT:

- New IoT architectures that consider the requirements of distributed intelligence at the edge, cognition, AI, context awareness, tactile applications, heterogeneous devices, end-to-end security, privacy, proven trustworthiness, and reliability.
- Development of digital twin concepts, technologies and standards that enable cross-domain collaboration and deployment.
- Augmented reality and virtual reality IoT applications.
- Autonomics in IoT towards the Internet of Autonomous Things.
- Inclusion of robotics in the IoT towards the Internet of Robotic Things.
- AI and ML mechanisms for automating IoT processes.

- Distributed IoT systems using securely interconnected and synchronised mobile edge IoT clouds.
- IoT systems architectures integrated with network architecture forming a knowledge-centric network for IoT.
- Intelligence and context awareness at the IoT edge, using advanced distributed predictive analytics.
- IoT applications that anticipate human and machine behaviours for social support including contextual behaviour and understanding.
- Stronger distributed and end-to-end holistic security solutions for IoT, preventing the exploitation of IoT devices for launching cyber-attacks, i.e., remotely controlling IoT devices for launching Distributed Denial of Service (DDoS) attacks.
- Pre-normative and standardisation activities to address IoT end-to-end security.
- Stronger privacy solutions, considering the requirements of the new General Data Protection Regulation (GDPR) [184] for protecting the users' personal data from unauthorised access, employing protective measures (such as Privacy Enhancing Technologies – PETs) as closer to the data generation source to perform anonymisation and thus guarantee the personal data protection.
- Cross-layer optimisation of networking, analytics, security, communication, and intelligence.
- IoT-specific heterogeneous networking technologies that consider the diverse requirements of IoT applications, mobile IoT devices, delay tolerant networks, energy consumption, bidirectional communication interfaces that dynamically change characteristics to adapt to application needs, dynamic spectrum access for wireless devices, and multi-radio IoT devices.
- Adaptation of software-defined radio and software-defined networking technologies in the IoT.
- Trusted software validation approaches supporting advanced firmware analysis and validation approaches before and during the firmware update process involving industrial IoT devices.
- Application-specific mechanisms installed on IoT devices to support quicker algorithmic implementations (encryption, etc.) and hardware solutions supporting distributed trust (e.g. Physical Unclonable Functions, etc.) to cope with limited processing power on IoT devices.
- Tactile IoT applications and supportive technologies.

3.3.1 Digitisation

Digitisation is entering many fields, and the influence of digital approaches and techniques is becoming more apparent as time passes. Buildings and cities are becoming smarter, vehicles are becoming self-driving, design processes are becoming highly efficient and objects and spaces can be visualised before being materialised. Devices with embedded sensors featuring complex logic are scattered everywhere to measure light, noise, sound, humidity, and temperature, and they are empowered to communicate with each other, forming an IoT ecosystem.

Digital transformation is pushing all market sectors to level up their digital capabilities to better serve customers and meet their expectations, requiring to quickly transform insights into the best user experience. This scenario is only feasible through the inherent core value of the information provided by IoT/IIoT applications and services.

The IoT technologies and applications are enablers and accelerators of digital transformation with IoT devices deployed in consumer sector and many several industrial domains. Of the enterprise segment in 2030, 34% of devices will be accounted for by "cross-vertical" use cases such as generic track-and-trace, office equipment and fleet vehicles, 31% by utilities, most prominently smart meters, 5% by transport and logistics, 4% by government, 4% for agriculture, and 3% each for financial services and retail/wholesale. The consumer Internet and media devices will account for one third of all devices in 2030. Smart Grid, including smart meters, will represent 14% of connections and connected vehicles, will represent 7% of the global installed base [4].

In the very near future, every company will either buy or sell information as this resource continues to increase its relevance in the value chain, creating strategic advantages in business models and empowering technology strategies for companies.

Digital technology is rapidly transforming the socio-economic fabric and the codes of ethics that rule algorithms-writing, access and exploitation of information have a critical impact on how the societies of the future evolve.

A common element in all these developments is that digitisation creates a great amount of information. A considerable part of this information reveals how objects work internally and as elements of more complex setups. Accordingly, many innovative technological installations offer creative solutions concerning how to collect and process this information and how to take necessary action.

The challenge with this information is related to how things interact with each other and with the environment while exhibiting behaviour that is often like human behaviour (also related to contextual processing). This behaviour cannot be accurately handled by robots, drones, etc., so this is where technologies, such as swarm logic and AI, come into play.

Security-perceived threats almost always trigger interactive installations equipped to sense and react to surrounding parameters. Changes in these parameters can be visualised, increasing the chances of real threats being detected and asserted.

Due to advanced visualisation techniques, the threat landscape is better defined and digested by the involved actors (being users, LEAs, or other involved agencies). While security used to be primarily about securing information, the landscape has widened considerably. The timely transfer of information, threat identification, isolation and correct and traceable actions all rely on security protection.

IoT ecosystems evolve together with security strategies, which have to account for the layered architecture, where all things, encryptions, communications and actions must be protected against a growing number of diverse attacks, whether via hardware, software or physical tampering.

The IoT system is a group of agents with non-coordinated individual actions that can collectively use local information to derive new knowledge as a basis for some global actions. The intelligence lies both in agents (AI) and in their interactions (collective intelligence). At the core of swarm logic is the sharing of information and interactions with each other and the surroundings to derive new information. However, this collective intelligence is prone to several attacks, especially related to malicious nodes sending false information to influence the decision-making system. Thus, reputation and trust management systems should be in place to be able to identify malicious or misbehaving system agents/nodes and remove them from the system until they behave normally again. These types of attacks can be easily identified and corrected at the edge of the network without having to move all the information to the cloud. Swarm agents can locate and isolate the threat and then converge towards a common point of processing. This is visualised by depicting the real-time state of the agent's movement.

Swarm-designed security is inspired by nature; hence, if IoT can uncover behaviour patterns (of birds, ants, etc.), it may also be capable of meeting security challenges with well-functioning solutions.

3.3.2 IoT/IIoT Platforms Evolution

An IoT platform is defined in [55] as an intelligent layer that connects the things to the network and abstract applications from the things to enable the development of services. The IoT platforms achieve a number of main objectives such as flexibility (being able to deploy things in different contexts), usability (being able to make the user experience easy) and productivity (enabling service creation to improve efficiency, but also enabling new service development). An IoT platform facilitates communication, data flow, device management, and the functionality of applications. The goal is to build IoT applications within an IoT platform framework. The IoT platform allows applications to connect machines, devices, applications, and people to data and control centres [55]. The functionality of IoT platforms covers the digital value chain of an end-to-end IoT system, from sensors/actuators, hardware to connectivity, edge, cloud, and applications. The platforms enable communication between IoT devices, manage the data flows, support application development and provide basic analytics for connected devices [56].

There are 620 IoT Platform companies on the world market, up from 450 IoT Platforms companies in 2017 according to [56]. The report analysis is based on the evaluation of 1,414 IoT projects used for tracking IoT platforms which highlights the importance and pervasiveness of IoT platforms in bringing IoT solutions to market.

The IoT platforms are classified based on their development origin in device-, connectivity-, edge-, cloud-, industrial-, or open-source centric IoT platforms.

The IoT device-centric platforms implement the functions for provisioning tasks to ensure connected devices are deployed, configured, and kept up to date with regular firmware/software updates. The IoT connectivity-centric platforms implement capabilities and solutions for connecting the IoT device, managing and orchestrating connectivity, and provisioning communication services for connected IoT devices.

The edge-based IoT platforms provide the computational and analytics AI-based capabilities close to the edge devices where data is generated. IoT edge platforms deliver the management capabilities required to provide data from IoT devices to applications while ensuring the availability of management services for the devices over their lifetimes. Several IoT edge platforms are used for supporting developers to create, test, and deploy IoT applications or services. The edge IoT platforms need to use hardware-agnostic scalable architectures to support the deployment of sets of functionalities across various types of IoT hardware without modifications.

Cloud-based IoT platforms are offered by cloud providers to support developers to build IoT solutions on their clouds. Infrastructure as a Service (IaaS) providers and Platform as a Service (PaaS) providers have solutions for IoT developers covering different application areas.

The advanced analytics AI-based platforms were initially developed separately to offer tools including ML and other AI techniques and streaming analytics capabilities to extract actionable insights from IoT data. The new IoT platforms include more and more the functions provided by the analytics platforms. However, the advanced analytics AI-based platforms are advancing towards providing AI training/learning tools for different AI-based algorithms.

The IIoT platforms have been developed based on the industrial manufacturing requirement to monitor IoT edge devices and event streams, support and/or translate a variety of manufacturer and industry proprietary protocols, analyse data at the edge and in the cloud [57]. The IIoT platforms integrate and combine IT and OT systems in data sharing and consumption, enhance and supplement OT functions for improved asset management life cycle strategies and processes and enable the application development and deployment.

The IIoT platforms embed AI-based components at different architectural layers and support the security, safety-, and mission-critical requirements associated with industrial assets and their operating environments.

Recent innovations in the field of IIoT, AI, and edge computing have accelerated the developments of IIoT platforms that provide a collection of functions for edge device management, IoT data AI-based analytics, modern sensor technologies and connectivity solutions that enhance industrial equipment and industrial operations with remote monitoring, predictive maintenance, and extensive information analytics. The IIoT platforms support the creation of intelligent, self-optimising industrial equipment and production facilities through the technology functions that are provided. These functions are device management that includes SW/HW that enables manual and automated tasks (e.g. to create, provision, configure, troubleshoot, and manage securely fleets of IoT devices and gateways remotely, in bulk or individually); integration that provides SW/HW, IIoT devices (e.g. communications modules, Supervisory Control and Data Acquisition (SCADA) systems, Programmable Logic Controllers (PLCs), etc.) tools and technologies (e.g. communications protocols, APIs and application adapters, etc.) addressing the processing, enterprise applications and IIoT ecosystem integration requirements at the edge, across cloud and on-premises implementations for end-to-end IIoT solutions. The IIoT platforms provide functions for data management that include capabilities to support the ingesting of IoT edge device data, storing data from the edge to enterprise platforms, data accessibility (e.g. for IIoT devices, IT and OT systems, external systems, etc.), tracking the flow of data and implementing enforcing mechanisms for data and analytics governance policies to ensure the quality, security, privacy and value of data.

The evolution of IIoT devices at the edge enable a new phase in the digital transformation providing more intelligent functions, seamlessly controlling and optimising the environment and systems, gathering data at every stage, together with being integrated with IoT/IIoT digital platforms, process automation and artificial/augmented intelligence. Gartner estimates that by 2025, 75% of data generated and processed by enterprises will exist at the edge rather than in the traditional centralised data centre or cloud with the capabilities of edge computing solutions ranging from event filtering to complex-event processing or batch processing that creates a need for enterprises to deploy computing power and storage capabilities at the network edge, or edge computing [3].

The new generation of IIoT platforms include AI-based analytics functions that integrate processing of data streams, (e.g. device, enterprise and contextual data), to deliver insights into asset state by monitoring use, providing indicators, tracking patterns and optimising asset use. Various AI techniques, rule engines, event stream processing, data visualisation and ML are applied and implemented to enhances the analytics capabilities.

The IIoT platforms have to integrate both the IT and OT security measures, and features and the IIoT platform security function includes the software, tools and practices facilitated to audit and ensure compliance, for establishing and executing preventive, detective and corrective controls and actions to assure privacy and the security of information across the IIoT solution on both IT and OT domains.

The evolution of edge computing for IIoT applications needs to address the secure data storage, efficient data retrieval and dynamic data collection. A data processing framework for IIoT by integrating the functions of data preprocessing, storage and retrieval based on both the fog computing and cloud computing was presented in [31]. The data processing system of IIoT consists of five entities (e.g. IIoT, edge server, proxy server, cloud server and data users). The IIoT continuously collects data from physical environments and then sends the data to the edge server. The time-sensitive data are extracted and processed by the edge server, and then the data is passed in the cloud to

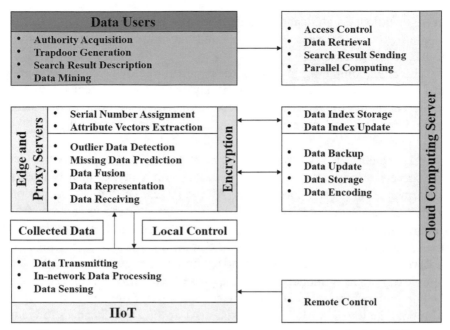

Figure 3.3   Framework of data collection, storage, retrieval, and mining [31].

be organised dynamically. A framework of data collection, storage, retrieval, and mining is presented in Figure 3.3 [31].

The IIoT devices are collecting data that is aggregated and fused at the data level and delivered to the edge server. After receiving the information, the edge server transforms the data into a unified representation framework and fuse the information at the feature level for convenient storage. The false data and missing data are also processed. To improve search efficiency, and support multiple search patterns, corresponding index structures need to be constructed. The data users and IIoT can communicate with the cloud server to execute specific instructions. Due to the amount of IIoT data, the data users likely employ the cloud server to mine the data. The cloud is used for its parallel computing capabilities.

In industrial platforms, IoT devices continuously monitor event-triggered information, which is further transmitted to a remote server, so apprehending monitoring of the industrial outcome.

Combining edge and cloud computing with IIoT data analytics can optimise the network traffic and latencies for ML tasks by employing the edge nodes and the evaluation of the degree of data reduction that can be achieved on edge without a significant impact on the ML task accuracy. Edge nodes act as intermediaries between IoT devices and the cloud, reducing the quantity of data sent to the cloud.

The edge computing in IIoT is focusing on deploying edge computing into different IIoT scenarios to reduce network traffic and decision-making delay. Therefore, the reference architecture of edge computing in IIoT needs to be improved and refined from the existing edge computing reference architectures.

Edge intelligence is applied on the edge of IIoT to enable edge devices and servers to perform more complex tasks with a higher data processing performance and lower latency. An AI model can be trained to perform predictions and make decisions with high-accuracy, using large amounts of training and verification data. For edge IIoT devices, training and leverage the AI model are hard due to the limited computing and storage resources. To resolve the conflict between the limited resources of edge devices and the high complexity of the AI model two basic approaches are used through enhancing the computing power of IIoT edge devices and simplifying or partitioning the AI model deployed on IIoT edge devices. AI at the edge enhances the range and computational speed of IoT-based devices in industries [34].

One of the advantages of IIoT is the massive amount of real-time data from multiple devices, sites, and infrastructures. Mining data values and making multi-dimensional business decisions improve significantly industrial production efficiency [32].

The traditional IIoT systems are represented in many cases by vertical, closed applications, focusing on maintaining the proper functioning of a machines/equipment or site, and the IIoT systems used constantly creates data islands. Adding edge computing to IIoT could support aggregation the data at lower processing levels and enhance its flexibility.

The complexity of the data security sharing problem is increased in IIoT applications. Opening data islands and sharing the same real-time data securely with any type and number of special applications and stakeholders is required in most of the edge-computing-based IIoT systems. There are two challenges that have to be addressed in edge data sharing: the inevitable increase of data interfaces that can lead to critical consequences (e.g. intrusion and destruction) and the limited performance of IIoT edge devices (e.g. strong/robust security algorithms are difficult to run on resource-constrained IIoT edge devices). The introduction of blockchain in edge computing in IIoT brings new challenges and opportunities for the secure sharing of data [33].

IIoT provides the network infrastructure for connecting IoT devices so that the monitoring and control of industrial manufacturing systems can be supported. From a cyber-physical system perspective, it is composed of both the physical subsystem and the cyber subsystem, which interact with each other so that the manufacturing process can be monitored and controlled with the aid of advanced information communication techniques.

By interacting with computing and networked objects in the physical subsystem, IoT devices (sensors, actuators, etc.) collect data, utilise the network subsystem to transmit the data to the operation centre, in which the data will be further analysed to assist system decision making and receive data to conduct actuation and modification of physical assets.

In IIoT, as numerous applications are time-sensitive, network performance is the key factor that affects the performance of IIoT applications. Nonetheless, to support automation and intelligence for IIoT applications, a large amount of data will be collected and analysed. While more data can provide better intelligence to IIoT applications, transmitting massive data through the network could lead to network congestion and further affect the monitoring and control performance of IIoT applications.

The integration of the IIoT platforms with other platforms in the industrial domains require addressing the application, collaboration and processes layers in the IoT architecture through the implementation of application enablement and management platform function that incorporates software/algorithms that enables business applications in any deployment model to analyse data and accomplish IoT-related business functions, interoperability and the flow of information across end-to-end systems. Application software components manage the operating system, standard input and output or file systems to enable other software components of the platform. The IIoT platforms with the integration of this function introduce application-enabling infrastructure components, application development, runtime management and digital twins that allows users to achieve edge and cloud federation while providing scalability and reliability to deploy and deliver IoT solutions seamlessly and in real-time. The future developments of IIoT platforms will accclerate the integration of AI-based components at all IoT architectural layers through a set of key capabilities, (e.g. data ingestion, processing and transmission) that optimise edge analytics by placing data and secure computing infrastructure closer to the factory floor that leads to improved industrial product quality, operational performance, prediction of downtime and automated operational flows.

Figure 3.4   Internet of Things Senses (IoTS) overview.

3.3.3 Internet of Things Senses (IoTS)

Internet of Things Senses (IoTS) is part of the IoT concept involving new sensing technologies to reproduce over the Internet the senses of sight, hearing, taste, smell, and touch, enabled by AI, VR/AR, intelligent connectivity, and automation. The IoTS developments are key for the IoT considering that the cognitive decision-making capabilities of the devices can be implemented by AI algorithms implemented into the intelligent IoT devices (e.g. robotic things) at the edge.

The IoTS complements and extends the capabilities of many IoT applications by including other senses (e.g. vision, hearing, touch, taste, smell, pain, mechanoreception-balance, temperature, etc.) and providing new perceptions and experiences by integrating augmented intelligence and information across senses, time, and space. A list of possible human senses is presented in [46].

The IoTS is bringing the technological advances for developing new sensing solutions as part of the intelligent things to address the implementation over the Internet of the senses of touch, sight, smell, taste, hearing, space.

Touch is discussed as part of tactile IoT and consists of several distinct sensations (e.g. pressure, temperature, light touch, vibration, pain, etc.). For humans, the sensations are part of the touch sense and are attributed to different receptors in the skin. For robotic and other types of things, the sensations are generated by sensors that mimic the reaction to pressure, temperature, touch, vibration, etc. Touch sense transmits different other messages in the case of human interaction, which can be transferred as well to things.

Sight is the capability of perceiving things through the vision system that can be represented by different types of cameras, AR glasses that support the navigation, searching for routes, identify places, recognising objects, persons, and sceneries. The sight will improve the capabilities of robotic things and enhance their functional abilities.

The smell is the ability to detect the different odours/scents/aromas (up to 1 trillion scents), that can be represented by different sensors that are sensitive to various odours. The remote smell integrated as online experience for humans and things can improve the capabilities of the robotic things to smell scents in different remote environments, provide new services and improve the perception in these environments.

Taste is the capability to sense different tastes like salty, sweet, sour, bitter, and savoury. The different tastes can be detected by various sensors that can provide a palette of tastes with a determined ranking. Information fusion from the different types of taste sensors is significant to experience a flavour. In many cases, other factors are needed to build the perception of taste in the cognition system of things.

Hearing is the sense to recognise the sound and decode the sound waves and vibrations. The detection of sound using different types of microphones, vibration sensors, etc. allows for enhancing the capabilities of the robotic things by developing techniques for voice control, automatically translate languages, voice biometrics, etc.

The sense of space is based on the information fusion from multiple sensors types and cognition process to comprehend where the things are in space. This is part of the proprioception includes the sense of movement and position of the movable parts of the items. This sense is critical in the future for autonomous intelligent robotic things operating in fleets in different cnvironments. Additional senses are used to sense movement to control balance and tilt the body of an object, sense the direction, acceleration, to attain and maintain equilibrium. Bio-and chemical sensors can be used to detect chemical substances and biologic materials (e.g. viruses, bacteria, etc.).

An Ericsson report [81] found that by the next decade digital sound and vision, complemented by touch, taste, smell and more, will transform the current screen based experiences into multi-sensory ones that are practically inseparable from physical reality. The report explores what that could mean for consumers, with AR glasses as the entrance point and presents what the consumers envisage as future developments driven by IoT sensory connectivity through AI, VR, AR, 5G and automation.

3.3.4 The Evolution of Tactile IoT/IIoT

The Tactile IoT/IIoT is a shift in the collaborative paradigm, adding sensing/actuating capabilities transported over the network to communications modalities so that people and machines no longer need to be physically close to the systems they operate or interact with as they can be controlled remotely.

Tactile IoT/IIoT combines ultra-low latency with extremely high availability, reliability and security and enables humans and machines to interact with their environment, in real-time, using haptic interaction with visual feedback, while on the move and within a certain spatial communication range.

Faster internet connections and increased bandwidth vastly expand the information garnered from onsite sensors within the industrial IoT network. This requires new software and hardware for managing storing, analysing, and accessing the extra data quickly and seamlessly through a Tactile IoT/IIoT applications. Hyperconnectivity is needed to take VR and AR to the next level for uniform video streaming and remote control/tactile internet (low latency).

The TIoT/TIIoT edge encompasses the sensors, actuators, computing, and communication resources deployed at the remote site where the tactile operation is controlled.

The TIoT/TIIoT can be classified based on the nature of the controlled environment (physical, digital, virtual) and based on the type of operator-teleoperator combination (e.g. machines, robotic things, humans).

The TIoT/TIIoT teleoperation in a physical environment (e.g. remote disaster management, cooperative autonomous systems, telesurgery) can be based on the nature of operator and teleoperator a 1) human operator-machine teleoperator, 2) machine operator-machine teleoperator (e.g. typical TIoT/TIIoT with minimal human intervention) or 3) human operator-human teleoperator (e.g. a human operator performs a physical, non-life-critical operation through a human teleoperator for the case of maintenance with the use of haptic-audio-video feedback from the controlled domain).

The TIoT/TIIoT teleoperation in a virtual/augmented reality involves controlling remote things in virtual environments using real-time interactions or simulations by employing digital twins. The teleoperators for these types of applications are robotic things and their digital/virtual representations. Robotic things and machines interacting remotely in virtual environments are used for testing and evaluation of different operational scenarios in the real world and optimise the design parameters for future physical devices. The case of human operator-machine teleoperator addresses the interactions between human operators with a virtual object via force feedback (e.g. immersive, multi-player, networked VR gaming, telemedicine, physiotherapy, etc.).

In the future, coworking with robots in IoT applications will favour geographical clusters of local production ("inshoring") and will require human

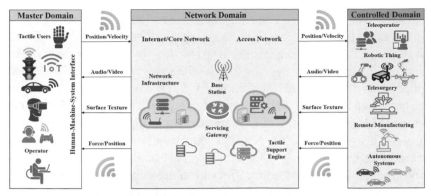

Figure 3.5 Tactile Internet of Things.

expertise in the coordination of the human-robot symbiosis with the purpose of inventing new jobs humans can hardly imagine or do not even know they want to be done. Fibre-wireless (Fi-Wi) enabled human-to-robot communications may be a stepping stone to merging mobile IoT/IIoT, and advanced robotics with the automation of knowledge work and cloud technologies, which together represent the five technologies with the highest estimated potential economic impact in 2025 [168, 169].

The tactile IoT/IIoT (TIoT/TIIoT) enables the real-time remote control and physical (haptic) experiences, and TIoT/TIIoT capabilities support the creation of a spatial safety zone that can interact with other nearby objects connected to robotic things that are part of different IoT applications.

The current network infrastructures are not able to support the emerging TIoT/TIIoT applications in terms of reliability, latency, sensors/actuators, access networks, system architecture and mobile edge-clouds. The design requirements of TIoT/TIIoT systems/devices to achieve real-time interactions are still dependent on the monitoring of the underlying system/environment based on human (or human-like) senses limited by the perception processes. TIoT/TIIoT applications need to adapt the feedback of the system to human reaction time.

TIoT/TIIoT applications have ultra-low end-to-end latency and ultra-high reliability design requirements and need to ensured data security, availability and dependability of systems without violating the latency requirements and considering the encryption delays and the end-to-end processing loop. The centralised architectures cannot meet these requirements, and more decentralised and distributed network architectures based on mobile-edge computing and cloudlets need to be developed to bring the TIoT/TIIoT applications closer to the end-users [170]. The wireless access networks used for these type of applications need to provide novel resource allocation techniques, feedback mechanisms, interference management and medium access control techniques in order to meet the stringent reliability and latency requirements of TIoT/TIIoT applications [157].

The enhancement in various aspects of physical and Medium Access Control (MAC) layers, emerging network technologies including Software Defined Networking (SDN), Network Function Virtualisation (NFV), network coding and edge/fog computing are part of the technologies that are supporting the TIoT/TIIoT applications in future connectivity networks.

The Tactile Internet standardisation is addressed in the IEEE P1918.X "Tactile Internet" [165] with the scope of defining a framework, incorporating the descriptions of its definitions and terminology, including the necessary functions and technical assumptions, as well as the application scenarios [162]. Within this framework, IEEE P1918.X defines the architecture technology and assumptions in Tactile Internet systems with IEEE P1918.X.1 dedicated for Codecs for the Tactile Internet, IEEE P1918.X.2 focussed on AI for Tactile Internet and IEEE P1918.X.3 addressing the MAC for Tactile Internet. The Industrial Internet Consortium (IIC) is working on developing a standard for low-latency TIoT/TIIoT for different smart cyber-physical systems including smart transportation systems, smart manufacturing, and smart healthcare systems [166, 167].

In the next generation of Internet of Robotic Things (IoRT) applications, it is expected that more network intelligence will reside closer to the robotic things. Several functions of IoRT applications can be implemented using TIoT/TIIoT ultra-low end-to-end latency and ultra-high reliability design to ensured data security, availability and dependability of IoRT systems across the end-to-end processing loop. This will lead to the rise of edge cloud/fog and Mobile Edge Computing (MEC) distributed architectures, as most data will be too noisy, latency-sensitive, or expensive to be transferred to the cloud. IoRT next intelligent generation requires to address the issues of unstable and intermittent data transmission via wireless and mobile links, efficient distribution and management of data storage and computing, interfacing between edge computing and cloud computing to provide scalable services and, finally, mechanisms to secure IoRT applications. To ensure the development of intelligent robotic things, it is necessary to reduce the amount of data processed and sent to the cloud. This requires to use a set of data functions for quality filtering and aggregation, and to fusion more functions into intelligent devices and gateways closer to the edge.

Moving from data centres and cloud computing into distributed edge computing infrastructures based on high-performance HW/SW platforms is opening the way for achieving the required latency constraint for TIoT/TIIoT implementation. The processing at the edge reduces the round-trip latency of transferred information, provides an efficient method for offloading information delivered to the core network, provides high bandwidth, offers new services and applications by accessing the network context information.

The master domain in TIoT/TIIoT includes the human and a human–system interface, together with machine and machine-system interface that transforms the human and machine operation into the control information by sensing technologies. The controlled domain consists of the teleoperators (e.g. remote things, such as robotic things) in remote environments that can be controlled by the master domain.

The network domain is an intermediary between the human/machine in the master domain and the remote environment. The sense of touch is generated by imposing an operation on the ambience and feeling the environment by a change or reaction force through the bidirectional haptic communication.

In TIoT/TIIoT, the human/machine delivers the command information to the remote thing. Then the remote thing interacts with the remote environment and feeds back the haptic information to the human/machine in the master domain. The haptic feedback information can contain two types of information (e.g. kinesthetic and tactile feedback information).

The kinesthetic feedback information is employed by the "things" in the master domain to estimate the force, location, speed, and torque in the target remote environment. The tactile feedback information is utilised to determine the roughness parameters of surface texture and friction information in remote environments.

The feedback information supports closing the global control over human/machine, communication network and remote environments allowing the master and controlled domains to exchange energy and tasks with each other based on the "real-time" interaction of commands and feedbacks.

The connectivity requirements of the TIoT/TIIoT/IoRT are matched by the capabilities of cellular networks. For example, 4G and 5G networks with dedicated radio base stations can be used to ensure that traffic remains local. In this case, on-site cellular network deployment with local data breakout ensures that critical production data do not leave the premises using quality of service (QoS) mechanisms to fulfil use case requirements and optimise reliability and latency. Critical applications can be executed locally and independently of the macro network through cellular network deployment with

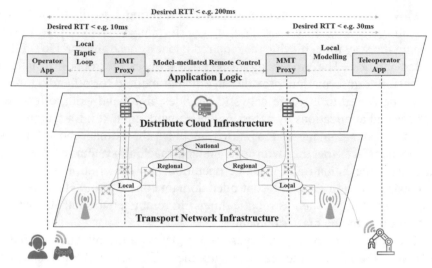

Figure 3.6 Long-distance haptic feedback loops through a national 5G network including MMT proxies deployed close to the operator and teleoperator [47].

edge computing. Edge computing, multi-access edge computing, processes data created around the network and the edge robotic things devices enable entry into core networks and computation is largely or completely performed on distributed device nodes, rather than primarily taking place in a centralised cloud environment.

An evaluation of the 5G ultrareliable and low-latency communications (URLLC) radio configurations in a reference urban macro network deployment, with 500 m distance in-between base stations sites for tactile internet devices located outdoors, was performed in [47]. The full outdoor coverage that was expected to be provided for latency was in the range 1–2 ms and 4–6 ms for the investigated NR-based and LTE-based URLLC configurations, respectively. The end-to-end latency for the tactile internet services depends on the service requirement, the latency components introduced by different domains (e.g. master, network, including multidomain network orchestration, controlled domain, etc.) along the end-to-end information path. Tuning and adapting the parameters to determine the optimal end-to-end latency can be implemented in an automated manner, to enable flexible, scalable, and cost-efficient service deployments.

The solution proposed in [47] uses a model-based approach for the service specification and configuration. It includes haptic control proxies instantiated within the 5G communication system tactile internet services

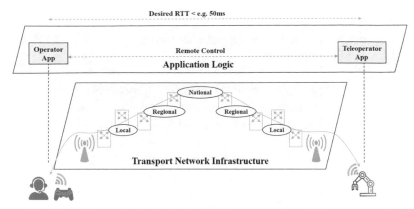

Figure 3.7 Long-distance haptic feedback loop through a national 5G network for the TDPA scheme, without MMT proxy [47].

when a specific end-to-end latency is exceeded. The haptics communication flow for a long-distance remote operation on a national level is illustrated in Figure 3.6. Long-distance haptic feedback loops through a national 5G network including Model-Mediated Teleoperation (MMT) proxies deployed close to the operator and teleoperator [47]. for teleoperating a drone-mounted robotic arm.

For this case, the MMT-proxy software is running on distributed cloud edge computing infrastructure located in vicinity to the remote "thing" (e.g. drone-mounted robotic arm) and operator. The corresponding haptics loop flow through the network for the time-domain passivity approach (TDPA) method (without a proxy) is shown in Figure 3.7. The authors in [47] used MMT as preferred method over the TDPA method only in case of a round trip time (RTT) that is larger than some threshold (e.g., 50 ms), implying that the selection of the method to use (and whether to use an MMT proxy or not) should ideally depend on the achievable RTT [47].

There are several challenges to be addressed for realising haptic communications over the wireless TI and TIoT/TIIoT as listed below [48–50].

Haptic Edge Sensors/Actuators – These edge devices are essential for collecting/capturing and transferring the information into correct actions. The haptic sensors sense the tactile information by interacting with the environment and are mounted both at the master domain (e.g. machines) and at the teleoperators end (e.g. remote things, such as robotic things) in remote environments. The sensed information is transmitted to the master domain in the form of the haptic feedback by the haptic actuators (also called haptic feedback devices). In the case of TIoT/TIIoT, the haptic sensors are mainly pressure sensors to detect the underlying pressure (e.g. capacitive, resistive pressure sensors). For IoTS, other types of sensors are used for mimicking the human and beyond senses.

The quality of the sensors, precision, range, sensitivity, response time, spatial resolution, cost, placement, temperature dependence and complexity are important design parameters for different TIoT/TIIoT use-cases in different environments. The information received from the sensors is converted in the different actions (that provide a feeling of touch similar to the feeling the receiver would get in the real-world context), by using haptic actuators (e.g. cutaneous – muscle type for force tension, kinaesthetic – skin type for vibration, pressure, pain, temperature, etc.). The implementation of lightweight energy-efficient, fast response time, low-cost, actuators providing capabilities of both the cutaneous and kinaesthetic feedback is one of the main challenges for TIoT/TIIoT deployments.

Surface Sensing and Actuation – The TIoT/TIIoT use-cases can require more than a single-point contact for the tactile and kinesthetic feedback as the master and controlled domains address applications requiring touch-based sensations across the surfaces (e.g. palm of the hand or other areas for humans or robotic things). Surface sensing implies the use of multiple arrays of sensors and techniques to identify the forces across the surface. Reading the sensing surface-based or distributed sensing and actuation requires technologies that increase the energy use and latency and impact the communication requirements due to the increased data rates and a different perception for the case of an information loss.

Multi-Modal Sensory and Information Fusion: Different types of sensing/actuation edge devices are used for collecting/providing the tactile information and other perception data (e.g. sound, visual, etc.) to increase the perception capabilities and performances. The use of multi-modal sensory and the fusion of the data from these sensors increases the various requirements for TIoT/TIIoT in terms of latency, fusion, transmission rate and sampling rate. Multiplexing and fusion the information from different types of sensing devices is another challenge for the implementation of TIoT/TIIoT real-time multiplexing schemes across different protocol layers for integrating the various modalities in dynamically varying wireless environments.

Collaborative Multi-User Haptic Communications - TIoT/TIIoT use cases integrate a multitude of edge humans/things that interact and collaborate in a shared remote environment, requiring the creation of peer-to-peer overlay to enable collaboration among multiple users. This overlay formation step brings new challenges in terms of meeting the requirements of TIoT/TIIoT applications as the overlay routing, and IP-level routing may further increase the end-to-end latency [49].

Efficient Resource Allocation – The different tasks in the master and controlled domains need to get specific communication channel parameters in the network domain based on the TIoT/TIIoT use case requirements. The efficient radio resource allocation in wireless/cellular networks brings new implementation challenges (e.g. the resources need to be shared among haptic, human-to-human, machine-to-machine, machine-to-human communications that have various and sometimes conflicting service requirements) due to the incorporation of haptic communications. Besides, as the haptic communications are bidirectional, symmetric resource allocation with the guarantee of·minimum constant rate in both the uplink and the downlink need to be ensured. These requirements bring new issues for managing and orchestrating the wireless/cellular networks parameters to provide priority for resources based on QoS, safety-, mission-critical features. Flexible resource allocation techniques across different protocols layers, including adaptive management and network slicing with on-demand functionality, are needed for future deployments.

Haptic Codecs – The transmission of the hepatic information in digital form requires sampling the signals at rates of 1 kHz [48], information that is then compressed and transmitted across the wireless/cellular networks. The future challenges are developing standardised groups of haptic codecs that can be integrated into the kinesthetic and tactile information, that are energy efficient and able to perform effectively in time-varying wireless environments.

Ultra-High Reliability – The TIoT/TIIoT applications require ultra-high reliability for wireless/cellular networks. Several factors, such as the lack of resources, uncontrollable interference, reduced signal strength and equipment failure, impact on network reliability. To provide the reliability requirements for the wireless/cellular networks the different layers of the protocol stack including the MAC layer, transport layer and session layer need to be reconsidered to enable ultra-high reliability in haptic communications. Trade-offs between reliability, latency, packet header to the payload ratio must be considered [48–50].

Ultra-Low-Latency – The future TIoT/TIIoT applications will require end-to-end latency below 1 ms across different protocols-layers, air interface, backhaul, hardware and core Internet. Optimising the transmission parameters at each layer and across the layers is required to achieve the overall end-to-end latency (e.g. upper bound fixed by the speed of light). At the physical layer providing shorter Transmission Time Interval (TTI) can lower the over-the-air latency, at the expense of higher needed bandwidth.

Stability for Haptic Control – The TIoT/TIIoT global control loop across the master, network and controlled domains need to be stable to avoid the degradation of the "continuum experience" to the remote environment. This is a challenge as the global loop integrates the humans/machines, the communication network, the remote environment, and the energy/tasks exchange among these components takes place via various commands and feedback signals [54]. The wireless environments can have time-varying delays and packet losses, and new techniques must be developed to reduce the instability due to communication channels parameters variations.

Performance Metrics – The TIoT/TIIoT requires evaluation methods beyond Technology Readiness Level (TRL) that include new QoS metrics, Quality-of-Experience (QoE) and Quality-of-Task (QoT), which ultimately will lead to Experience Readiness Level (ERL) classification. This will allow identifying new suitable performance metrics for analysing and comparing the performance (e.g. information fusion, connectivity features, data processing techniques, data reduction/control, compression, etc.) of various haptic systems over the TIoT/TIIoT [47]. The introduction of experience evaluation allows to analyse the difference of the physical interaction across a network and the same manipulation carried out locally and measure the accuracy by which a tactile user can perform a task [47].

The research challenges across different domains of TIoT/TIIoT are summarised in Table 3.2.

A generalised framework for TI beyond the 5G era that can be applied to TIoT/TIIoT is presented in Figure 3.8 [51].

The basic architecture of TIoT/TIIoT is composed of a master domain, a network domain and a controlled domain and this is reflected in the generalised framework [51] comprising various aspects of TI including key technical requirements, main application domains, a basic architecture and enabling technologies. The key technical requirements of TI include ultra-responsive connectivity, ultra-reliable connectivity, intelligence at the edge network, efficient transmission and low-complexity processing of tactile data as presented in the previous subsections.

Table 3.2 Research challenges across different domains of TIoT/TIIoT

TIoT/TIIoT Domain	Research Challenges
Connectivity	• Ultra-low latency ($<$ 1ms) • Ultra-high reliability ($>$ 99,999%) • Very-high data rates (Gbps-Tbps) • Very-high backhaul bandwidth • Support for communication overhead with cloud/edge/fog networking infrastructures
Computation	• Online processing of haptic feedback for near real-time interactions • In field processing to reduce ingress transmission • Support high-computation AI processing and training/learning at the tactile edge
Artificial Intelligence	• Predict reliable end-to-end communication • Movement and action prediction to compensate physical limitations of remote latency • Inferring movements and actions techniques • Acquisition and replication of skills for optimising the TIoT/TIIoT global loop
Haptics	• Codecs to acquire remote interactions • Methods to vary modalities for interactions • Machine-to-machine and machine-to-human coordination mechanisms and interfaces • Techniques for increasing the stability of haptic control • Surface-based sensing and actuation • Sensing fusion methods (e.g. multiplexing of multi-modal sensory information) • Performance metrics and ERL methods
AR/VR	• Localisation and tracking precision and efficiency • Scalability and heterogeneity • Quality-data rate-latency trade-offs • In-network vs. in-VR (edge) computation • Information processing theoretical advancements
Collaborative autonomous mobility systems	• Dynamic rout selection for autonomous systems • Traffic management for IoRT • AI/ML/DL techniques for prediction in real-time movements/actions • Multi-autonomous things perception/control for safety (fail-operational), traffic/movements efficiency • Ultra-low latency and ultra-high reliability design techniques for connecting (V2X) multiple collaborative autonomous mobility systems in real-time.

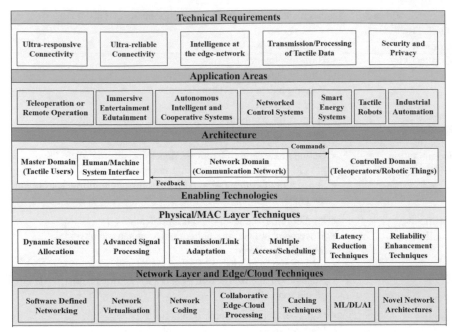

Figure 3.8 A generalised framework for TI. Adapted from [51].

3.3.5 IoT Digital Twins

Digital twins (DTs) are virtual representations of physical assets and things across their life cycle using real-time sensor data. The DTs can be utilised to expose a set of services allowing to execute certain operations and produce data describing the physical thing activity.

The concept was introduced by NASA as part of its spacecraft monitoring mission and is now a mainstream approach as technologies such as sensors, data analytics and edge computing fuel a new generation of IoT/IIoT devices, industrial assets, buildings and intelligent city infrastructure. In IIoT applications, DTs can be used in digital factories that consist of multi-layered integration of the information related to various activities along with the factory and associated resources.

A digital twin is comprised of a virtual object representation of a real-world item in which the virtual is mapped to physical things in the real-world such as equipment, robots, or virtually any connected business asset. This mapping in the digital world is facilitated by IoT platforms and software that is leveraged to create a digital representation of the physical asset. The digital twin of a physical asset can provide data about its status, such as its physical state and disposition. Conversely, a digital object may be used to manipulate and control a real-world asset by way of teleoperation of DT modelling solution.

Digital twins use AI, ML and software analytics with data to render real-time digital simulation models that can update and change as their real, physical counterparts, or "twins" change. These IIoT systems can be used to optimise the operation and maintenance of physical assets, systems, and processes in real-time. Operational intelligence can be used towards building digital twins. The insights produced by real-time intelligence enable operators to understand the performance of distributed infrastructure, make predictions, improve efficiency, and even prevent disasters. Operational intelligence is one catalyst for the DT concept as it supports to digitise infrastructure, monitor operations in real-time, predict events, take actions based on intelligence, and engage with different stakeholders.

There are different definitions for Digital Twins across industry and academia. As described by the Digital Twin Consortium [138], a digital twin is an abstraction of something in the real world. It may be physical (a device, product, system, or other assets) or conceptual (a service, process, or notion). A digital twin captures the behaviour and attributes of its physical sibling with data and life cycle state changes potentially moving in either, or both, directions.

As per CIRP Encyclopedia of Production Engineering [139], a digital twin is a digital representation of a unique active (real device, object, machine, service, or intangible asset) or unique product-service system (a system consisting of a product and a related service) that comprises its selected characteristics, properties, conditions, and behaviours by means of models, information, and data within a single or even across multiple life cycle phases.

The advances in IoT/IIoT technologies accelerate the development of tools and technologies that can support the design of digital models for the IoT digital devices twins integrating 3D modelling tools, computer-aided engineering (CAE) software with IoT/IIoT platforms. The diversity and heterogeneity of IoT/IIoT devices is reflected as well in the data formats used for representing, designing the digital twin. The next-generation IIoT platforms are designed to include capabilities for building digital twins and simulate the interactions between them as well as simulate different what-if scenarios that will support preparedness, impact, and mitigation management.

The capabilities of these platforms include key elements such as data integration, the accuracy of the physics simulation software and the capacity to update physics models in digital twins with real-time data streaming from IoT devices placed in the field.

Thanks to technologies, such as blockchain, swarm logic and AI, digital twins now have these capabilities. In the pursuit of better security, digital twins can trigger and simulate threat scenarios in the digital world, as well as optimise the security strategy to handle such scenarios should they occur in the real world.

In the context of IoT, digital twins are the representation of physical IoT devices that offer information on the state of the physical twin, respond to changes, improve operations, and add value. The digital twin, as a virtual representation of the IoT's physical object or system across its lifecycle, using real-time data to enable understanding, learning, and reasoning is a one-element connecting the IoT and AI.

The digital twin represents the virtual replica of the IoT physical device by acting like the real thing, which helps in detecting possible issues, testing new settings, simulating all kinds of scenarios, analysing different operational and behavioural scenarios and simulating various situations in a virtual or digital environment, while knowing that what is performed with that digital twin could also happen when it is done by the "real" physical "thing".

Digital twins as part of IoT technologies and applications are being expanded to more applications, use cases and industries, as well as combined with more technologies, such as speech capabilities, AR for an immersive experience and AI capabilities, enabling to look inside the digital twin by removing the need to go and check the "real" thing. The digital twins' evolution is shown in Figure 3.9

Digital twins for IoT must possess a minimum of attributes:

- Correctness – give a correct replication of the IoT ecosystem and its devices.
- Completeness – updated vis a vis the functionality in the real-world system.
- Soundness – exhibit only the functionality available in the real-world system.
- Abstractness – free from details specific to implementations.
- Expandability – adapt easily to emerging technologies and applications.
- Scalability – must be able to operate at any scale.
- Parameterisation – accessible for analysis, design, and implementation.

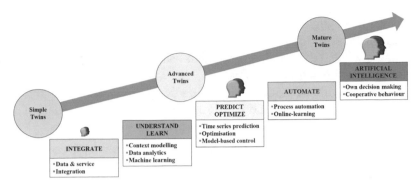

Figure 3.9 Digital Twins evolution. Adapted from IBM.

- Reproducibility – be able to replicate the same result for the same input as the real system.

For IoT, digital twins can expand the interface between man and machine through their virtual representation and advanced technologies on levels, such as AI and speech, which enable people and devices/machines to take actions based on operational data at the edge (provided by IoT devices and edge computing processing).

To fully exploit the potentials of IoT's digital twins several gaps need to be addressed to foster the market uptake and adopt the technology in several application scenarios.

There is a need to put more effort into the validation and quantification of the benefits perceived using DTs against exiting processes and systems. Both the scale and the type of benefits and improvements need to be formalised, and the selection of the type of digital twins should be justifiable. There is an emerging need for an evaluation framework that can categorise the various levels of sophistication of digital twins. These levels should help the industry to use a common language when describing a digital twin and its capabilities. An evaluation framework that is based on 5 distinct levels has been proposed for the built environment [140]:

- Level 1: A digital model linked to the real-world system but lacking intelligence, learning or autonomy.
- Level 2: A digital model with some capacity for feedback and control, often limited to the modelling of small-scale systems.
- Level 3: A digital model able to provide predictive maintenance, analytics, and insights.

- Level 4: A digital model with the capacity to learn efficiently from various sources of data, including the surrounding environment. The model will have the ability to use that learning for autonomous decision making within a given domain.
- Level 5: A digital model with a wider range of capacities and responsibilities; ability to autonomously reason and to act on behalf of users (AI); interconnected incorporation of lower-level twins.

Another aspect that will further boost DT innovation is the consideration of solutions that span across the entire product life cycle. Current frameworks and implementations focus on specific use cases and specific life cycle phases. It is critical to understand the requirements of the DTs at each phase of the life cycle, as well as the required number and interdependencies of digital twins required for example in the production phase in contrast to the operational and maintenance phase.

Other important aspects to be addressed that may accelerate future DTs solutions are (i) standardisation and interoperability aspects such that virtual entities can communicate and interoperate, (ii) the data ownership of Digital Twin data (iii) aspects related to data privacy and personal data protection.

Several standardisation activities have emerged in recent years. The International Organisation for Standardisation (ISO) covers industrial data in TC 184 SC 4 [141]. The standard for a Digital Twin Manufacturing Network is currently under development (ISO/DIS 23247) [142]. ISO/IEC JTC1 provided a technology trend report by its joint advisory group on Emerging Technology and Innovation (JETI). In the report, "Digital Twin" was identified as the number one area needing in-depth analysis [143]. Within the ISO/TC 184 a working group for Digital Twins has been created ISO/TC 184/AHG 2 [144]. The IEEE Standards Association initiated a project IEEE P2806 that aims to define the system architecture of digital representation for physical objects in factory environments [145]. In ITU-T SG 13 study group [146], requirements and capabilities of a digital twin system for smart cities is under study. The recently established Digital Twin Consortium is a program of Object Management Group dedicated to the widespread adoption of digital twin technology and the value it delivers [138]. It aims at contributing to standards by deriving requirements that will be submitted to international standards' development organisations, such as Object Management Group and ISO/IEC.

BuildingSMART International (bSI) is another initiative, committed for creating and developing open digital ways of working for built asset environment and their standards help asset owners and the entire supply chain work more efficiently and collaboratively through the entire project and asset lifecycle. They recently published a position paper "Enabling an Ecosystem of Digital Twins" [147], where three areas identified to focus further developments. They are closely related to the topic of standardisation: (i) standards for data models, (ii) standards for data management and integration and (iii) data security and privacy.

Projects funded under the H2020 work programme such as IoTwins project [148], which aims to build a reference architecture for developing and deploying distributed and edge-enabled digital twins of production plants and processes will act as facilitators for Digital Twin implementations. More initiatives in this direction are required.

The idea of the implementation of National Digital Twin infrastructures on local and/or national level has also gained attention. A good example is "Virtual Singapore" [149] a research and development programme initiated by the National Research Foundation of Singapore, which is a dynamic three-dimensional (3D) city model and collaborative data platform, including the 3D maps of Singapore. When completed, Virtual Singapore will be the authoritative 3D digital platform intended for use by the public, private, people and research sectors. It will enable users from different sectors to develop sophisticated tools and applications for test-bedding concepts and services, planning and decision-making, and research on technologies to solve emerging and complex challenges for Singapore.

Another example is the National Digital Twin Programme of the Centre for Digital Built Britain [150]. The programme seeks to deliver a smart digital economy for infrastructure and construction, and to transform the UK construction industry's approach to the way we plan, build, maintain and use our social and economic infrastructure for the future. The programme's objectives are to deliver the Information Management Framework, to enable the National Digital Twin and to align industry, academia, and Government on this agenda.

A DT life cycle and supporting tools and functionalities to exploit the approach are presented in Figure 3.10 [35]. The model can be applied to DTs used in IoT applications together with the functions associated. The DTs for IoT will be represented in the design phase by a logical object that is the first software representation of the physical object or thing. In the production phase, the digital twin is released, and the software representation is used to test and experiment with the future physical product/thing. The software aspects of the DT support optimising the physical item and in carrying out

Figure 3.10   Life cycle of a DT – Functions and tools. Adapted from [35].

the validation, testing and optimisation. In the operation phase, the DT runs in the digital domain as a one to one representation of the real thing. In some cases, several physical things can have a DT representation that represents the capabilities of the class of instantiated physical things. In many IoT applications, the production and operation phases are used to manufacturing physical items or products and later use the DT to mirror the properties of the real thing to measure, simulate and optimise the behaviour and performance of products.

Mirroring the characteristics of the real thing into the DT requires that there be a continuous link between these entities that allows for updates/upgrades. Depending on the type of application, the link/connection can be real-time, permanent/intermittent, resilient, etc. The flow of information is predominantly from the physical thing to the DT, with specific situations when the DT sends data and information to the physical thing. In all IoT/IIoT applications, the DT must be continuously synchronised with the production system and the IoT device that it represents. The synchronisation supports the use of the DT to simulate and predict the behaviour of the real thing in a new scenario and use case. Using the bidirectional exchange of information based on real data measurement and events flowing from real to virtual can improve the accuracy of the simulations and predictions.

The DT can be used in the TIoT/TIIoT to support the optimisation of the global loop (master, network, slave domains) by identifying through simulations the parameters of the physical thing to perform in specific environments and conditions.

A representation of an architectural model and general framework for DT is illustrated in Figure 3.11 [35]. The description is following a layering concept that can be mapped to the 3D IoT reference architecture. The first layers are mapped to the physical and network layers in the 3D IoT reference architecture. The upper layers integrate the properties of the DT, into the processing, storage, abstraction, and service IoT architecture layers where different components implement the functions for modelling of objects, instantiation, self-management, orchestration, entanglement, and other. Collecting and contextualising the data as well as to execute data analysis and information inferring for the DT must be integrated into the IoT data processing and abstraction.

Semantics and ontologies need to be aligned with the ones defined for IoT devices and integrated into the service layer. The DT simulation functions must be implemented as part of the IoT abstraction and service layer components. The use of open APIs that can be programmable at different levels and with different abstraction capabilities support the interaction of IoT applications with different IoT platform functions using structured data.

Applying the digital twins to different sensors, is exemplified by the work on IEEE 1451 smart sensor digital twin federation for cyber-physical systems (CPS) [36]. The IEEE 1516 high-level architecture (HLA) is a standard for the modelling and simulation of distributed, heterogeneous processes. The digital twin developed is a digital simulator or digital replica of a real IEEE 1451 smart sensor. The DT emulates both desired, non-linear behaviours and failure modes to simulate an actual sensor in the field [36].

Several features of the DTs in IIoT applications are the following [37]: connectivity (e.g. the ability to communicate with other entities and DTs), autonomy (e.g. the possibility for the DT to live independently from other entities), homogeneity (e.g. the capability to allow the use of the same DT regardless of the specific production environment), customisation flexibility (e.g. ability to modify the behaviour of a real thing by using the functionalities exposed by its DT), and traceability, (e.g. capability to follow the traces of DT's activity of the corresponding physical item). One of the challenges with DTs in IIoT applications is data interoperability as data moves from physical devices and equipment in the field (e.g. manufacturing floor), to the software used in data IoT twin modelling systems. Different IIoT applications can use several different digital twins, at different levels (e.g. physical component, asset, system, process, etc.). The hierarchy of IoT digital twins can produce different perspectives and generate different types of data and complex relationships between data sets that can result in data interoperability issues.

An architectural model, which incorporates digital twins into edge networks for real-time data analysis and network resource optimisation for smart manufacturing, is presented in Figure 3.12 [38]. The framework contains three layers the user layer, edge layer, and digital twin layer. The user layer consists of client devices in IIoT such as smart machines, vehicles, IoT devices and can be mapped to the physical layer in the 3D IoT reference architecture.

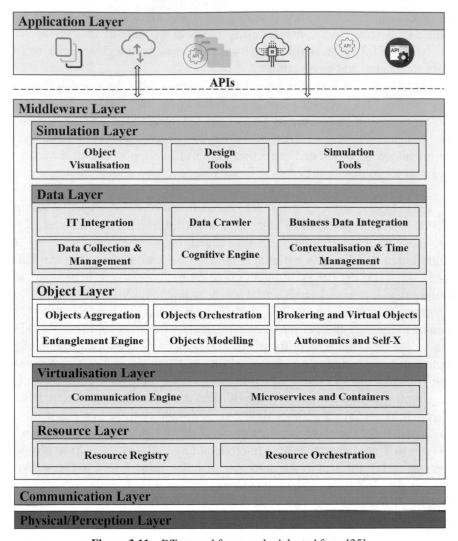

Figure 3.11   DT general framework. Adapted from [35].

The edge layer is composed of base stations that are equipped with MEC servers. The edge layer can be mapped to the network and processing layer in the 3D IoT reference architecture. The base stations are connected with user devices under their coverage via wireless communications. The digital twin layer can be mapped to the abstraction and service layers in the 3D IoT reference architecture.

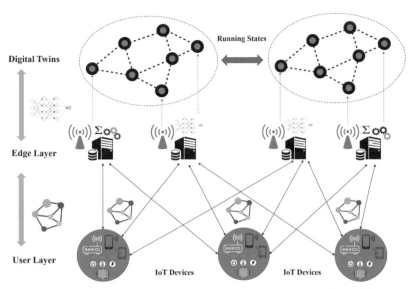

Figure 3.12 Digital Twin Edge Networks for IIoT. Adapted from [38].

To model digital twins, the authors used federated learning to build digital twins from the historical running data of devices. The raw data transmission is avoided, and data privacy is enhanced by federated learning. An optimisation problem was formulated that aimed at reducing the communication cost of federated learning and provided the solution by decomposing it and using DNN for communication resource allocation. Numerical results on the benchmark real-world dataset corroborated that our proposed mechanism could improve the communication efficiency and reduce the overall energy cost [38].

It is evident that standardisation efforts, research, and development funding programmes, as well as reference implementations on local, national, and international level with public-private funding schemes, will be the main drivers towards acceleration and market up-take of DT technology.

3.4 Internet of Things Augmentation

The advances in IoT technologies enabled by heterogeneous integration of functions to sense, collect, act, process, infer, transmit, create notifications of/for, manage and store information allows the interactions between autonomous systems and humans in a continuum of physical, digital, virtual and cyber environments.

IoT augmentation is a blend of methods, techniques and technologies that are applied to improve the sensing, action, or cognitive abilities of IoT devices. The concept is based on the integration of the capabilities offered by augmented reality (AR), virtual reality (VR), digital twins, and AI, to generate and visualise virtual 3D models of the real world that are evolving into smart and interactive environments related to the context of things for physical objects. The concept expands to humans by providing interactive digital extension of human capabilities (e.g. replication, supplementation) by using IoT sensing and actuation technologies, AI, fusion and fission of information, AR, VR, and digital twins to improve human productivity and capabilities.

IoT devices act as bridge between physical assets, digital and cyber infrastructure while the AR/VR brings the digital content and the digital twins by interacting with the physical objects in real-time in the virtual environment.

These new developments allow the technicians accessing real-time IoT data on the shop floor and augmented reality IoT devices aware of the spatial configuration of the environment of the operator can sense what the operator is looking/focusing at to intuitively display only the data needed for the operation to be performed.

IoT augmentation expands its potential in healthcare for virtual monitoring and simulations of the living environment integrating the information from various IoT devices connected in real-time for measuring vital health parameters and information combined with environment and context monitoring.

In retail and the smart, connected supply chain, IoT augmentation bridges the online and offline environments providing the users with the tools to model, simulate and visualize the products and how those products can be customised to fit the user's needs or interact with other products.

From vehicles to mobile devices, the manufacturing processes require putting together hundreds of various components in a precise and predetermined order. Using a 3D design superimposed onto the actual process provides more straightforward access and stepwise guidance. Implementing an IoT-based connected supply chain assures that machines parts are always stocked and ready for use.

IoT augmentation extends to augmented senses, augmented actuation and augmented cognition that applies both to humans and machines/things and enhance the human-machines interactions and collaboration.

As the IoT augmentation develops, the need for standardisation is evident as the new IoT applications encompass various systems and technologies that deliver real-time information, make inferences, provide analytics and take decisions that have to be interoperable with other systems with which they are interacting and collaborating.

The standardisation needs to address the fail operational functions for IoT augmentation that are important features that are required to be integrated into the system to provide the functional safety and eliminate the sources of error and ensures that the right parts, processes and sequences are always followed.

The next wave of IoT/IIoT applications will include a form or another of IoT augmentation that need to scale, dynamically adapt to the context and seamlessly adopted for multiple functions and in diverse environments, and operation conditions to enhance the human and machines capabilities.

3.5 Edge Computing

The model of centralised cloud computing, including data analysis, and storage for IoT/IIoT is limiting the capabilities of IoT applications, creating silos, challenges regarding interoperability and data spaces. To further develop their capabilities and provide new opportunities for enterprise data management, IoT/IIoT applications (e.g. real-time) and services are moving the developments toward the edge and, as a consequence, most of IoT data generated and processed by enterprises will exist at the edge rather than in the traditional centralised data centre in the cloud. Edge computing provides significant benefits such as reduced latency, data analysis close to the data source, lower bandwidth, and reduced energy consumption in networks.

IoT/IIoT technologies including IoT devices, advanced autonomous IoT systems, IoT intelligent servers providing gateways capabilities, end-point-enabling networks form the foundation of edge computing, accelerating the move of IoT applications to edge in industries where IoT is deployed, such as energy, utilities and manufacturing. Edge computing allows the easy deployment of private and local area networks with all the capabilities associated to the IoT/IIoT systems.

IoT systems deliver disconnected or distributed capabilities into the embedded IoT world and edge computing is part of the technological fabric across the industrial sector that deliver these capabilities empowered with advanced and specialised processing resources, data storage, and analytics. The IoT applications powered by edge computing allow to keep the traffic and processing local, to reduce latency, exploit the capabilities of the edge and enable higher autonomy of the IoT devices at the edge.

Edge computing is redefining the IoT/IIoT, embedded, and mobile processor landscape, accelerating the development of high-performance circuits using AI/ML techniques and embedded security for addressing the edge and deep edge data analysis processing. Edge computing provides mechanisms for distributing data and computing on the edge, which makes IoT applications much more resilient to malicious and no malicious events. Distributed deployment models are expected to address more efficiently connectivity and latency challenges, bandwidth constraints and higher processing power and storage embedded at the edge of the network. In addition, they preserve privacy as raw data is processed locally and only aggregated data is shared in the cloud. Using efficiently the edge computing layer of the IoT architecture move most of the data traffic and processing closest to the end-user applications and devices that generate and consume data. The use of IoT edge, near edge and deep edge capabilities and diverse edge systems, centralised cloud services will enhance the functionalities of cloud technology to provide, manage and update software and services on edge and near edge devices. Centralised cloud services could become hubs in coordinating and federating operation across highly distributed edge devices, and in aggregating and archiving data from the edge or intermediate gateways and servers. The centralised cloud services for intelligent IoT applications will be used as robust and additional scalable ML and sophisticated processing capabilities linked to traditional back-office processing.

The edge computing through different forms (e.g. mobile edge, fog, dew computing, etc.) creates multi-dimensional architecture consisting of a wide range of heterogeneous "things" with different sensing/actuating, connectivity, processing and intelligence capabilities connected to services/applications in a dynamic mesh linked by platforms and distributed services located at the edge/cloud level.

From an architecture perspective, the computing landscape is represented by a tier model as represented in [25] with many alternative implementations of hardware and software at each tier, but all of them are subject to the same set of design constraints. In this model the Tier-1 represents the data centres and the cloud, Tier-2 is organised into small, dispersed data centres called cloudlets representing the essence of edge computing, Tier-3 represents the gateway-base processing devices, and Tier-4 represented by the edge physical devices. The future computing landscape is evolving, including new computing paradigms such as distributed AI in the future IoT and the emergence of new programmable quantum architectures, new compilers that target emerging quantum machines. Future decentralised and distributed architectures and computing paradigms for IoT/IIoT inspired by neuromorphic computing with hybrid implementations based on architectures for neural DL that could provide optimising computing in such systems leading computing paradigm.

Embedding sensors/actuators, storage, compute and advanced AI capabilities in IoT edge devices and integrate them within mesh architectures, will enable dynamic, intelligent, responsive, peer-to-peer IoT end-devices that exchange information with enterprise IoT platforms and conduct peer-to-peer exchanges with other IoT devices operating across different industrial sectors.

The integration of data analysis technologies in IoT devices is raising edge security challenges that have to be addressed to set and enforce security, privacy and compliance standards for the vastly increased number of devices on the edge, dealing with expanded network types and connections, and various software deployments supporting key features such as low latency, the ability to perform deterministic real-time computing, the support for mission-critical or safety-critical use cases, and the ability to extend computing beyond humans to the extremes of the environment and IoT devices.

IoT devices at the edge need to embed privacy-by-design mechanisms (e.g. encryption of static and dynamic/moving data, anonymisation and credential protection against physical intrusion), together with an increased authentication process governing device participation and data exchange in the cloud, in order to ensure the full authentication and encryption of all traffic between cloud resources and edge IoT devices.

Edge computing is filling the gap between the centralised cloud and the need for a decentralised processing medium for IoT devices and paves the way for IoT connectivity (e.g. 5G and beyond), and AI convergence. Mobile edge computing (MEC), and other edge computing paradigms such as Mobile Cloud Computing (MCC), fog computing, and cloudlets are complementary, possibly competing, options for filling the gap. MEC offers new opportunities for network operators, service, and content providers to deploy versatile and uninterrupted services on IoT applications. MEC supports IoT by providing IoT devices with significant additional computational capabilities through computation offloading.

Edge computing provides low latency connectivity by processing the information closer to its source and to the end user. Edge computing supports the implementation of 5G and AI to intelligently and efficiently manage the network edge by provisioning load balancing, supporting multiple levels of nodes for hierarchical networking, allowing for resource pooling, universal orchestration and management, multiple access modes providing each edge network node with the resource applications it requires, improving reliability, security, resiliency, supporting virtualisation, mobile IoT applications, providing agility with a horizontal platform and supporting all vertical markets and becoming more scalable by moving computation, networking, or storage capabilities across or through levels of hierarchy.

3.5.1 Edge Computing Architectures

As the IoT technologies have evolved and moved from centralised to decentralised and distributed computing, the architectures needed to perform data analysis at the edge, and the development of the edge cloud computational capabilities as applied to distributed node IoT edge have generated new concepts such as edge cloud, edge computing with specific capabilities (e.g. fog, mobile edge computing, dew computing, etc.) that offer complementary and additional features than the ones provided by cloud computing. The evolution was possible due to the technology progress in the network field (e.g., 1000 times bandwidth increase and significant cost reduction), in the computing processing power (e.g., increased computing power and reduced cost), and in the storage capacity (e.g. 10 000 times capacity increase of a single disk followed by cost reduction).

AI edge processing today is focused on moving the inference part of the AI workflow to the device, keeping data constrained to the device. There are several different reasons why AI processing is moving to the edge device, depending on the application. Semiconductors companies are providing dedicated microcontrollers and microprocessors with hardware dedicated to accelerating DL processing coupled with tools for easy mapping of algorithms in those architectures [41]. Privacy, security, cost, latency, and bandwidth all need to be considered when evaluating cloud versus edge processing [29]. Tractica's report [29] provides a quantitative and qualitative assessment of the market opportunity for AI edge processing across several consumer and enterprise device markets. The device categories include automotive, consumer and enterprise robots, drones, head-mounted displays, mobile phones, PCs/tablets, security cameras, and smart speakers. The report includes segmentation by processor type, power consumption, compute capacity, and training versus inference for each device category, with unit shipment and revenue forecasts for the period from 2017 to 2025. The report predicts that AI edge device shipments will increase from 161.4 million units in 2018 to 2.6 billion units worldwide annually by 2025.

Cloud computing is characterised by a high-processing and computer power, centralised architecture, high latency, centralised data analytics, and AI capabilities, implementing centralised cybersecurity mechanisms and providing very large storage capacity. Edge cloud computing offers a decentralised architecture, medium-low latency, dedicated bandwidth based on the computing needs, AI processing capabilities, cybersecurity mechanisms and networking effect through connectivity with the other IoT edge site nodes. Edge computing is further extending the decentralised architecture and moving towards distributed computing, offering low/ultra-low latency, efficient use of bandwidth, local networking, processing, and storage capabilities. These new architectures provide more efficient in energy and cost solutions.

Edge computing is used to process real-time data storage and computation on the device or data source, rather than sending it to a remote data centre, thus decreasing latency and reducing the bandwidth used by IoT devices. When necessary, the centralised cloud can serve as storage facility for large amounts of data and additional processing. The IoT devices where data analysis and control occur act as nodes. Specialised AI circuits used in IoT devices can process more data on the edge in real-time, which combined with the fact that the devices are closer to the data source, results in faster responses and actions while decreasing latency and saving on bandwidth and equipment costs. Edge computing is addressed in the standards setting communities at ISO/IEC JTC-1 SC-41 from a network centric perspective.

In summary, edge computing provides the functions of a distributed open platform at the network edge, close to the IoT devices and other data sources, integrating the capabilities of collecting information, processing information, networking and exchange of information, storage, analytics, services and applications. Edge computing and IoT technologies are enablers for digitisation of industry offering dynamic connectivity, real-time services, data optimisation, application intelligence, security, privacy protection and delivering IoT edge intelligence services and applications.

The term *edge cloud* is used to describe the decentralisation of the traditional, large cloud data centres with moving cloud storage and compute closer to the edge source while also scaling down the size. Edge locations may connect to each other or to a central cloud for added data inputs and processing or storage capabilities, or isolated in instances of a data breach or service compromise. Edge cloud requires additional remotely administered data centres that are called edge sites, close to the end users. The edge sites are placed at specific locations where increased compute and processing is needed beyond what can be completed at the edge in conjunction with low-latency, time-sensitive IoT operations.

3.5.2 Deep Edge Computing

In many new IoT applications, the data is collected and processed at deep edge (such as sensor nodes) and send via a communication channel to the gateways and processing units at the edge. Learning and inference will happen on the edge and cloud and the decisions are communicated back to the nodes at the edge. The system architecture including the deep edge depends on the envisioned functionality and deployment options considering that at the core of these devices are cognitive sensor modules that can acquire, understand, and react to data.

Deep edge computing architecture rely on the principles of distributed computing with computing units having limited resources and relying on high efficiency energy consumption and connectivity functions.

IoT devices are becoming more intelligent, equipped with various sensors using new computing paradigms for processing the information and providing the "intelligent deep edge", allowing the IoT applications to become ubiquitous and merge into the environment where various IoT devices can sense its environments and react intelligently to it. Providing AI capabilities to IoT devices significantly enhance their functionality and usefulness, especially when the full power of these networked devices is harnessed – a trend that is often called AI on the edge.

To support connected or collaborative operation, the intelligent sensor modules include at least one communication channel that supports the necessary bandwidths and latencies. Using the channel, the local controller can inform other edge devices or an edge-based service of its current context.

3.5.3 Fog Computing

Fog computing, a term created by Cisco, refers to extending computing capabilities to bring cloud computing capabilities to the edge of the network. Fog enables repeatable structure in the edge computing concept, so enterprises can push compute out of centralised systems or clouds for better and more scalable performance. A Fog computing implementation is a virtualised platform, located between cloud data centres (hosted within the Internet) and end user devices, providing strong support for IoT and complementary to cloud computing platforms.

Fog computing is defined in [1] as an horizontal, system-level architecture that distributes computing, storage, control and networking functions closer to the users along a cloud-to-thing continuum. This horizontal architecture provides support for multiple industry verticals and applications. domains, delivering intelligence and services to users and business. The cloud-to-thing continuum of services assure that services and applications are distributed closer to the IoT devices or things, and anywhere along the continuum between cloud and things. The system level concept is covering the cloud-to-thing continuum over the network edges and across multiple protocol layers without depending on specific communication and protocol layer.

An important component of the Fog computing layered architectures is the Fog service orchestration layer that provides dynamic policy-based life-cycle management of Fog services and the distributed orchestration functionality as the underlying Fog infrastructure and services.

Fog computing has the processing capabilities at the LAN end while data is gathered, processed, and stored within the network, using IoT gateways or fog computing nodes (FCN). Information is transmitted to gateways from various sources in the network and the information is processed in FCN, and the processed data and additional commands are transmitted to the other devices. Using these mechanisms, the fog computing implementation enables a single, processing device to process data received from multiple edge devices and send information where it is needed, with lower latency than centralised cloud computing processing. The Fog computing architecture is scalable and can be integrated with the different cloud computing architectures.

Fog computing has several key differences compared with edge computing as it works with the cloud and has a hierarchical structure. Fog addresses computation, networking, storage, control, and acceleration. Fog computing is an extension of the traditional cloud-based computing model where implementations of the architecture can reside in multiple layers of a network's topology. Fog architectures selectively move compute, storage, communication, control, and decision making closer to the network edge where data is being generated in order solve the limitations in current infrastructure to enable mission-critical, data-dense use cases [1].

Fog computing is described in [2] as a huge number of heterogeneous (wireless and sometimes autonomous) ubiquitous and decentralised devices communicate and potentially cooperate among them and with the network to perform storage and processing tasks without the intervention of third-parties. These tasks can be for supporting basic network functions or new services and applications that run in a sandboxed environment. Users leasing part of their devices to host these services get incentives for doing so.

Fog computing is based on a decentralised computing infrastructure, where computing resources and application services are distributed in the most logical, efficient place, at any point along the continuum from IoT data source to the cloud. The fog provides high data processing efficiency due to the reduction of amount of data to be transported to the cloud for data processing, analysis, and storage, increasing the security and optimising the data transfer.

In many IoT applications Fog can be used to offload the storage and computations from the IoT devices by using a network edge. The Fog-based IoT network offers a number of design improvements compared with the cloud infrastructure [44, 45] as listed below:

- Better offloading and reduced server strain considering that Fog computing allows the network to offload the data processing to its cloudlets, which reduces data traffic and server strain and allow that more users can be managed more efficiently with edge computing.
- Scalability through parallelism by using Fog computing architecture, decentralisation and adding edge devices to the network, while the cloud computing scales a cloud server by increasing its size in terms of data capacity.

For IoT applications, fog computing offers significant amount of storage at or near the IoT nodes data collection (avoiding primarily to store in large-scale data centres), efficient communication close to the IoT nodes (avoid routing through the backbone network) and optimised management, including network measurement, control and configuration, performed close to the IoT devices.

3.5.4 Mobile Edge Computing

Mobile edge computing relates to computing at the edge of a network and the edge can be seen as a distributed cloud with proximity close to the end user that delivers ultra-low latency, reliability, and scalability.

The aim of Mobile Edge Computing (MEC) is to provide an IT service environment and cloud-computing capabilities at the edge of the mobile network, within the Radio Access Network (RAN) and near mobile subscribers. MEC is a new development in the evolution of mobile base stations and the convergence of IT and telecommunications networking. The main expected goal of MEC is to reduce latency, to ensure highly efficient service delivery, with the result of providing an improved user experience. This goes together with improved and much more efficient and transparent network operation.

Based on MEC parameters such as characteristics, actors, access technologies, applications, objectives, computation platforms, and key enablers a MEC taxonomy is presented in Figure 3.13 [7]. The implementation of MEC in real applications is support by several key enablers, which contribute to provide context-aware, low latency, high bandwidth services to the mobile subscribers at the RAN close proximity as listed below [7]:

- Cloud and Virtualisation: Virtualisation allows to create variant logical infrastructure in the same physical hardware, with the computing platform at the edge of the network creating different virtual machines using virtualisation technology to provide different services of cloud computing (e.g. Software-as-a-Service (SaaS), Platform-as-a-Service (PaaS), and Infrastructure-as-a-Service (IaaS)).
- High Volume Servers: Mobile Edge Servers are deployed in each mobile base station of the edge network to perform network traffic forwarding and filtering and executing the offloaded task by the edge devices.
- Network Technologies: Multiple small cells are deployed in Mobile Edge Computing environment with Wi-Fi and cellular networking as main networking technologies used to connect the mobile devices with the edge sever.
- Mobile and IoT Devices: Mobile and IoT devices at the edge network compute low intensive task, and hardware related tasks which are non-off loadable to the edge network. Mobile devices perform peer-to-peer computing within edge network through Device-to-Device communication.
- Software Development Kit: Software Development Kit (SDK) with standard Application Programming Interface (API) supported in adapting existing services and foster on expediting the development of new elastic edge applications. These standard APIs can be easily integrated in application development process.

MEC can create opportunities for new use cases and complementary roles for mobile operators as well as for application and content providers by enlarging their business models to better monetise the mobile broadband experience.

MEC enables the creation of new and innovative services over the mobile network for consumers, enterprise customers as well as industries with

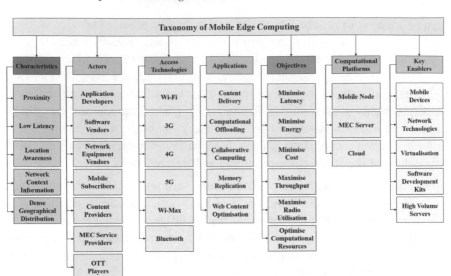

Figure 3.13   Taxonomy of Mobile Edge Computing. Adapted from [7].

mission-critical applications. To this extent, MEC must rely on a standardised, open environment to allow the efficient and seamless integration of applications across multi-vendor Mobile Edge Computing platforms. This was a major driver for the creation of a standardisation effort in support of MEC.

3.5.4.1 The Industry Specification Group (ISG) on Multi-access Edge Computing

Multi-access edge computing technology is currently being standardised in an ETSI Industry Specification Group (ISG) of the same name launched in 2015 [1, 5]. The ISG MEC has already published a set of specifications (including a "Framework and Reference Architecture" with a first release in 2016 [6] and an updated one in 2018 [8]) focusing on management and orchestration (MANO) of MEC applications, and a large range of APIs for application enablement, service deployment and the User Equipment (UE) application. Multi-access edge computing is shifting compute functions from a centralised location to the edge of the network closer to the end user that enables real-time analytics of video surveillance, vehicle-to-vehicle communications, traffic management and other public safety functions.

Multi-access edge computing as it is deployed currently in the 4th generation LTE networks, is connected to the user plane via one of the options

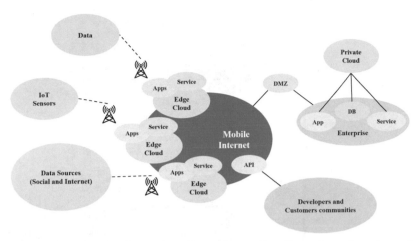

Figure 3.14 Mobile Edge Computing – Improving quality of experience through proximity with end users.

described in [9, 10]. Considering that LTE networks have been already deployed for several years, it was necessary to design the MEC solution as an add-on to a 4G network to offer services in the edge. Consequently, the multi-access edge computing system and its interface specifications, as defined in [8], is to a large extent self-contained, covering everything from management and orchestration down to interactions with the data plane.

As a result, multi-access edge computing servers can be deployed at multiple locations, such as at the LTE macro base station (eNodeB) site, at the 3G Radio Network Controller (RNC) site, at a multi-Radio Access Technology (RAT) cell aggregation site, and at an aggregation point (which may also be at the edge of the core network).

3.5.4.2 A new paradigm for the development of applications
Multi-access edge computing offers to application developers and content providers cloud-computing capabilities and an IT service environment at the edge of the network. Consequently, Multi-access edge computing introduces a standard for supporting an emerging cloud paradigm for software development communities.

Multi-access edge computing in 5G gives service providers and wireless carriers the opportunity to offer distributed sites, including cloud radio access networks, aggregation points, central offices and cell towers as real points for the end users to compute data and store them as part of a virtualised

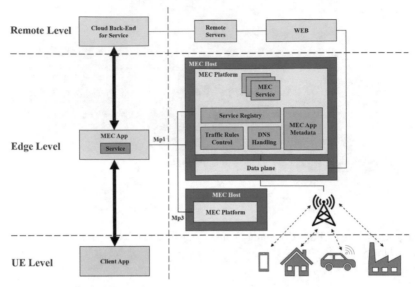

Figure 3.15   The MEC application deployment approach.

and automated cloud network. The role of the edge and multi-access edge computing used for 5G is to reduce network congestion, increase speed, efficiency, scalability and improve application performance by achieving related task processing closer to the user. Edge computing provides more throughput by storing and processing data closer to where it is generated or used.

3.5.4.3 The challenge of network transformation

The deployment of complex, large-scale, and heterogeneous 5G networks is posing a major challenge to the telecom industry. One of the most critical aspects is the issue of management with two main requirements regarding the possible solutions. Firstly, they need to be highly automated and not to require lots of complex (and possibly manual) configuration. Secondly, they need to collect large amounts of relevant data, process them and act on them in an automated fashion with the support of AI and ML.

To this extent, a lot of work has been done in emerging NFV-based architectures and solutions, and there is an opportunity for multi-access edge computing to benefit from the large developments (e.g., the Management and Orchestration (MANO) framework) and deployments done around NFV (the NFV operationalisation) and to extend NFV in order to accommodate the

MEC architecture. To that end, ETSI MEC has defined a "MEC-in-NFV" reference architecture in ETSI Group Specification MEC 003 [6].

3.5.5 Cloudlet

The *Cloudlet* was introduced in [11] as a framework to overcome the overhead of Virtual Machines (VMs) while benefiting of the other features of VMs (i.e. management and reliability issues). The Cloudlet is an edge computing technology that enables new classes of mobile applications that are both compute-intensive and latency-sensitive in an open ecosystem based on cloudlets.

A definition of the cloudlet is given in [14] as a trusted, resource-rich computer or cluster of computers that is well-connected to the Internet and is available for use by nearby mobile devices. For IoT this means the existence of a resource rich computing infrastructure with high-speed Internet connectivity to the cloud that can be used by the IoT mobile devices to augment its capabilities and to enable real-time applications.

Several optimisations in Cloudlet were proposed to reduce the amount of outer data transfer (i.e. the amount of data sent across the network) by processing data within the physical node and provide locality aware global execution.

The Open Edge Computing [23] addresses the Cloudlet based on Open-Stack [24]. Open Edge Computing considers that any edge node can offer computational and storage resources to any user in proximity using a standardised mechanism. Edge computing technologies are characterised by openness, as operators open the networks to third parties to deploy applications and services, while their differences enable edge computing technologies to support broader IoT applications with various requirements.

A cloudlet represents a small-scale cloud that provides services like cloud computing with limited resources closer to users. The users are directly connected to other devices in a cloudlet and the latency of the response to user requests is significantly reduced. The cloudlet can ensure better security and privacy. The cloudlet concept forms an initial prototype of fog computing and has four key attributes [22]:

- Maintains only soft state: It is built for microservices and containers and may buffer data originating from a mobile device to a cloud service. Each cloudlet adds close to zero management burden after installation and can be entirely self-managing.

- Powerful, well-connected, safe and secure: It has sufficient compute power (i.e., CPU, RAM, etc.) to offload resource-intensive computations from one or more mobile devices with proper connectivity to the cloud and is not limited by finite battery life. Its integrity as a computing platform is assumed and, in a production, quality implementation this will have to be enforced through some combination of tamper-resistance, surveillance, and run-time attestation.
- Located at the edge of the network: It is logically close to the mobile devices. "Logical proximity" is defined as low end-to-end latency and high bandwidth (e.g., one-hop Wi-Fi).
- Builds on standard cloud technology: It encapsulates offload code from mobile devices in VMs, and thus resembles classic cloud infrastructure such as OpenStack. Each cloudlet has functionality that is specific to its cloudlet role. The cloudlet term is driven primarily by the ISO / IEC JTC-1 SC38 standardisation subcommittee working group.

The cloudlet is enabling resource-intensive and interactive mobile applications such as IoT by providing computing resources to mobile devices with lower latency.

The clouds and cloudlets need a strong isolation between untrusted device-level computations, mechanisms for authentication, access control, and metering, dynamic resource allocation for device-level computations; and the ability to support a wide range of device-level computations, with minimal restrictions on their process structure, programming languages or operating systems. The clouds and cloudlets achieve device and user isolation through virtual machines.

In the case of cloudlets as the mobile and IoT device moves from one physical area to another, its current cloudlet has to hand off the device/user's virtual machine to the new cloudlet.

3.5.6 Dew Computing

Dew computing is defined in [26] as an on-premises computer SW/HW organisation paradigm in the Cloud computing environment where the on-premises computer provides functionality that is independent of cloud services and is also collaborative with cloud services. The Dew computing aims to maximise the potential of on-premises computing and cloud services and use the resources at the lowest level as self-organising systems solving the processing in these environments. Two features describe the nature of dew computing applications: independence, which indicates that the application is inherently distributed and collaboration that indicates that the application is inherently connected. Collaboration implies that the dew computing application automatically exchange information with cloud services during its operation through synchronisation, correlation, or other kinds of interoperation. Independence represents the ability of the on-premises computing units to provide the requested functionality without cloud services connection meaning that an dew application is not a completely-online application or cloud service [26].

Another definition [28] is considering the Dew computing as a programming model for enabling ubiquitous, pervasive, and convenient ready-to-go, plug-infacility empowered personal network that includes Single-Super-Hybrid-Peer P2P communication link with the aim to access a pool of raw data equipped with meta-data, which can be rapidly created, edited, stored, and deleted with minimal internetwork management effort (i.e. offline mode) on local premises. The Dew computing model is composed of six essential characteristics such as. Rule-based Data Collection, Synchronisation, Scalability, Re-origination, Transparency, and AnyTime Any How Accessibility; three service models such as Software-as-a-Dew Service, Software-as-a-Dew Product, Infrastructure-as-a-Dew; and two identity models (e.g. Open, Closed) [28]. IoT can be one major application for Dew where heterogeneous devices act together to perform a set of tasks and do not need to be connected to the Cloud services all the time creating the so call Dew of Things (i.e. DoT). The Dew computing technical challenges include power management, processor utility, data storage, viability of existing operating system, network model, communication protocols, programming principles, database security, and data efficiency exchange.

The Dew computing paradigm is different from the Fog/Cloud paradigms, as the edge and low-level devices cannot be used in a "conventional" programmable way, but they have to cooperate on the lowest level to solve processing needs, and be able to pass (and consume) information from all hierarchical levels.

Dew computing uses on-premises processing units to provide decentralised, cloud-friendly, and collaborative micro services to end-users. Dew Computing is complementary to Cloud computing, and the Dew computing processing units provide on-premises functionality independent of cloud services and exchange information and collaborate with cloud services.

The Dew computing architecture includes modules for addressing the hybrid and extremely heterogeneous information collectors, information distributors, information processors, information presenters and information consumers, at the edge level directly connected to processing units and IoT devices which are part of the common physical environment, and at the highest level interconnected into the global information processing and distribution system. In the Dew computing paradigm, the individual IoT devices are collecting/generating the raw data and are the components of the computing ecosystem that are aware of the context the data were generated in, therefore dew devices must produce and exchange information.

Dew computing is context-aware, giving the meaning to data being processed with data that is context-free, while information is data with accompanying meta-data and the meta-data places the data in a specific context.

A Dew computing architecture for cyber-physical systems and IoT is proposed in [27]. The dew computing implementation in cyber-physical systems allows autonomous devices and smart systems, that can collaborate and exchange information with the environment, and be independent of external systems or act as processing element in connected complex cyber-physical system of systems.

3.6 Artificial Intelligence IoT

AI has developed over time several (sometimes nature-inspired or human-behaviour mimicking) computational methodologies that address complex real-world problems, in particular when mathematical modelling is not able to provide effective solutions. AI is one of the modalities by which a machine is able to perform logical analysis, acquire knowledge, and adapt to an environment that varies over time or in a given context by making use of abilities that allow a machine (i.e. computer, robot, or intelligent IoT device) to perform functions such as learning, decision making, or other intelligent human behaviours. Applications where IoT technologies converge with AI are growing in depth and functionalities, creating new markets and opportunities that, in turn, require clarifying the future technology requirements and the effective ways to integrate more AI techniques and methods in IoT applications.

The basic concepts behind AI have been around since the 1950's but thanks to modern programming techniques (such as python), the availability of huge quantities and qualities of data, open source tools for neural network training, powerful computing centres and ever improving embedded process-ing systems, AI is taking off as a world-changing technology today with a special role for ML in general and DL in particular.

ML is a subset of AI that refers to techniques which enable machines to recognise underlying patterns and learn to make predictions and recommen-dations by analysing data and experiences, rather than through traditional explicit programming instructions. The system adapts with new data and experiences to improve prediction performance over time. DL is a subset of ML. It is based on an ability to learn data patterns and dependencies by using a hierarchy of multiple layers that mimics the neurons connections of the human brain and make up any deep neural network. DL techniques can work with very large data sets by analysing data, recognizing patterns and making prediction on next data. With DL, a computer can train itself with a large set of data collected for this purpose. If the Training stage, in which the neural network learns to classify different patterns, use datasets labelled in advance the process is referred to as Supervised Learning. In the case of unlabelled datasets, the learning process is called Unsupervised Learning and the neural network tries to cluster the dataset into groups with similar patterns. In both cases the result is an Artificial Neural Network (ANN) that contains all the information necessary to carry out the task. The ANN uses the knowledge acquired in the training to infer data features from new incoming data. This is called Inference stage and can be deployed in embedded devices with memory and processing capabilities orders of magnitude smaller than the servers used to train the ANN itself.

AIoT is enabling and accelerating the developments of IoRT applica-tions to achieve more efficient IoRT operations, improve human-machine interactions and enhance data management and analytics.

The edge computing paradigm is a key element in the evolution of IoT platforms towards IoT decentralised and distributed architectures, intelli-gence at the edge and optimising the use of resources in IoT applications (i.e. communication, processing, energy consumption at IoT devices, sub-systems and systems level, etc.). In many applications, the processing is done at the edge, close to the real-time processes (such as in factories and industrial plants). Many of these processes are time-critical, business-critical, privacy-critical, and part of connectivity bandwidth-intensive, and bandwidth-void applications.

In edge computing, the data generated by different types of IoT devices can be processed at the network edge instead of being transmitted to the cen-tralised cloud infrastructure thus creating bandwidth and energy consumption concerns. Edge computing can provide services with faster response and greater quality, in comparison with cloud computing. The integration of edge computing with IoT provides efficient and secure services for many end-users, and the development of intelligent edge computing-based architectures can be considered a key element for the future Intelligent IoT infrastructure [42].

That real-time system properties are essential in mobility applications such as autonomous driving applications is obvious, but also new applications within manufacturing, predictive maintenance and robotics are real-time dependent to ensure increased quality, efficiency, reliability, and safety. Moving computing capabilities from the cloud to the edge by implementing AI methods, for instance in IoT devices that operate directly or close to the actual operation and independently of external input from the cloud or other computer centre, will in many cases be the preferred or only possible solution. Enabling technologies like IoT are developing fast, including increasing computing and wireless communication facilities for a range of applications, although interoperability issues are still needed to overcome.

Bringing AI in IoT devices will require, on the one hand, to define the adequate processing unit and, on the other hand, to ensure the easy porting of AI apps on the target platform in an effective way. Regarding the processing unit, either it is a microcontroller or a microprocessor, and in many IoT domains the inference engine usually does not require high computational power. Application domains such as activity monitoring, acoustic event detection, predictive maintenance deal with data (in term of sampling rate and size) that can be analysed by a software base inference engine.

More complex application domains, as for example those based on computer vision, deal with huge amount of data to be analysed in real-time. Inference engines for these domains require dedicated hardware accelerators to be integrated in the processing unit. In both cases tools should be designed and adopted to easy the porting of AI solutions in embedded devices, enabling system designers to concentrate on the building blocks where they have the core IP.

The advantages of bringing AI-enhanced decision-making at the edge (edge-based AI) include the following [39]:

- Edge-based AI is highly responsive and closer to real-time than the typical centralised IoT model deployed to date. Insights are immediately delivered and processed, most likely within the same hardware or devices.

- Edge-based AI ensures greater security. Sending data back and forth with Internet-connected devices exposes data to tampering even without anyone being aware. Processing at the edge minimises this risk, and preserve privacy, with an additional plus: edge-based AI-powered devices can include enhanced security features.
- Edge-based AI is highly flexible. Smart devices support the development of industry-specific or location-specific requirements, from building energy management to medical monitoring to predictive maintenance.
- Edge-based AI does not require highly qualified personnel to operate. Since the devices can be self-contained, AI-based edge devices do not require data scientists or AI experts to maintain. Required insights are either automatically delivered where they are needed, or visible on the spot through highly graphical interfaces or dashboards.
- Edge-based AI provides for superior customer experiences. By enabling responsiveness through location-aware services, or rerouting travel plans in the event of delays, AI helps companies build trust and rapport with their customers.
- Edge-based AI seems to currently provide the required processing power and speed to execute tasks relating to data security or may identify points of altered data or intrusions (data smoothness check, data abnormality checks etc.).
- Edge-based AI reduces data transfers back and forth and thus supports lower bandwidth investments and backhaul costs.
- Edge-based AI supports more efficient and real-time decision making and requires reduced latency as ML triggers real-time actions and decision making. Having data closer to the decision-making point is an added value.

The intelligent edge allows humans to tackle multi-faceted processes by replacing the manual process of sorting and identifying complex data, key insights, and actionable plans. This can help humans gain a competitive edge by having better decision-making, improved return of investment (ROI), operational efficiency, and cost savings. At the same time, there are several challenges facing ML based edge computing such as [40]:

- The cost of deploying and managing an edge.
- The evaluation, deployment, and operation of edge computing solutions.
- Security challenges for processing at the edge and the fear of data breach.
- The type and number of operations performed on the data.

• The size of the data to be processed during the learning phase when the inference engine is generated.

The concept of incorporating AI into edge computing is evolving and more research is needed to get intelligent edge-based solutions to be fully set up, functional and running smoothly in production.

As smart devices require access to data for optimised performances and analysis accuracy, data access must be granted to a technology or solution provider, a new scenarios must be regulated to prevent restriction of data access to only a few cloud service providers able to collect and restrict access to new data. Industry 4.0 and IIoT are examples where there will be an exponential use of AI solutions. The IoT paradigm has evolved towards intelligent IoT applications which exploit knowledge produced by IoT devices at the edge using AI techniques and methods. Knowledge sharing between IoT devices is a challenging issue and the way in which IoT devices effectively produce, cumulate, and share the self-taught knowledge with other IoT devices at the edge in the vicinity to form the distributed intelligence in the IoT ecosystem is a research priority for the future.

Advances in AI and ML combined with the sensing/actuating, processing, connectivity capabilities of the IoT devices is driving transformation across industries and workstreams, across various industrial vertical sectors.

AI and ML processing happened in the past mostly in the cloud. As the enabling technologies are evolving AI computing is increasingly moving onto the IoT devices themselves, reducing dependence upon the cloud. This goes together with the emergence of a new kind of architecture that supports AI at the edge as depicted in Figure 3.16.

The intelligent IoT devices at the edge generate information that can be processed locally or close to the source by AI techniques and methods and exchange the information among the edge computing based IoT platforms setting the stage for AI-enabled IoT applications and devices at the edge.

Edge computing is putting the data processing power to the edge by providing energy efficient computing processing with guaranteed performance for real-time operations to manage the data generated by IoT devices, while preserving end-to-end security and privacy. In this context, edge AI means that AI software algorithms are processed locally on the hardware of IoT devices and the algorithms are using data (sensor data or signals) that are created on the IoT device, the device using edge AI software can process data and take decisions independently without a centralised cloud.

Running the edge AI software on IoT devices, gateways and edge infrastructure is critical in the new IoT/IIoT system architecture. The software

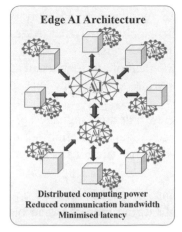
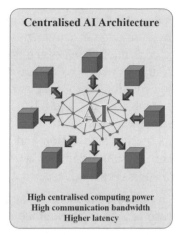

Figure 3.16   Edge AI vs centralised AI architecture (Source: STMicroelectronics).

works in real-time, reports conditions and behaviours on the device based on real-time sensor data. The AI based IoT devices at the edge integrate hardware and software for capturing sensor data, software for training the AI model for different application scenarios and binary software that runs the AI models and algorithms on the IoT devices and gateways. The use of virtualisation is extending the capabilities of edge computing and edge AI.

Edge computing is enabling edge AI as a distributed computing paradigm that brings computation and data storage closer to the location of the edge devices. Edge computing and edge AI are providing the processing of data at the edge allowing to deliver AI-enabled personalised features by deploying different distributed architectures with increased security mechanisms and features for different devices. Edge computing and edge AI create more distributed services at the network and device level implementing connectivity techniques with low-latency when exchanging information across networks and edge devices.

The use of edge computing and edge AI enables the load and processing balancing and increases the applications end to end resiliency when running on distributed systems architectures.

Combining the edge AI and cloud AI many applications in the different industrial sectors will operate in a hybrid manner with a part of AI processing on-device, at the edge and other at the cloud level. The type of AI processing needs determines the optimal federation between edge AI and cloud AI in any given use case depending on the application requirements.

The edge AI processing circuits are focusing on AI workloads that are deployed in edge environments, which include IoT/IIoT edge devices, gateways, edge micro servers and on-premise servers.

IoT continues to evolve and the development of affordable and accessible AI-enabled solutions across industries provide new concepts and integrated solutions combining IoT and AI.

In this context, new reference architectures, design languages, application generators, design automation and respective standardisation are obviously constituents of such engineerable new IoT, AI solutions. Further work is needed to address challenges such as:

- Device-centric AI that runs and trains DL models on edge devices and the use of AI in autonomous manufacturing processes and blockchain networks.
- Development of AI based processor chips for edge devices, PLCs, Distributed Computing Systems (DCSs), and integration of distributed intelligence for wireless and Ethernet-based connectivity on new processor platforms.
- Embedded advanced algorithms. AI based algorithms that can learn with less data and are applied to the edge devices and systems.
- The scale of the IIoT requires that DL networks capabilities are pushed out from the cloud to the edge, and into edge mobile and fixed devices. The edge devices in the industrial sector require intelligent control and coordination, based on DL and AI systems that are collecting data locally, process it and make decisions at the edge in real-time.
- Device-centric AI that runs and trains ML models on edge devices and the use of AI in autonomous manufacturing processes and blockchain networks.
- Security related involving the execution of cryptographic algorithms or other security related firmware at the edge.

The next generation IoT systems require distributed architectures connecting heterogeneous intelligent IoT devices having a context based dynamic behaviour. AI and ML combined with the sensing/actuating, processing, connectivity capabilities of the IoT devices affect the decisions on how to distribute the ML algorithms across device, edge and cloud to create a continuum of knowledge transfer and exchange across various applications domains.

Appling and implementing AI and ML techniques and methods at the edge at the device level requires the semiconductor technology providing the following capabilities resources low energy consumption, high integration (size, weight, volume), real-time operation, high processing power, integrated connectivity, tools to easy develop new algorithms features, security features, and cost.

The energy efficiency and battery lifetime (e.g. low-power design requirements) for mobile and IoT devices at the edge require implementing AI methods and techniques that take into consideration the duty cycle between the different power modes (active, standby, off) and the energy consumption of the sensors/actuators, CPU and communication interface, while optimising data processing/exchange to the edge/cloud that affect the power consumption of the communication interface in relation to processing the data locally.

The implementation of AI and ML at the edge is highly dependent on the real-time requirements as the IoT systems demand real-time or near real-time responses of 1–2 seconds delay that can be associated with sending the data to the cloud and waiting for the response, while other demand real-time responses of milliseconds or below a millisecond that can be implemented only with edge computing processing.

In many cases loading the communication channels by sending gigabits of data to the cloud, in real-time as in the case of autonomous vehicles applications is not feasible due to the cost, latency and due to the fact that such loads can block all other traffic. Furthermore, the cloud solutions are very costly energy hungry.

One challenge for developing AI and ML for IoT devices at the edge is addressing the multiple platforms and in multiple languages as the AI and ML algorithms are implemented most likely of heterogenous environments with different operating systems, a different toolchain and a different software language. Splitting an AI and ML algorithm and migrating it from one environment to another is a resource-intensive process, which could affect the performance of the AI and ML algorithm, the performance of the IoT application as well as introduce errors and bugs.

Updating and maintaining the AI and ML algorithms on the IoT devices deployed in the field is critical to keep the functionality and the security of the IoT systems. The distribution of configuration and updates to IoT devices need new solutions to orchestrate the operation of semi-autonomous and autonomous IoT nodes to provide higher level functions. To achieve the full functionality expected of an IoT system, research should be done in advanced network reorganisation and dynamic function reassignment.

Considering that the AI and ML algorithms on the IoT devices are dynamic and require to evolve based on learning, all updates, debug/troubleshoot need to be considered in real-time. The IoT edge devices will experience continuously new scenarios where the AI and ML algorithms will need to be adapted, based on failure or the need to improve their performance. Identifying the issues and improving accuracy, requires parallel simulation and modelling of the raw data considering different scenarios.

AI and ML implementation at the edge for different heterogenous IoT devices requires flexibility, agility and scaling at the device, network, edge processing and application levels. In this context, considering that the edge resources are agile and support flexible architecture that make use of the intelligent connectivity infrastructure capabilities and cloud environments, performance and memory can scale up and down depending on context and analytics requirements which allows data to improve and upgrade the ML algorithms or transfer the processing across this continuum.

The edge processing and transfer of AI and ML at the device level can improve privacy, ethical and security concerns as data or commercially sensitive data is restricted to the local use. Data collected locally can be processed locally and only information filtered by the user can be sent to the cloud. When large amount of sensitive data needs to be processed the edge, this can go together with the capability to transfer only filtered results to the cloud.

Developing IoT/IIoT applications integrating AI end-to-end solutions require addressing the components illustrated in Figure 3.17.

The IoT/IIoT devices and other data generators represent the structured and unstructured data sources that provide the raw input that is fed into a data conditioning step in which they are fused, aggregated, structured, accumulated, and converted to information. The information is processed by different AI algorithms. The components are implemented in different layers of the 3D IoT architecture. The computing capabilities required for performing AI functions are integrated with the other IoT/IIoT functions and provide acceleration mechanisms for processing the AI algorithms at the edge or in the cloud. To support robust AI implementations of IoT/IIoT applications elements like training/learning, metrics, bias assessment, verification, validation, security, safety, policy rules and ethics must be considered when designing complete solutions.

3.6.1 Training/Learning – Federated Learning

The development of AI technologies has to address the continuum between deep edge, edge, cloud and data centres, with the AI technology stack addressing the infrastructure and developer environment to cover the

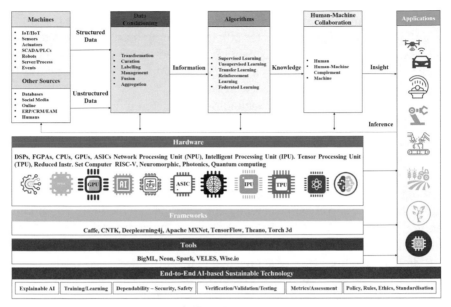

Figure 3.17   Edge AI ecosystem.

hardware, interfaces, platforms, training/learning, applications and services. The intelligent infrastructure at the edge refers to the tools, platforms, and techniques used to run store data, build, and train AI/ ML algorithms, and the algorithms themselves. This infrastructure already exists and is run today by large corporations and AI cloud service providers.

The developer edge environment is a key element for the transition of AI processing to the edge and refers to the tools that assist in developing code to bring out AI capabilities. The developer edge environment must cover all the layers of the AI technology stack offering end-to-end (E2E) development solutions.

Training is a key part of AI technology stack. AI training is defined as the process of creating ML algorithms and involves the use of a learning framework (e.g., TensorFlow) and training datasets provided by static or real-time sources (e.g. databases, IoT/IIoT edge devices, etc.). The data collected is the source of the training data that can be used to train AI-based models for (e.g. ML, DL, etc.) a different use cases, from pattern recognition, object detection, failure detection to consumer intelligence.

AI training systems need to store large volumes of data as the systems refine the algorithms. AI inference systems store only input data that could be useful in future training.

The AI training requires hardware platforms optimised for the type of AI neural networks and algorithms for efficient training used to address how the processing of neural networks is being performed on the platforms, and how application-specific accelerators are designed for specific neural networks for optimising the throughput and energy efficiency.

The emergence of new AI technologies has brought several problems, especially regarding communication efficiency, security threats and privacy violations. To this end, Federated Learning (FL) has received widespread attention. In contrast to centralised training, Federated Learning is a ML setting where the goal is to train a high-quality centralised model while training data remains distributed over a large number of clients each with unreliable and relatively slow network connections that facilitate a distributed learning process [151]. FL allows a ML to be synchronously dispatched to distributed data source locations to be locally trained. The resulting updated ML models are subsequently aggregated at a central location *i.e.,* the trained model parameters are transferred instead of the data, delivering a new updated global model ready for subsequent dispatching cycles. As a result, sensitive data are not exposed to the entity that maintains the global ML model. In certain cases, this approach further yields network resource savings, where training data transfer volume exceeds that of the ML model e.g., video stream data.

ML frameworks such as TensorFlow and PyTorch have already taken steps recently towards privacy with solutions that incorporate federated learning. Instead of gathering data in the cloud from users to train data sets, federated learning trains AI models on mobile devices in large batches, then transfers those learnings back to a global model without the need for data to leave the device. With open source initiatives such as OpenMined [152], AI models can be governed by multiple owners and trained securely on an unseen, distributed dataset.

In the IoT ecosystem, federated learning is proposed to train a globally shared model by exploiting a massive amount of user-generated data samples on IoT devices. Although this is very promising approach the heterogeneities of IoT device as well as the complexity of IoT environments pose great challenges to traditional federated learning. Potential concerns such as man-in-the-middle attacks, model poisoning, bandwidth and processing limitation need to be carefully addressed. Enabling technologies such as 5G and DLTs will play an important role towards this direction.

Cross-silo applications have also been proposed or described in several domains such as finance, health, and smart manufacturing. Cross-silo setting can be relevant where several companies or organisations share incentive to train a model based on all their data but cannot share their data directly. This could be due to constraints imposed by confidentiality or due to legal constraints [153].

3.7 Distributed Ledger Technologies (DLTs)

A Distributed Ledger is a record of transactions or data that is maintained in a decentralised form across different systems, locations, organisations, or devices. It allows data (e.g. funds) to be effectively sent between parties in the form of peer-to-peer transfers without relying on any centralised authority to broker the transfer. A distributed consensus mechanism allows members of the network (nodes) to establish "trust" and thus maintain a common "distributed trust machine". Adding trust in IoT improves the system capabilities and enhance the IoT platforms capabilities [60, 62, 63].

The blockchain refers to Distributed Ledger Technology (DLT) solution where data from different transactions is linked, hashed, and organised per unit, one block at a time and each block is cryptographically "sealed". The unique seal is the start of the next block of transactions that creates the blockchain structure. Examples of DLTs that are classified as blockchains' applications are Bitcoin, Ethereum, Neo, Stellar, Hyperledger. In this context, the blockchain is the mechanism that allows the implementations (e.g. Bitcoin, Ethereum, Neo, Stellar, Hyperledger, etc.) to work, and the implementations are applications that uses blockchain. The main characteristics of blockchains are the decentralised architecture, "trust less" system properties (e.g. the fact that the system can operate without the need for participants involved to know or trust each other or a third party), the existence of consensus mechanisms, the maintenance of history of transactions and the insurance of immutability.

A blockchain is constituted of a digital DLT that is immutable, noneditable and shared among all participants in a blockchain network. The blockchain is constructed as a data structure constituted of time-stamped and cryptographically linked blocks. In this context, the individual blocks have a cryptographic hash, a list of validated transactions, and a reference to the previous block's hash, Using this mechanism the nodes can verify that a participant owns an asset without the need for a central governing authority.

The blockchain allows for participants to engage in trust less peer-to-peer transactions and the decentralised, trust less transactions are the key innovation of the blockchain. The following element are part of the common

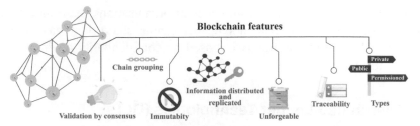

Figure 3.18   Blockchain features.

vocabulary of the blockchains applications and are used to describe the different functions performed by the blockchain.

- Consensus refers to consensus algorithms that represent the mechanism by which all nodes in the network agree on the same version of the truth. Consensus algorithms permit nodes on the system to trust that a specific part of data is valid and that it has been synchronised with all other nodes.
- Cryptographic keys refer to the use of symmetric keys and asymmetric (public-private) key pairs for the use of signing and verifying transactions.
- Decentralised Application (DAPP) refers to decentralised applications that are built on top of a blockchain-based system.
- Ledger represents a shared and distributed history of all transactions and balances.
- Merkle Tree Root (also called binary hash tree) is the result of all leaves hashed together to a single hash.
- Mining/Miners is the process of generating a new legitimate block by applying proof-of-work (e.g. the case of Bitcoin). In specific applications the nodes are dedicated to "mine" new blocks and these nodes are defined as "miners".
- Nodes are represented by any computer or device (e.g. IoT device) connected to a blockchain network.
- Secure Cryptographic Hash Function is defined as a secure cryptographic hash function that preserves one-way, easy computation, and makes impossible to reverse engineer.

The blockchains are classified in the following different categories [67] as presented in the following:

- **Blockchain 1.0** is represented by using blockchain in digital currency (e.g. Bitcoin and other cryptocurrencies payment systems) applications for the decentralisation of money or payment systems.

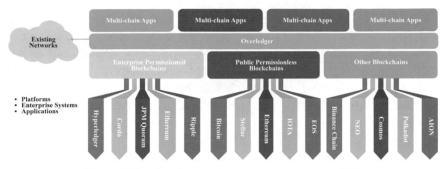

Figure 3.19 Overleger enterprise operating system [69].

- **Blockchain 2.0** is represented by the technology known as contracts, which goes beyond peer-to-peer payment systems and includes the transfers of other property such as stocks, bonds, smart property, and smart contracts.
- **Blockchain 3.0 is** found in the other applications beyond currency and markets, including the use of blockchain in areas like healthcare, governments, and commercial settings.

In the last years, the blockchain developments have addressed the creation of blockchain operating system (OS) that inter-connects blockchains and existing enterprise platforms, applications and networks to blockchain and facilitates the formation of internet scale multi-chain applications known as mApps [69]. Overledger provides interoperability with the full range of DLT technologies including Enterprise Permissioned blockchains like Hyperledger, R3's Corda, JP Morgan's Quorum, permissioned variants of Ethereum and Ripple (XRPL) and Public Permissionless blockchains / DAGs such as Bitcoin, Stellar, Ethereum, IOTA, EOS and the most recent blockchain like Binance Chain. The overledger is positioned as an enterprise operating system that interconnect enterprise platforms and networks providing interoperability to connect to any networks and hyper decentralised applications that run and store data on multiple blockchains is illustrated in Figure 3.19.

Blockchain interoperability is defined as "a composition of distinguishable blockchain systems, each representing a unique distributed data ledger, where atomic transaction execution may span multiple heterogeneous blockchain systems, and where data recorded in one blockchain are reachable, verifiable, and referable by another possibly foreign transaction in a semantically compatible manner" [70].

The emergence of several blockchain platforms brings challenges for IoT applications in particular with respect to the solutions and the interoperability features provided considering the multitude of IoT platforms and blockchain platforms custom-made for specific purposes, (e.g. public, private, or consortium), that adds overhead to manage workflows. Public and private blockchains scalability is a challenge and new techniques for implicit consensus and data sharing (e.g. cross-blockchain transaction routing and retrieval, and asset referencing/discovery) are needed to provide improvements in transaction throughput and storage. Another interoperability issue is related to the fact that different blockchains have different architectures, use different protocols, service discovery, access control, and use different transaction mechanisms. For the integration of IoT and blockchains interoperability, security, privacy, and scalability remain priority research challenges in the near future for the emergence of efficient IoT blockchain integration.

The usage of ledgers (for record-keeping) has been applied in the past but have been recently enhanced to distributed forms, thus significantly improving their capabilities based on easy to understand architectures.

DLTs are intertwined with IoT platforms and used to provide efficient data management in terms of security, privacy, and safety [66].

There are several benefits from using DLTs in IoT applications, as summarised below:

- Increased security: Avoiding centralised networks, increased autonomy of devices, immutability over injection or penetration vulnerabilities, advanced cryptography, prevention of data augmentation and overall increased trust on the IoT nodes.
- Improved secure data management and reliability: devices communicating with each other directly, consensus on data propagation, reduced junk data processing, reduction of costs associated to centralised architectures setup and maintenance.
- Lower bandwidth: reduction of connectivity bottlenecks, cost efficiency, improved latency for decisions.
- Increased auditability: Through ledger immutability and auditing features.
- New types of contracts: ability to automate decision making once particular conditions are met (smart contracts).

DLTs have still to demonstrate their high value and applicability in modern IoT systems being supported by PKI (public key infrastructure) as well as Physical Unclonable Function (PUF) technologies. Combinations of several of these technologies have strong potentials of increasing even further the safety, security, and privacy capabilities of DLTs. However, this combination may in most cases require support from the hardware devices, especially when providing more processing power is required at the edge.

3.8 Intelligent Connectivity

Next generation IoT will make use of an intelligent connectivity that rely on the consolidation of communication technologies. Whereas the Internet is today the largest world-wide communication network, Intelligent connectivity for IoT still requires further development to allow high demand in bandwidth and quality in signal in support of more important content and data exchange. The requirements of real-time response may be, amongst other, safety and mission critical (like in telemedicine for example) and, consequently, IoT communications solutions are numerous and diverse. An important characteristic of the next generation IoT is the requirements range from high reliability and resilience in the communication network to ultra-low latency and high capacity at the communication channel. IoT intelligent connectivity solutions are also very dependent on the context in which they are applied and whether it is necessary to respond to strict energy efficiency constraints or cover large outdoor areas, deep indoor environments or vehicles moving at high speeds. As they are highly dependent upon both the creation of new technologies and the deployment of new communication networks (e.g. 5G mobile network), many of the next generation IoT capabilities may not be largely available before 2025.

3.8.1 Wireless and Cellular Communication Protocols Used for IoT Applications

The next generation IoT will rely on new communication network technologies. An important development is 5G because of its wider broadband capacity, and other technological enhancements to the 5G network that will allow the new connectivity to be a catalyst for the next-generation IoT services through advanced modulation schemes for wireless access, network slicing capabilities, automated network application lifecycle management, software-defined networking and network function virtualisation. In addition, support will be provided for edge- and cloud-optimised distributed network applications.

Next generation IoT builds on the technologies that have been developed and deployed over time (as summarised in Figure 3.20) providing constant evolution of the available data rates over time.

Similarly, next generation IoT takes advantage of the evolution on the spectrum availability and potential usage as summarised in Figure 3.21.

The next generation IoT must address the convergence between different cells and radiation and develop new management models to control roaming

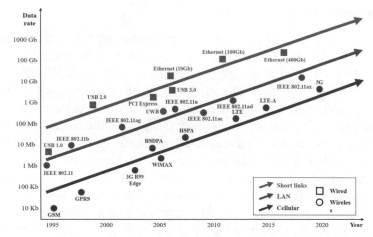

Figure 3.20   Evolution communication technologies (Source: IEEE).

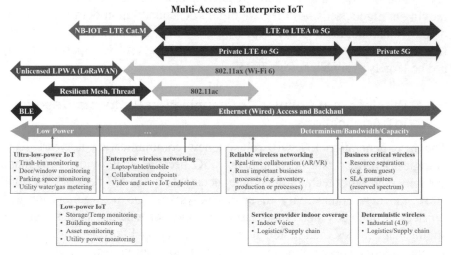

Figure 3.21   Spectrum use by IoT communication protocols. Adapted from GSMA.

while exploiting the coexistence of different cells and radio access technologies. New management protocols to control user assignment regarding cells and technology will have to be deployed in the mobile core network to access network resources more efficiently. Satellite communications must be considered a potential method of radio access, especially in remote areas. With the emergence of safety applications, minimising latency and various protocol translations that will benefit end-to-end latency. It is part of this

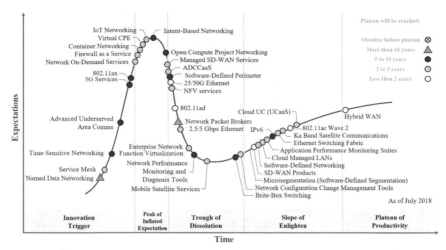

Figure 3.22 Enterprise networking and communications (Source: Gartner Hype Cycle).

latency where tactile technologies as part of the tactile Internet and the capacity to exchange data in real-time where the next generation IoT are involved.

Intelligent connectivity covers the networks and communications technologies as part of the IoT Network Communication Layer covering various communication protocols to provide seamless connectivity for heterogenous IoT devices with different levels of intelligence and connectivity needs.

Intelligent connectivity creates a scalable mobile platform and at the same demands a modular integration of technology (modems for 2G/3G/4G LTE), enabling high-speed data and voice, the implementation of various onboard protocols (i.e. LoRa, Sigfox, On Ramp Wireless, NWave/Weightless SIG, 802.11 Wi-Fi/Wi-Fi Aware, Bluetooth, ZigBee, 6LowPAN, Z-Wave, EnOcean, Thread, wMBus) and the simultaneous use of multiple ISM radio bands (i.e. 169/433/868/902 MHz, 2.4 GHz and 5 GHz).

Connectivity modules are based on integrated circuits, reference designs and feature-rich software stacks created for flexibility and modularity so that they could be implemented in various application domains.

As a result of the progress in data rates, the evolution of the available spectrum and, most importantly, the very large and diverse work in standardisation, the IoT protocols landscape is very large, with varying degrees of development and maturation as shown in Figures 3.22 and 3.23.

Communication technologies used by different IoT devices have various power consumption, bandwidth used, range, data rate, frequency

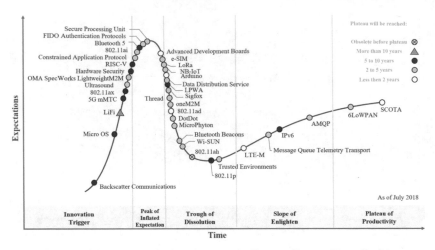

Figure 3.23 IoT standards and protocols (Source: Gartner Hype Cycle).

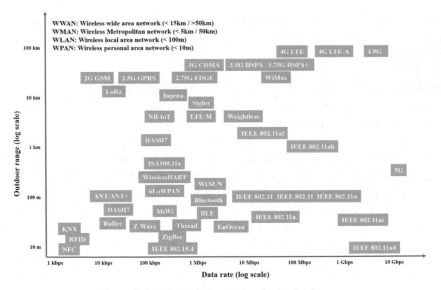

Figure 3.24 IoT wireless technologies landscape.

requirements to ensure seamless communication with other IoT devices and with infrastructure in the operating environments to connect to Anything (any device), to transfer information from/to Anyone (anybody), located Any place (anywhere), at Any time (any context), using the most appropriate physical path from Any path (any network) available between the sender and the recipient based on performance and/or economic considerations, to provide Any service (any business).

The protocol landscape has evolved and has been structured according to two complementary dimensions. The dimensions in the organisation of the protocol landscape regards the available data rate and the range as shown in Figure 3.24.

The following sub-sections will detail the available protocols with a specific sub-section dedicated to the application protocols.

3.8.2 Wireless Personal Area Network (WPAN)

The development of IoT and mobile devices support various connectivity protocols for different applications and industrial sectors. Different types of wireless area networks are defined to cover the various requirements such as range, data rates, power consumption, and applications.

The wireless personal area network (WPAN) is defined as a personal, short distance area wireless network for interconnecting devices e.g. IoT devices, mobile/cellular phones, tablets, PCs, wireless wearable devices, pagers consumer electronics, PCs, etc.) centered around an individual person's workspace.

The IEEE 802 LAN/MAN Standards Committee develops and maintains networking standards and recommended practices for local, metropolitan, and other area networks, using an open and accredited process, and advocates them on a global basis. The standards covered are used for Ethernet, Bridging and Virtual Bridged LANs Wireless LAN, Wireless PAN, Wireless MAN, Wireless Coexistence, Media Independent Handover Services, and Wireless RAN. IEEE created an individual Working Group that provides the focus for each area.

The WPAN standard protocols for IoT devices are covered under IEEE P802.15 Wireless Personal Area Network (WPAN) Working Group that has 10 major areas of development: IEEE 802.15.1: WPAN / Bluetooth, IEEE 802.15.2: Coexistence, IEEE 802.15.3: High Rate WPAN, IEEE 802.15.4: Low Rate WPAN, IEEE 802.15.5: Mesh Networking, IEEE 802.15.6: Body Area Networks, IEEE 802.15.7: Visible Light Communication, IEEE P802.15.8: Peer Aware Communications, IEEE P802.15.9: Key Management Protocol and IEEE P802.15.10: Layer 2 Routing.

The discussions in this section will focus mostly on the IEEE 802.15.1 and IEEE 802.15.4. IEEE 802.15.1 will be further analysed for the Bluetooth and Bluetooth Low Energy. The IEEE 802.15.4 standard defines the functions of the Physical and Media Access Control (MAC) layers and is the foundation for different protocol stacks, (e.g. Zigbee, Zigbee RF4CE, Zigbee Pro, WirelessHART, ISA 100.11a, etc.). In the 802.15.4 network there are two types of devices, one that is the full-function device (FFD) implementing all of the functions of the communication stack, which allows it to communicate with any other device in the network and the other that is the reduced-function device (RFD), with very reduce resource and communication capabilities. FFD can relay messages, and it is dubbed as a personal area network (PAN) coordinator, which oversees its network domain by allocating the local addresses and acting as a gateway to other domains or networks. RFDs can only communicate with FFDs and they cannot act as PAN coordinators as their rationale is to be embedded into the "things".

Different network topologies (e.g. star, mesh, or cluster tree topology) formed from clusters of devices separated by suitable distances can be built and every network needs at least a single FFD to act as the PAN coordinator. For example, the star topology is represented by a hub-and-spoke model where all devices communicate through a single central controller, namely, the PAN coordinator. The PAN coordinator (typically main powered) is represented by the hub, and all other devices (battery operated) form spokes that connect only to the hub. Several IoT applications use this type of network configuration (e.g. home automation, personal health monitors, wearables, etc.) where each star network selects a PAN identifier, that is in use by any other network within the radio range, allowing each star network to operate independently of other networks.

3.8.2.1 6LoWPAN

6LoWPAN combines the Internet Protocol (IPv6) and Low-power Wireless Personal Area Networks (LoWPAN) applied to devices by providing encapsulation and header compression mechanisms that allow IPv6 packets to be sent and received over IEEE 802.15.4 based networks. IoT on the Internet layer is the adaptation of the layer's functions to Link layer technologies with restricted frame size.

The base maximum frame size for 802.15.4 is 127 bytes, out of which 25 bytes need to be reserved for the frame header and another 21 bytes for link layer security. IEEE 802.15.4 g increased the maximum frame size to 2047 bytes, that make possible to compress IPv6 packet headers over the Link layer.

IETF defined in RFC6282 IPv6 over low-power wireless personal area networks, and 6LowPAN provides three main functions, IPv6 header compression, IPv6 packet segmentation and reassembly, and layer 2 forwarding. 6LowPAN, allows to compress the IPv6 header into 2 bytes, as most of the information is already encoded into the Link layer header.

6LowPAN uses a mesh topology, operates in the 2.4 GHz frequency band providing a data rate of 250 kbps with a coverage range of 100 m. The security implemented is AES-128 link layer security defined in IEEE 802.15.4 protocol providing link authentication and encryption. Security features defined in RFC 5246 standard are enabled by the transport layer security mechanisms over TCP. The RFC 6347 standard defines the transport layer security mechanisms over UDP.

3.8.2.2 ANT/ANT+

ANT is an ultra-low power wireless protocol designed for low data rate sensor network topologies (e.g. peer-to-peer, star, mesh, broadcast, ANT-FS, shared cluster) in personal area networks and for local area networks using Gaussian frequency-shift keying (GFSK) modulation. The IoT applications include sports, fitness, wellness, home health, homes, and industrial automation applications.

The protocol uses the 2.4GHz frequency and provides ultra-low power, network flexibility and scalability (e.g. self-adaptive and able to do practical mesh). ANT devices may use any RF frequency from 2400MHz to 2524MHz, except for 2457MHz, which is reserved for ANT+ devices. ANT devices may use the public network key, a private network key, or a privately-owned managed network key. The ANT+ network key is reserved only for ANT+ devices. The ANT protocol provides device profiles that are tied to a specific use case. The device profiles are shared among all the ANT+ adopters, enabling any ANT+ adopter to create a specific device for the specific use case (e.g. heart monitor) that will operate interchangeably with one another. ANT+ branded on a device assure the interoperability with other ANT+ branded devices [104, 106].

Each ANT node can operate as a slave or master and can transmit and receive as well as function as a repeater. ANT uses a very short duty-cycle technique and deep-sleep modes to ensure very low power consumption and a single 1-MHz channel for multiple nodes due to a time-division-multiplex technique. Each node transmits in its own time slot, with basic message length of 150 μs, and message rate ranging from 0.5 Hz to 200 Hz with an 8-byte payload per message. ANT ensures management of physical, data link, network, and transport layers of OSI stack and ANT+ manages session, presentation, and application layers to provide data and devices interoperability.

ANT devices that communicate with each other are part of the same network and the ANT channel (e.g. independent, shared and scan channel) between two devices uses the same frequency, message period, device type and transmission type (i.e. slave or master). The ANT devices are establishing a channel by pairing for communication, process in which an ANT slave device gets the complete unique channel ID of the master, that plans to communicate with, and agrees upon same frequency and message period. Establishing a channel can be permanent, semi-permanent or transitory. The ANT nodes representing a wireless sensor device, include an ANT protocol engine, a Micro Controller Unit (MCU) and can be configured as master and slave nodes to participate in one or more networks [104, 106, 107].

The range provides is approximate 30 m and the data rates are 12.8 Kbit/s – 60kbit/s. ANT supports a 8-byte (64-bit) network key and 128-bit AES encryption for ANT master and slave channels. Authentication and encryption can be further implemented through the application level.

3.8.2.3 Bluetooth and Bluetooth Low Energy WPAN – IEEE 802.15.1

Bluetooth was designed for point-to-point or point-to-multipoint, star (up to seven slave nodes) network configurations and data exchange among mobile devices. Bluetooth is designed as an open wireless protocol used in the unlicensed Industrial, Scientific, and Medical (ISM) 2.4 to 2.483 GHz short-range radio frequency bandwidth. The protocol is utilised for exchanging data over short distances from fixed and mobile devices, creating personal area networks (PANs). Bluetooth uses frequency-hopping spread spectrum, which chops up the data being sent and transmits chunks of it on up to 79 frequencies. The Bluetooth modulation used in the basic mode is GFSK. Bluetooth devices must operate in one of four available modes from mode 1 (insecure mode) to mode 4 -where security procedures are initiated after link setup. Secure Simple Pairing uses Elliptic Curve Diffie Hellman (ECDH) techniques for key exchange and link key generation in mode 4.

Bluetooth Low Energy (BLE) was optimised for power consumption to address small-scale consumer IoT applications. BLE is integrated into several IoT devices such as fitness and medical wearables (e.g. smart-watches, glucose meters, pulse oximeters, etc.), smart home devices (e.g. door locks) and IoT tracking devices (e.g. objects, animals, etc.) whereby data is conveniently communicated to and visualised on smartphones. Bluetooth Mesh specification aims to enable a scalable deployment of BLE devices (e.g. retail contexts. logistics, etc.). BLE is providing versatile indoor localisation features and IoT beacon networks are used for different IoT service applications (e.g. in-store navigation, personalised promotions, content delivery, etc.). BLE is incompatible/non-interoperable with Bluetooth and to achieve interoperability a dual-mode device needs to be implemented.

BLE uses different set of technical and radio techniques to ensure very low power consumption implementing the data protocol to create low-duty-cycle transmissions or a very short transmission burst between long periods, combined with very low-power sleep modes.

BLE uses a specific frequency-hopping spread-spectrum (FHSS) scheme that employs forty 2 MHz-wide channels to ensure reliability over longer distances. Bluetooth offers data rates of 1, 2, or 3 Mbits/s, while BLE's rate is 1 Mbit/s with a net throughput of 260 kbits/s. BLE has a 0 dBm (1 mW) power output, latency of 6 ms, uses 128-bit AES security and provides a range of 50 meters. The adaptive frequency-hopping technique to avoid interference, a 24-bit cyclic redundancy check (CRC), and a 32-bit message integrity check improves BLE link reliability. In BLE (e.g. version 5.0), the new waveforms and coding techniques are implemented in order to achieve longer ranges, less power consumption and latency, better robustness and support for higher number of subscribers in a single Bluetooth network.

3.8.2.4 UWB

UWB is using wireless connectivity at wide bandwidth for short-range applications. UWB technology transmits information spread over a large bandwidth (more than 25% of the centre frequency or at least 500 MHz) with very low power levels therefore not interfering with other narrower band devices nearby. The receiver translates the pulses into data by listening for a familiar pulse sequence sent by the transmitter. As the data is moving on several channels at once, it can be sent at high speed, up to 1 gigabit per second. UWB technology can penetrate walls. Frequency regulations limit UWB to low power levels to keep interferences below the level of noise produced unintentionally by electronic devices (e.g. TV sets). UWB is limited to short-range applications, enabling wireless connectivity. IEEE 802.15.3 – standard addresses the high-data-rate WPAN designed to provide sufficient quality of service for the real-time distribution of multi-media. Apple launched the three phones with ultra-wideband capabilities in 2019 (e.g. iPhone 11, iPhone 11 Pro, iPhone 11 Pro Max).

3.8.2.5 EnOcean

EnOcean is a wireless technology based on the wireless EnOcean radio standard (ISO/IEC 14543-3-10/11) in sub 1GHz is optimised for use in buildings, with a radio range of 30m indoors and 300m outdoors. The standard covers the OSI (Open Systems Interconnection) layers 1-3 which are the physical, data link and networking layers. EnOcean wireless data packets are relatively small, with the packet being only 14 bytes long and are transmitted at 125 kbit/s. RF energy is only transmitted for the 1's of the binary data, reducing the amount of power required.

The transmission frequencies used for the devices are 902 MHz, 928.35 MHz, 868.3 MHz, and 315 MHz. EnOcean technology uses energy harvesting techniques to harvest the energy from mechanical motion, indoor light, temperature differences. Energy converters are used to transform energy fluctuations into usable electrical, electromagnetic, solar, and thermoelectric energy.

Th authentication method offers field-proven secure and reliable communication in building automation. The unique 32-bit identification number (ID) of the standard EnOcean modules cannot be changed or copied. For additional data security, the security mode protects battery less wireless communication with enhanced security measures to prevent replay or eavesdropping attacks and forging of messages. A specific feature is a maximum 24-bit rolling code (RC) incremented with each telegram which is used to calculate a maximum 32-bit cypher-based message authentication code (CMAC). The CMAC uses the AES 128 encryption algorithm. Another security mechanism is the encryption of data packets by the transmitter. The data is encrypted using the AES algorithm with a 128-bit key.

3.8.2.6 ISA100.11a

ISA100 is developed by ISA as an open-standard wireless networking technology based on IEEE 802.15.4 (MAC and physical layer), TDMA (Time Synchronized Mesh Protocol), utilising DSSS with channel/frequency hopping (with blacklists of noisy channels) and mesh routing. The official description is "Wireless Systems for Industrial Automation: Process Control and Related Applications".

The standard provides reliable and secure wireless operation for non-critical monitoring, alerting, supervisory control, open-loop control, and closed-loop control applications. ISA100.11a specification was approved as IEC 62734.

The standard adds features and functions that give support to comply with ETSI EN 300328 v1.8.1 (e.g. country codes to identify device location and the capacity to attenuate output power levels). The IEC 62734 standard incorporates Annex V, that presents multiple scenarios and approaches to assure ETSI compliance.

ISA100.11a uses mesh and star topologies, operates in the 2.4 GHz frequency band providing a data rate of 250 kbps with a coverage range of 100 m. The security implemented in ISA 100.11a standard is embedded with integrity checks and optional encryption at data link layer of the OSI reference model. Security mechanisms are provided in transport layer and 128 bits keys are used in both transport and data link layers. To join a ISA 100.11a network, a sensor node needs to use a shared global key, a private symmetric key or certificate.

3.8.2.7 NFC

The Near Field Communication (NFC) protocol operates at high frequency band at 13.56 MHz and supports data rate up to 424 kbps. The applicable range is up to 10 cm where communication between active readers and passive tags or two active readers can occur. NFC is a short-range technology and protocol that allows two devices to communicate when they are placed into the touching distance. NFC enables sharing power and data using magnetic field induction at 13.56MHz (HF band), at short range, supporting varying data rates from 106kbps, 212kbps to 424kbps.

A key feature of NFC is that it allows two devices to interconnect. In reader/writer mode, an NFC tag is a passive device that stores data that can be read by an NFC enabled device (e.g. smart poster, for which a technical specification was developed). NFC devices are designed to exchange data in Peer-to-Peer mode.

Bluetooth or Wi-Fi link set up parameters can be shared using NFC and data like digital photos, and virtual business cards can be transferred. The NFC device itself acts as an NFC tag in Card Emulation mode, resembling an external interrogator as a traditional contactless smart card. This facilitates contactless payments and e-ticketing. NFC standards are acknowledged by major standardisation bodies like ISO (e.g. ISO/IEC 18092).

Security is implemented in the NFC using mechanisms such as Digital Signature and Trusted Tag. The Digital Signature (defined in the NFC Forum Signature RTD 2.0) uses asymmetric key exchange and the Digital Signature is a part of the NFC Data Exchange Format (NDEF) message, which includes also a Certificate Chain and a Root Certificate. Each NFC device has a private and a public key. The Trusted Tag method is fully compliant with NFC Forum Tag Type 4 and works with any NFC Forum compatible devices. The Trusted tag is protected from cloning and embedded with cryptographic code that is generated by every "tap" or click on NFC button. This cryptographic code protects the content of the transmitted information.

3.8.2.8 RuBee

RuBee is a peer to peer communication protocol designed for active or passive tags operating at low frequency (using magnetic induction), suitable in environments containing water and/or metal. IEEE is standardising it as P1902.1 "IEEE Standard for Long Wavelength Wireless Network Protocol". The standard offers a "real-time, tag searchable" protocol using IP and subnet addresses associated with asset taxonomies that operate at speeds of 300 to 9600 Baud. RuBee Visibility Networks are operated by Ethernet-enabled routers. RuBee enables tag networks and applications. RuBee tags and tag data may be seen as a stand-alone via web servers from anyplace in the world. RuBee tags that are properly programmed can be discovered and monitored over the Internet using search engines.

3.8.2.9 DASH7 (D7A)

The DASH7 protocol complies with the ISO/IEC 18000-7 standard. ISO/IEC 18000-7 is an open standard for the license-free 433 MHz ISM band air-interface for wireless communications. Using 433 MHz frequency provides D7A devices with long propagation distance and better penetration in objects (e.g. buildings).

The protocol defines four different device classes such as blinker that only transmits and does not use a receiver, endpoint that can transmit and receive the data and supports wake-up events, sub controller that is a full featured device using wake on scan cycles similar to end points and gateway that connects D7A network with the other networks and is always online and always listens unless it is transmitting.

DASH7 uses AES-CBC for authentication and AES-CCM for authentication and encryption.

3.8.2.10 RFID (WPAN – IEEE 802.15)

RFID standards include ISO 11784/11785, ISO 14223, ISO/IEC 14443, ISO/IEC 15693, ISO/IEC 18000, ISO/IEC 18092, ISO 18185, ISO/IEC 21481, ASTM D7434, ASTM D7435, ASTM D7580, ISO 28560-2.

Passive RFID tags have no power source/battery and the power is provided by the interrogator and operate in short read range (from mm to m). The passive RFID tags consist of an integrated circuit attached to an antenna and use either the magnetic field or the electric field to transmit their signals.

Semi passive tags are RFID devices which use a battery to maintain memory in the tag and sometimes provide power to the processing unit or power the electronics that enable the tag to modulate the reflected signal, or to support sensors. The radio aspects remain passive in that they only react to received signals and use the power from the carrier signal.

Active RFID systems usually operate at 433 MHz, 868MHz (no standardised RFID protocol) 2.45 GHz, or 5.8 GHz and have a read range of up to 100 meters. Low frequency (LF) and very high frequency (VHF) systems are also available. Some active tags now include GPS positioning capability.

Active RFID beacons are used in real-time locating systems (RTLS), where the precise location of an object is tracked. In an RTLS, a beacon emits a signal with its unique identifier at pre-set intervals. The beacon's signal is picked up by at least three interrogator antennas (triangulation or trilateration) positioned around the perimeter of the area where assets are being tracked. RTLS are usually used outside, in a container base or inside large facilities to track parts or personnel. Active tags can be read reliably because they broadcast a signal to the interrogator.

The implementation of security mechanisms in RFID technology is based on confidentiality, integrity, and availability. Confidentiality is the information protection from unauthorised access. Integrity is related to data protection from modification and deletion by unauthorised parties. Availability represents the capability for data access when needed.

An overview of the RFID technologies and the related standards is given in Table 3.3.

One functional area of great relevance to many supply chain applications is the ability to monitor environmental or status parameters using an RFID tag with built in sensor capabilities. Parameters of interest may include temperature, humidity, and shock, as well as security and tamper detection.

3.8.2.11 Thread

Thread is a mesh networking low-power wireless protocol, based on the IPv6, designed to address the interoperability, security, power, and architecture challenges of the IoT. Thread utilises 6LoWPAN that employs the IEEE 802.15.4 wireless protocol with mesh communication. Thread is IP-addressable, with cloud access and AES encryption. Thread 1.2 release

Table 3.3   RFID technologies and related standards

Frequency	LF Low Frequency	HF High Frequency	UHF Ultra-High Frequency	SHF Super High Frequency
Frequency range	30kHz to 300kHz	3MHz to 30MHz	300MHz to 3GHz	3GHz to 30GHz
RFID Frequency Air Interface	30-50kHz 125/134kHz[1] 131/450kHz	6.78MHz[2] 7.4-8.8MHz 13.56MHz 27MHz	433MHz 840-960MHz 2.45GHz	3.1-10,6GHz 5.8GHz 24.125GHz
Standard/Protocol Air Interface	ISO/IEC 18000-2 USID (Ultrasound) ISO 11784/5 ISO 14223 IEEE P1902.1/ RuBee EM4100, Sokymat UNIQUE	ISO/IEC 18000-3 ISO/IEC 15693 ISO/IEC 18092/NFC ISO/IEC 10536	ISO/IEC 18000-7 ISO/IEC 18000-6 Type A, B, C EPC C1G2 ISO/IEC 18000-4 IEEE 802.11 IEEE 802.15 WPAN IEEE 802.15 WPAN Low Rate IEEE 802.15 RFID	ISO/IEC 18000-5 (withdrawn) IEEE 802.15 WPAN UWB
Availability	> 30 years RuBee -2 years	> 10 years	US > 9 years, EU > 7 years, (4-channel plan is from 2008)	> 10 years
Multiple tag reading	Limited	Up to 50 tags/sec	Up to 200 tags/sec	Up to 300 tags/sec
Reading distance	0.001–1 m	0.001–0.5m	US ~ 0-6 m, EU ~ 0-4 m Passive tags ~ 0–100 m, Active tags	~ 0–1000 m, Active tags
Data transmission rate	Low	Medium	Fast	Very fast
Power Source	Passive – inductive coupling Active – using battery RuBee protocol considers both active and passive tags.	Passive tags, using inductive or capacitive coupling	Active tags with embedded battery Passive tags using capacitive, E-field coupling, Backscattering	Active tags with embedded battery Passive tags using capacitive, E-field coupling

[1,2] According to Annex 9 of the ERC Rec 70-03, inductive RFID Reader systems primarily operate either below 135 kHz or at 6.78 or 13.56 MHz. Therefore, the correlated transponder data return frequencies reside in the following ranges:
LF Range Transponder Frequencies: f_C = < 135 kHz, f_{TRP} = 135 to 148.5 kHz
HF Range Transponder Frequencies: f_C = 6.78 MHz f $_{TRP}$ = 4.78 to 8.78 MHz
f_C = 13.56 MHz f_{TRP} = 11.56 to 15.56 MHz.

includes commercial extensions, the integration of enterprise device life-cycle management and Bluetooth Low Energy Extensions. Enterprise-level security is handled using information technology authentication, authorisation, accounting, and the Thread 1.2 release expands the IP connectivity and end-to-end security for BLE devices. Applications using the protocol can consolidate multiple Thread networks into an extensive virtual Thread network with thousands of nodes, including predictable, stable addressing and management even as devices migrate within that virtual network.

Thread uses mesh network topology, in the 2.4GHz frequency spectrum, providing data rates of 250 kbps with a coverage range of 30 m. The security implemented uses a 128-bit AES encryption system. The encryption cannot be disabled.

Thread utilises a network-wide key that is used at the Media Access Layer (MAC) for encryption. This key is used for standard IEEE 802.15.4 authentication and encryption. IEEE 802.15.4 security protects the Thread network from over-the-air attacks originating from outside the network. Each node in the Thread network exchanges frame counters with its neighbours via an MLE handshake. The frame counters help protect against replay attacks. Thread allows the application to use any internet security protocol for end-to-end communication. The protocol can connect up to 250 devices in a wireless mesh network.

3.8.2.12 WirelessHART

WirelessHart is an open standard wireless networking technology developed by HART Communication Foundation. The protocol utilises a time synchronised, self-organising, and self-healing mesh architecture. WirelessHart currently supports service in the 2.4 GHz ISM Band using IEEE 802.15.4 standard radios. WirelessHART was developed as an interoperable wireless standard specifically for the requirements of Process field device networks. The HART Protocol uses Frequency Shift Keying (FSK) standard to superimpose digital communication signals at a low level on top of the 4-20mA.

WirelessHart uses mesh and star topologies, providing data rates of 250 kbps with a coverage range of 200 m. The security implemented uses a 128-bit AES encryption system. The encryption cannot be disabled.

The security manager in the WirelessHART gateway administers the Network ID, Join key and Session key. A common network key is shared among all devices on a network to facilitate broadcast activity in addition to individual session keys. A 128-bit join encryption key is used to keep sent and received data private, during the joining process.

3.8.2.13 Z-Wave

Z-Wave is a mesh network low-energy wireless communications protocol used mainly for wireless control of home appliances, lighting control, security systems, swimming pools, garage door openers, thermostats, windows, locks, etc. Z-Wave systems can be controlled via the Internet and locally through devices or using a Z-Wave gateway or central control device serving as hub controller and portal. Z-Wave's interoperability at the application layer assures that Z-Wave devices share information and allows all Z-Wave hardware and software to work together.

Z-Wave uses the unlicensed industrial, scientific, and medical (ISM) band and operates at 868.42 MHz in Europe and 908.42 MHz in the US. Z-Wave provides data rates of 9600 bps and 40 kbps, with output power at 1 mW.

The Z-Wave range between two nodes is up to 100m in an outdoor, unobstructed setting. For in-home applications, the range is 30 m for no obstructions and 15 m with walls in between. Z-Wave Alliance requires the mandatory implementation of Security 2 (S2) framework on all devices receiving certification.

Z-Wave mesh networks become more reliable as more devices are added (e.g. a Z-Wave network with 50 devices is more reliable than a Z-Wave network of 25 devices).

Z-wave provides packet encryption, integrity protection and device authentication services. End-to-end security is provided on application level (communication using command classes). It has in-band network key exchange and AES symmetric block cipher algorithm using 128-bit key length.

3.8.2.14 ZigBee WPAN – IEEE 802.15.4

ZigBee is a short-range, low-power, wireless standard (IEEE 802.15.4), deployed in mesh topology to extend coverage by relaying IoT sensor data over multiple sensor nodes. ZigBee/ZigBee Pro is based on a specification that adds application profile, security, and network layers to IEEE 802.15.4 standard for wireless low-rate personal area networks. It operates in the UHF/microwave bandwidth with battery-powered tags that communicate with each other. Zigbee protocol features include support for multiple network topologies such as point-to-point, point-to-multipoint and mesh networks, low duty cycle that provides long battery life, low latency, direct sequence spread spectrum (DSSS), 128-bit AES encryption for secure data connections, collision avoidance, retries and acknowledgements.

ZigBee protocols are designed to be used in embedded applications requiring low data rates and low power consumption, enabling devices to form a mesh network of up to 65 000 nodes, covering a very large area target general-purpose, inexpensive, self-organising, mesh networks which are deployed for building/home automation, industrial control, embedded sensing, medical data collection, smoke and intruder warning, domotics, etc. ZigBee networks use low power for communication, and individual devices can run for a few years on the installed battery. ZigBee is a perfect complement to Wi-Fi in many of these applications. ZigBee provides data rates at medium power-efficiency due to mesh configuration. The protocol is designed for physical short-range (< 100m) mesh configurations best-suited for medium-range IoT applications with an even distribution of nodes in proximity.

The Zigbee PRO Specification adds child device management, improved security features, and new network topology options to Zigbee networks. Commissioning devices into networks has also been improved and made more consistent through Base Device Behaviour (BDB). The specification furthermore requires Green Power Basic Proxy functionality in all devices to further support Green Power capabilities and compiles all profile clusters into a single specification.

The Zigbee standard operates on the IEEE 802.15.4 physical radio specification and operates in unlicensed bands including 2.4 GHz, 900 MHz and 868 MHz Zigbee 3.0 supports the increasing scale and complexity of wireless networks, and copes with large local networks of greater than 250 nodes. The data rates provided are 250 kbps (2.4 GHz) 40kbps (915 MHz) 20kbps (868 MHz). Zigbee also handles the dynamic behaviour of these networks (with nodes appearing, disappearing, and re-appearing in the network) and allows orphaned nodes, which result from the loss of a parent, to re-join the network via a different parent.

The self-healing nature of Zigbee Mesh networks also allows nodes to drop out of the network without any disruption to internal routing. Zigbee's supports over-the-air (OTA) upgrade for software updates during device operation and provides enhanced network security using methods such as centralised security by employing a coordinator/trust centre that forms the network and manages the allocation of network and link security keys to joining nodes or distributed security where there is no coordinator/trust centre. Any Zigbee router node can subsequently provide the network key to joining nodes.

ZigBee is a secure wireless communication protocol, with security architecture built in accordance with IEEE 802.15.4 standard. Security mechanisms include authentication – authorised access to network devices, integrity protection and encryption with key establishment and transportation.

Device authentication is the procedure of confirming a new device that joins the network as authentic. The new device must be able to receive a network key and set proper attributes within a given time frame to be considered authenticated. Device authentication is performed by the Trust Center.

Integrity protection is realised on the frame level using message integrity checks (MIC) to protect the transmitted frames and ensure they are not accessed and manipulated. A 128-bits symmetric-key cryptography is implemented in ZigBee's security architecture.

JupiterMesh is a robust, low-power industrial IoT wireless mesh network for Neighbourhood Area Network (NAN) with flexible data rates that enables neighbourhood and field area communications for utilities and municipalities deploying intelligent grid and smart city solutions. JupiterMesh is supported by ZigBee Alliance and is built on open IETF and IEEE standards uses parts of the IEEE 802.15.4g standard used by the Wi-SUN Alliance, as well as IEEE 802.15.4e and the IETF's Ipv6, 6LoWPAN, UDP, TCP, RPL, and CoAP protocols.

The protocol includes advanced technologies such as IPv6, frequency hopping, multi-band operation, authentication, encryption, and key management to drive industry realisation of interoperable multi-vendor implementations that scale and that are secure and easy to manage. JupiterMesh can operate in the sub-GHz ISM bands, as well as the 2.4GHz band, using FSK, O-QPSK, and OFDM modulation schemes.

3.8.3 Wireless Local Area Network (WLAN)

3.8.3.1 Wi-Fi WLAN – IEEE 802.11

IEEE 802.11 wireless technologies (Wi-Fi) address many of the requirements of IoT and have a number of challenges related to power consumption for end devices, due to the need for client devices to wake up at regular intervals to listen to AP announcements, waste cycle in contention processes, etc. and the frequency bands (e.g. 2.4–5 GHz), which are characterised by short transmission range and high degree of loss due to obstructions.

Wi-Fi provides high-throughput data transfer for both industrial and home environments. Wi-Fi has high energy requirements and is used for IoT applications and services that do not require large networks of battery-operated IoT sensors. For such applications Wi-Fi major limitations are in coverage, scalability, and power consumption. Many IoT devices that are connected to power outlet (e.g. smart home IoT monitoring devices and appliances, digital signages or security cameras, etc.) are using Wi-Fi as connectivity protocol.

Wi-Fi – specification (ISO/IEC 8802-11) is an international standard describing the characteristics of a wireless local area network (WLAN). The name Wi-Fi (short for "Wireless Fidelity") corresponds to the name of the certification given by the Wi-Fi Alliance, formerly WECA (Wireless Ethernet Compatibility Alliance), the group which ensures compatibility between hardware devices that use the 802.11 standards. Wi-Fi networks are networks that comply with the 802.11a-x specifications. The different frequency bands used by the different Wi-Fi implementations is presented in Figure 3.25 [91].

The new Wi-Fi generation is represented by the Wi-Fi 6 protocol that enhances network bandwidth (i.e. <9.6 Gbps) to improve data throughput per user in congested environments.

Wi-Fi 6E brings the technology into 6 GHz and features more contiguous spectrum, wider channels, less interface, gigabits speed, very low latency and high capacity [213]. The increased network capacity and the improved simultaneous communication between access points and multiple endpoints allows Wi-Fi 6 (IEEE 802.11ax) to offer improved performance in crowded areas. Several technologies facilitate these improvements and the most important are [214]; firstly building upon the Orthogonal Frequency Divisional Multiple Access (OFDMA) technology, already available in previous Wi-Fi standards, gives access points the ability to divide channels into many sub-channels, i.e. that the access points can communicate with multiple devices at the same time at a lower data rate; secondly, utilising the Multi-User Multiple-Input-and-Multiple-Output (MU-MIMO) which already are used by Wi-Fi 5 for downlinks.

Wi-Fi 6 utilises MU-MIMO for uplinks and enable access points to simultaneously receive communication from multiple clients. The combination of these technologies and some others (e.g. 1024-QAM) results in more efficient networks and reduces the overall network latency. The power management for connected IoT devices is also improved. Wi-Fi 6 supports a so called "target wake time" feature, which allow the Wi-Fi access points to put IoT devices in sleep mode for a given period [214]. The possibilities to reduce the power consumption of battery driven IoT devices can be very useful.

Figure 3.25   Wi-Fi frequency bands [91].

Today, Wi-Fi dominates as the infrastructure from small companies to large enterprises at the expense of the mobile network operators and their cellular network. The technologies are fundamentally different, primarily due to their differences in unlicensed and licenced environments. With the upcoming Wi-Fi 6 and 5G technologies, Wi-Fi and cellular becoming closer to each other, and Wi-Fi 6 is the first generation of WLAN technology that promises to seamlessly interact with the cellular solutions [215]. Wi-Fi 6 access points are already commercialised and are being built into networks [214]. Wi-Fi 6 phones are available together with Wi-Fi 6 adapters and routers. There are still early for private cellular IoT, but it is moving in that direction. Early findings indicate that unlicensed and licensed spectrum will continue to exist as complements [215], but it is a matter of quality of service (QoS) and costs for the users.

Wi-Fi 6 is a private-network technology with similar capabilities to 5G. The majority of IoT devices connect to Wi-Fi or an IoT-specific protocol designed for local ranges. Wi-Fi 6 lacks roaming capability, which has a significant impact on mobile sensor applications. Both 5G and Wi-Fi 6 are based on Orthogonal Frequency Division Multiple Access (OFDMA).

Wi-Fi 6 is less prone to interference than previous Wi-Fi standards, requires less power consumption and has improved spectral efficiency. 5G offers fully private 5G networks based on either licensed or unlicensed, shared spectrums and could offer alternatives to Wi-Fi 6. Specific IoT applications will require the capabilities offered by 5G and Wi-Fi 6 and the factors

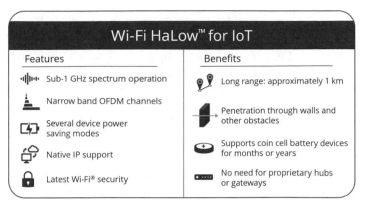

Figure 3.26 Wi-Fi HaLow features [91].

for deciding between the technologies are availability, range, the interplay of mobility and roaming capability, overall costs, etc.

3.8.3.2 Wi-Fi HaLOW

Wi-Fi HaLow wireless technology is based on the IEEE 802.11ah protocol standard. The technology augments Wi-Fi by operating in the spectrum below one gigahertz (GHz) to offer longer range and lower power connectivity. Wi-Fi HaLow meets the requirements for the IoT to enable a variety of use cases in industrial, energy (e.g. smart metering), smart home/building, agricultural, and smart city environments [91]. Some features and benefits are illustrated in Figure 3.26.

Wi-Fi HaLow utilises 900 MHz license-exempt bands to give extended range Wi-Fi networks, compared to Wi-Fi networks working in the 2.4 GHz and 5 GHz bands. The protocol has low energy consumption, allowing for IoT implementations with large groups of stations or sensors that cooperate to share data. The protocol's low power consumption competes with Bluetooth and has the benefit of higher data rates and wider coverage range.

3.8.3.3 White-Fi

White-Fi wireless technology is based on the IEEE 802.11af protocol standard that supports WLAN operation in the he VHF and UHF bands between 54 and 790 MHz TV white space spectrum, designed for ranges up to 1 km. The data rate for IEEE 802.11af per spatial stream reaches 26.7 Mbit/s for 6 and 7 MHz channels and 35.6 Mbit/s for 8 MHz channels. The maximum data rate can reach 426.7 Mbit/s for 6 and 7 MHz channels and 568.9 Mbit/s for the 8 MHz channels by using four spatial streams and four bonded channels. In order avoid that the system does not create any undue interference with existing television transmissions, the White-Fi can utilise cognitive radio to detect transmissions and move to alternative channels or geographic sensing by using a geographic database and a knowledge of what channels are available to avoid interference with used channels. The White-Fi technology has benefits like propagation characteristics by using frequencies below 1 GHz that allow for greater distances or additional bandwidth. To achieve the required data throughput rates is necessary to aggregate several TV channels to provide the bandwidths that Wi-Fi uses on 2.4 and 5.6 GHz. Unused channels in any geographic area can vary in frequency, and special techniques need to be used for managing the data sharing across the different channels (e.g. as in technologies such as LTE).

3.8.3.4 Li-Fi

Light-based Li-Fi (light fidelity) is a wireless communication technology that uses light to transmit data and position between devices. The technology addresses some of the short comings of radio-based wireless communications and has applications for industrial IoT connectivity by improved solutions for security, scalability, bandwidth, interference, latency. Li-Fi uses the modulation of light intensity to transmit data at high speeds over the visible light (e.g. LED devices), ultraviolet, and infrared spectrums. Li-Fi can transmit at speeds of up to 100 Gbit/s. A unique feature of Li-Fi is that it combines illumination and data communication by using the same device to transmit data and to provide lighting [92]. Figure 3.27 shows the concept of a Li-Fi attocell network.

The illumination in the room is provided by several light fixtures with each light driven by a Li-Fi modem that serves as an optical base station or access point (AP). Each optical base station is connected to the core network by high speed backhaul connection. The light fixtures have an integrated infrared detector to receive signals from the terminals. The illuminating lights are modulated at high rates so the high frequency flickers are much higher than the refresh rate of a computer monitor are not visible to the occupants of the room [92].

Li-Fi's can safely function in areas susceptible to electromagnetic interference (e.g. aircraft cabins, hospitals, military, industrial). Li-Fi and visible light communication (VLC) are used in different IoT applications. Several companies offer uni-directional VLC products, that is not the same as Li-Fi. IEEE 802.15.7 supports high-data-rate visible light communication up to

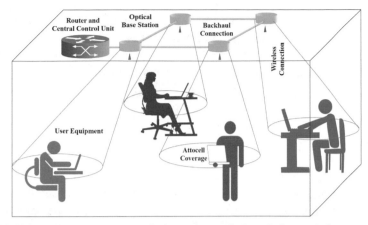

Figure 3.27 Li-Fi concept. Attocell networks applied to indoor wireless networking. Adapted from [92].

96 Mb/s by fast modulation of optical light sources which may be dimmed during their operation. IEEE 802.15.7 provides dimming adaptable mechanisms for flicker-free high-data-rate VLC. VLC technology using Li-Fi was demonstrated and data rates of 8 Gbit/s can be achieved over a single light source (e.g. LED) and complete cellular networks based on Li-Fi can be created.

3.8.3.5 Wi-SUN
Wi-SUN protocol is used for applications like the industrial grade utility, smart cities field area networks, agriculture, and asset monitoring.

Wi-SUN provides the communications profile definitions based on open standards for field area, IoT wireless networks, interoperability testing and certification for peer to peer wireless mesh networks based on IEEE 802.15.4g and IPv6. IEEE 802.15.4g is designed for Smart Utility Networks (SUN) supported by Wi-SUN Alliance. The 802.15.4g variation is aimed at deployments that can have millions of endpoints, deployed across large geographic scales, using minimal infrastructure in a peer-to-peer self-healing mesh. Wi-SUN devices may securely discover and join an existing Wi-SUN network. For this purpose, Wi-SUN relies on EAP and 802.1x security standards. Trickle timers are used to reduce interference and battery consumption while still maintaining responsive connectivity.

Wi-SUN is delivering star and mesh-enabled field area networks (FAN) to provide resilient, secure, cost-effective connectivity with ubiquitous coverage in a range of topographical environments, from dense urban neighbourhoods to rural areas, with minimal additional infrastructure. Wi-SUN protocol characteristics include a coverage in the range of 1 km, high bandwidth of up to 300 kbps, low latency (e.g. 20 ms), resilient and scalable mesh routing, low power consumption (e.g. less than 2 uA when resting, 8 mA when listening), scalable networks of up to 5000 devices and security using public key certificates (e.g. AES, HMAC, dynamic key refresh, hardened crypto). Wi-SUN protocol is a viable solution as an implementation choice for LPWAN.

The Wi-SUN security is specified by implementation of the x.509 certificate-based, public-key infrastructure to authenticate devices, as well as Advanced Encryption Standard (AES) encryption and message integrity check. Devices protect their digital credentials either by storing them in hardened cryptographic processors that are resistant to physical tampering or by using physically unclonable function (PUF) technology.

3.8.4 Wireless and Cellular Wide Area Networks (WWAN)

3.8.4.1 WiMax IEEE 802.16

IEEE 802.16 technology has been put forward to overcome the drawbacks of WLANs and mobile networks. It provides different QoS scheduling for supporting heterogeneous traffic including legacy voice traffic, VoIP (Voice over IP), voice and video streams and the Internet data traffic. The prominent features of WiMAX include quality of service, high-speed Internet, facility over a long distance, scalability, security, and mobility.

IEEE 802.16 (WiMax) uses PMP network topology, in 2.3 GHz, 3.5 GHz, 5.8 GHz frequency spectrum, providing data rates of 40 Mbit/s for mobile, 1 Gbit/s for fixed networks with a coverage range of 50 km.

Different security solutions are enabled in WiMax networks, like Advanced Encryption Standard (AES) with 128-bit key: Rivest, Shamir and Adelman (RSA) with 1024-bit key and Triple Digital Encryption Standard (3-DES). Both Advanced Encryption Standard (AES) and Triple Digital Encryption Standard (3-DES) are symmetric encryption algorithms using a block-cipher method. Rivest, Shamir and Adelman (RSA) is an asymmetrical algorithm.

The air interface in IEEE 802.16 networks is secured by authentication procedures, secure key exchange and encapsulation. With encapsulating data from authorised users, the base station limits the access of unauthorised users. Besides, it supports the Privacy Key Management (PKM) protocol for secure two-layer-key distribution and exchange and real-time confirmation of subscribers' identification, which ensures secure wireless data transport.

3.8.4.2 2G (GSM)

IoT applications rely on data transfer over *cellular technologies* such as 2G (GSM, D-AMPS, PDC), 2.5G (GPRS), 2.75G (EDGE), 3G (UMTS/WCDMA, HSPA, HSUPA , EvDO), 4G (i.e. LTE, LTE-A), 5G. M2M (*Machine-to-Machine*) connectivity is referred within the cellular context or MTC (*Machine-type Communication*) within 3GPP (3rd Generation Partnership Project). The approximate range is 35km for GSM and 200km max for HSPA. The data rate for typical download is in the range of 35–170 kps (GPRS), 120–384 kbps (EDGE), 384Kbps–2 Mbps (UMTS), 600 kbps–10 Mbps (HSPA), 3–10 Mbps (LTE).

3.8.4.3 3G (GSM/CDMA)

3G and 4G technologies such as 3GPP LTE are enabling technologies that offer wide area cover- age, QoS support, mobility and roaming support, scalability, billing, high level of security, the simplicity of management as well as connectivity of sensors through a standardised API [127] LTE-A (Long Term Evolution – Advanced) and Mobile WiMAX Release 2 (*Wireless MAN – Advanced* or *IEEE 802.*16 m) enabling higher speeds, more scalability, and low costs.

3.8.4.4 4G (LTE)

3GPP specified technologies such as eMTC (enhanced Machine-Type Communication), NB-IoT, and EC-GSM-IoT. The eMTC brings some LTE enhancements for MTC such as a new Power Save Mode (PSM). Release 14 brings new eMTC feature enhancements such as support for positioning and multicast, mobility for inter- frequency measurements, and higher data rates [131]. It brings enhancements such as lower costs, reduced data rate/bandwidth, and some other protocol optimisations.

Release-14 delivers new enhancements for the NB-IoT technology such as support for multicast, power consumption and latency reduction, mobility, and service continuity enhancements, etc. EC-GSM-IoT delivered EGPRS enhancements, which in combination with PSM makes GSM/EDGE systems IoT ready.

This technology brings improvements such as extended coverage, support for massive number of devices: at least 50 000 per cell, improved security compared to GSM/EDGE, etc. The eMTC, NB-IoT, and EC-GSM-IoT has are described in the report on progress on 3GPP IoT [130]. QoS and network congestion are very challenging issues due to a huge number of deployed nodes (devices) [128]. In 4G LTE enhanced security was added such as unique identifiers (ID) for end-mobile device (UE), secure signalling between the UE and MME (Mobile Management Entity) and security for interworking between 3GPP networks and trusted non-3GPP users (e.g. using EAP-AKA) the UMTS Authentication and Key Agreement protocol.

3.8.4.5 5G

Cellular next generation 5G with high-speed mobility support and ultra-low latency is positioned to be the future of autonomous vehicles and augmented reality. 5G is also expected to enable real-time video surveillance for public safety, real-time mobile delivery of medical data sets for connected health, and several time-sensitive industrial automation applications in the future.

5G networks are expected to support the new IoT technologies, enabling IoT/IIoT device producers to develop and deploy new IoT devices and systems across multiple industries and provide IoT/IIoT applications globally.

The cell radius in 2 G systems is 35 km, in 3G systems 5 km, in 4G systems 100 m, and in 5G about 25 m to reuse the available RF spectrum more efficiently and to achieve higher data densities.

To implement the ultra-low latency and very high bit-rate applications, the connectivity technologies require more sizable contiguous blocks of the spectrum than those available in frequency bands that have been previously used. As the IoT/IIoT and the underlying connectivity technologies aim for worldwide coverage, there is a need for harmonised worldwide bands to facilitate global roaming and the benefits of economies of scale.

Technology advancements enabling 5G deployments can be divided into radio deployments in new bands in the sub-6GHz range, deployments in millimetre wave frequency bands and deployments in existing LTE bands. The first two categories combined can provide a forecasted population coverage of 55 percent in 2025, while the third category, where the networks can be upgraded to support 5G services in existing LTE bands by utilizing spectrum

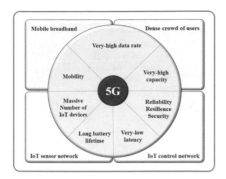 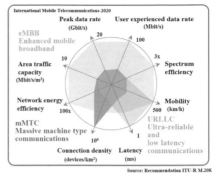

Figure 3.28 5G key capabilities of IMT-2020 defined by ITU.

sharing can provide a 10 percentage population coverage is achievable, creating a potential of up to 65 percent coverage in 2025 [82].

The ITU-R M.2083-0 define the capabilities of 5G to use cases such as mobile broadband, massive-machine communication, and mission-critical communication that include the full deployment of IoT solutions. 3GPP Releases 15 and 16 addresses the set of 5G standard specifications starting from LTE-Advanced Pro specifications. The functional specifications include mMTC (massive Machine Type Communications) requirements, specifications for eMBB (enhanced Mobile Broadband) and URLLC (Ultra-Reliable and Low Latency Communications), etc. A survey of the 5G cellular network architecture and key emerging technologies like interference management, spectrum sharing with cognitive radio, cloud computing, SDN, etc. is described in [119]. An overview of unique characteristics and characteristics of IoT and 5G technologies are given in [129].

The 5G technologies spectrum is distributed within three key frequency ranges to deliver the required coverage and support different use cases across various applications. The three ranges are: below-2 GHz, 2-8 GHz and above 24 GHz as presented in Figure 3.29.

Below 2 GHz low frequencies – low bands, support widespread coverage across urban, suburban, and rural areas and accelerate the IoT services for massive machine-type communications. Low-frequency bands extend the 5G mobile broadband to more extensive areas and deeper indoor environments. URLLC and mMTC type services benefit from enhancing coverage at the low frequency-bands.

The 2-8 GHz medium frequencies – mid bands offer a mixture of coverage and capacity benefits that includes spectrum within the 3.3-3.8 GHz range

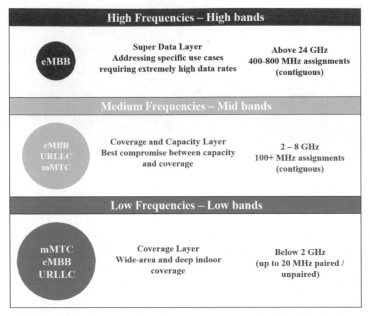

Figure 3.29   Ranges of the frequency spectrum for 5G technologies.

expected to form the base for many 5G services. The mid bands include the 2.3 GHz and 2.6 GHz frequencies. The unpaired (TDD) bands at 3300-4200, 4400-5000, 2500-2690 and 2300-2400 MHz deliver the best compromise between wide-area coverage and high capacity.

Above 24 GHz high frequencies – high bands are needed to ensure the ultra-high broadband speeds. The 26 GHz and/or 28 GHz bands have the most international support in this range. High-frequency bands are essential for providing additional capacity and delivering the extremely high data rates required by some 5G eMBB applications at specific locations ("hotspots"). The 400-800 MHz of contiguous spectrum per network operator is recommended from higher frequencies to achieve good return on investment and meet service requirements.

The frequency bands 24.25–27.5 GHz (global), 37–43.5 GHz (global), 45.5–47 GHz (regional/multi-country), 47.2–48.2 GHz (regional/multi-country) and 66–71 GHz (global) were identified for International Mobile Telecommunications (IMT) for the deployment of 5G networks by the World Radiocommunication Conference 2019 (WRC-19), that took place in Egypt, 28 October to 22 November 2019. WRC-19 took measures to ensure

Group 30 (GHz)	Group 40 (GHz)	Group 50 (GHz)	Group 70/80 (GHz)
24.25-27.5 31.8-33.4	37-40.5 40.5-42.5 42.5-43.5	45.5-47 47-47.2 47.2-58.2 50.4-52.6	66-71 71-76 81-86

Figure 3.30 Frequency bands identified by the WRC-19 for the deployment of 5G networks.

appropriate protection of the Earth Exploration Satellite Services, including meteorological and other passive services in adjacent bands.

The rage for 5G depends on the frequency bands used. The low band 5G has a range of tens of kilometres (similar range to 4G), mid-band 5G has several kilometres range, and high band 5G has hundreds of meters up to 1.5 km range. The data rates for the different bands are 30–250 Mbps for low-band 5G (600–700 MHz), 100–900 Mbps for mid-band 5G (2.5–3.7 GHz) and downloading speeds of 1–3 Gbps for high-band 5G (25–39 GHz and higher frequencies up to 80 GHz).

5G aims to integrate different portions of the unlicensed spectrum, to work in concurrence with licensed bands or independently. 5G NR-U (unlicensed) will support operations on 5 GHz spectrum used by Wi-Fi and new frequencies in the 6 GHz spectrum (e.g. future support includes the 3.5-4.2 GHz, 6-7 GHz, 37.37.6 GHz (US) and 57-70 GHz bands). 5G NR-U supports wideband carriers, flexible numerologies, beamforming, and dynamic TDD, where the uplink-downlink allocation could change to adapt to traffic conditions. The Licensed Assisted Access NR-U (LAA NR-U) aggregates carriers of both licensed and unlicensed spectrum, using NR and LTE carriers in the licensed band as anchors combined with NR-U carriers in the unlicensed spectrum to increase network capacity and provide broadband services at a lower cost. Stand-alone NR-U enables stand-alone operation in the unlicensed spectrum and can be deployed by different stakeholders. The use scenarios include local private networks, for specific uses industrial IoT applications or mobile enterprise broadband. Other potential applications are the delivery of broadband connectivity to public locations such as in stadiums or shopping malls. For private networks, 5G NR-U can deliver improved coverage capacity, mobility, increased reliability and precise timing due to the integration with time sensitive networking (TSN) that provides deterministic services over IEEE standard 802.3 Ethernet wired networks, guaranteeing low-latency packet transport, low packet delay variation and low packet loss.

The encryption is enhanced compared with 4G and the level of 5G security is not defined by the number of specified security mechanisms. A new approach is necessary for addressing 5G security to provide the security baseline of trustworthy, cost-efficient, and manageable 5G networks for various IoT/IIoT applications.

The Next Generation Mobile Network (NGMN) [121] introduced the concept of 5G network slicing and defined as an end-to-end (E2E) logical network/cloud running on a common underlying (physical or virtual) infrastructure, mutually isolated, with independent control and management that can be created on demand. The NGMN slice capabilities consists of several layers [120], the 5G Service Instance Layer (5GSIL) that represents different services which are to be supported, the 5G Network Slice Instance (5GNSI) that provides network characteristics which are required by a 5GSI, and the 5G Resource Layer (5GRL) that consists of physical resources (e.g. assets for computation, storage or transport including radio access) and logical resources (e.g. partition of a physical resource or grouping of multiple physical resources dedicated to a Network Function (NF) or shared between a set of NFs).

A network slice may be expressed by of cross domain components from separate domains in the same or different administrations, or components applicable to the access network, transport network, core network, and edge networks. Network slices are manageable, and programmable, self-contained, mutually isolated, to support multi-service and multitenancy.

Standardisation organisations have defined the network slicing considering various perspectives. The 3GPP [123] described network slicing as a "technology that enables the operator to create networks, customised to provide optimised solutions for different market scenarios which demand diverse requirements (e.g., in terms of functionality, performance, and isolation)" [124]. The ITU-T designated the network slicing as the a concept of network softwarisation that facilitates a Logical Isolated Network Partitions (LINP) composed of multiple virtual resources, isolated and equipped with a programmable control and data plane [123].

The network slicing overview and programmability are illustrated in Figure 3.31 and Figure 3.32. Network slicing concept can support multiple logical and self-contained networks on top of a shared physical infrastructure platform [122]. Network slicing present various operational challenges [125] such as configuration capability to allow users to adjust and modify the network functions as well as underlying resources within the network slice instance provided for them, monitoring capability for following the traffic

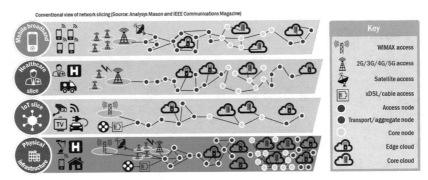

Figure 3.31 Network slicing overview. (Source: Analysis Mason and IEEE Communications Magazine).

characteristics and performance (e.g., data rate, packet drop, and latency), end user's geographical distribution, etc., in addition to per session/user/slice instance-based monitoring, etc. and control capability to enable the customers to utilise application programming interfaces (APIs) provided by the operator to control network service. Considering different applications where the network slicing concept is used, a slice encompasses a combination of relevant network resources, functions, and assets required to fulfil a specific business case or service, including operations support system/business support system (OSS/BSS), and DevOps processes.

Internal and external slices are used. The internal slices are partitions used for internal services of the provider, retaining full control and management of them and the external slices are partitions hosting customer services, appearing to the customer as dedicated networks/clouds/datacentres.In the initial phase of network slicing, operators are likely to launch a handful of slice types (e.g., eMBB, URLLC, and mMTC) with multiple tenants per slice. Over time, the number of slice types should increase and become more service specific (e.g., video gaming, smart meter connectivity, autonomous vehicles, specific IoT/IIoT use cases, etc.). For mobile network operators (MNOs) to capture value beyond basic connectivity, they must change their business models to address industry vertical-specific ecosystems. This will require a fundamental change in the way operators manage monetisation to address the multitude of use cases that network slicing will offer, including fixed wireless access, augmented reality/virtual reality (AR/VR) broadcasting, and industrial IoT [125].

Network programmability and slicing offer flexibility and control and the implementation steps are illustrated in Figure 3.32 [74]

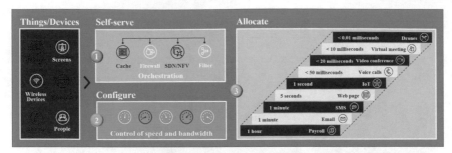

Figure 3.32 Network programmability and slicing. Adapted from Telestra [74].

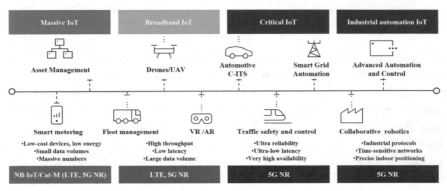

Figure 3.33 Cellular IoT use case segments. Adapted from [79].

Cellular IoT use cases have differing connectivity requirements and can be divided into four segments as presented in Figure 3.33. Verticals using Massive IoT include utilities with smart metering, healthcare in the form of medical wearables and transport with tracking sensors. NB-IoT and Cat-M will be able to fully co-exist in spectrum bands with 5G NR. Broadband IoT includes wide-area use cases that require higher throughput, lower latency, and larger data volumes (e.g. peak data rates in the multi-Gbps range and radio interface latency as low as 10ms) such as smart watches, drones/UAV, etc.

Critical IoT includes both wide-area and local-area use cases that have requirements for extremely low latency and ultra-high reliability such as such as interactive transport systems in the automotive industry, smart grids with real-time control and distribution of renewable energy in the utilities industry, and real-time control of manufacturing robots in the manufacturing industry. Industrial automation IoT includes very specific use cases, with the most demanding requirements coming from the manufacturing and industrial sites.

Table 3.4	IoT connections (billions) [82]		
IoT	2019	2025	CAGR
Wide-area IoT	1,6	5,5	23%
Cellular IoT	1,5	5,2	23%
Short-range IoT	9,1	19,1	13%
Total	**10,7**	**24,6**	**15%**

Note: The figures for cellular IoT are also included in the figures for wide-area IoT.

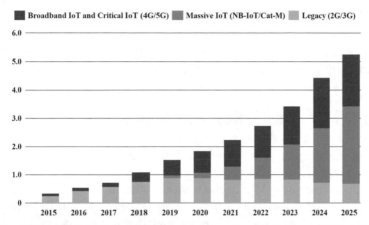

Figure 3.34 Cellular IoT connections by segment and technology (billion) [82].

Time-sensitive networks, industrial protocols running over ethernet, and very precise positioning will be needed [79].

It is projected that more than five billion active cellular IoT connections will be available by 2025 which are split between LTE-M and NB-IoT as shown in Table 3.4. The cellular IoT connections by segment and technology are presented in Figure 3.34.

The massive IoT technologies NB-IoT and Cat-M1 are rolled out at a slightly slower pace in 2020 than previously forecast due to the impact of the pandemic crisis. 2G and 3G connectivity enable the majority of IoT applications. massive IoT connections reached around 100 million connections at the end of 2019.

NB-IoT and Cat-M technologies complement each other, and it is projected to account for 52 percent of all cellular IoT connections [82]. Cat-M includes both Cat-M1 and Cat-M2. Only Cat-M1 is being supported today. Commercial devices for massive IoT include different types of meters, sensors, trackers, and wearables.

The deployment of 5G New Radio (NR) will increase data rates and the broadband IoT that includes wide-area use cases that utilise higher through-put, lower latency, and larger data volumes. Critical IoT that is used for time-critical communications in wide- and local-area use cases requires guaranteed data delivery with specified latency targets, will be introduced in 5G networks with the advanced time-critical communication capabilities of 5G NR in 2021. The use cases for critical IoT include cloud-based AR/VR, robotics, IoRT, autonomous vehicles, edge computing, advanced cloud gaming, and real-time coordination and control of machines and processes.

The 5G devices are initially supporting mobile broadband capabilities, and performance is expected to evolve towards time-critical communication capabilities where needed, via software upgrades on devices and networks [82].

3.8.4.6 NB-IoT (LTE Cat NB1 and LTE Cat NB2)
The 3GPP has started the standardisation of a set of low cost and low complexity devices targeting machine type-communications (MTC) in Release 13. The standardisation addresses the IoT market from the enhanced machine type communications (eMTC), the narrow band IoT (NB-IoT) and the EC-GSM-IoT. eMTC is an evolution of the work developed in Release 12 that can reach up to 1 Mbps in the uplink and downlink and operates in LTE bands with a 1.08 MHz bandwidth.

In Release 14 3GPP introduced five new FDD frequency bands for NB-IoT: 11 (central frequencies – UL 1437.9 MHz, DL 1485.9 MHz), 21 (central frequencies – UL 1455.4 MHz, DL 1503.4 MHz), 25 (central frequencies – UL 1882.5 MHz, DL 3962.5 MHz), 31 (central frequencies – UL 455 MHz, DL 465 MHz), and 70 (central frequencies – UL 1702.5 MHz, DL 2007.5 MHz).

The NB-IoT, CAT-NB1, use the existing 4G/LTE network [216]. NB-IoT and LTE coexist, and re-use of the LTE physical layer and higher protocol layers benefits the technology implementation. NB-IoT has been designed for extended range, and the uplink capacity can be improved in bad coverage areas. NB-IoT devices support three different operation modes [216, 217]:

- Stand-alone operation (in other spectrum/non-LTE) – utilizing one or more GSM carriers (bandwidth of 200 kHz replacements) utilizing for example the spectrum currently being used by GERAN systems as a replacement of one or more GSM carriers.

- Guard band operation (in the guard band of an LTE carrier) – utilizing the unused resource blocks within an LTE carriers' guard-band (frequency bands to prevent interference).
- In-band operation (within an LTE carrier) – utilizing resource blocks within a normal LTE carrier.

NB-IoT and LTE-M defined in 3GPP Release 13, has two user network equipment categories: Cat-NB1 for NB-IoT networks and Cat-M1 for LTE-M networks [217].

The coverage enhancement modes introduced as part of LTE-M can also be optionally supported by ordinary LTE user equipment categories. Ten years battery lifetime and low-cost devices are available and support a high number of low throughputs in IoT for different applications.

NB-IoT is designed to exist in independently licensed bands, in unused 200 kHz bands that have previously been used for GSM or CDMA, or on LTE base stations that can allocate a resource block to NB-IoT operations or in their guard bands (where regulations allow it).

The LTE Cat NB1 provides data rates of 66 kbps (multi-tone), or 16.9 Kbit/s (single tone), while LTE Cat NB2 data rates of 159kbps.EC-GSM-IoT

EC-GSM-IoT is an evolution of EGPRS towards IoT, operates in the frequency band of 850-900 MHz (GSM bands), uses a star topology, provides a data rate of 70 kbps (GSMK), 240 kbps (8PSK) with a coverage approximate range of 15km.

The EC-GSM-IoT has improved security, compared to the existing GSM/GPRS networks – offers integrity protection, mutual authentication and implements stronger ciphering algorithms. The EC-GSM-IoT technology implementation is based on software upgrades of the existing GSM networks.

3.8.4.7 LTE-M

LTE-M is a LPWAN cellular radio technology standard developed by 3GPP to enable a wide range of cellular devices and services IoT applications.

LTE-MTC Cat 0 uses the technology frequency bands (700 MHz, 800 MHz, 900 MHz, 1700 MHz, 1800 MHz, 1900 MHz, 2300 MHz, 2400 MHz, 2500 MHz, 2700 MHz). The technology uses a star topology, provides a data rate of 1 Mbps with a coverage range of 10km that is variable and depends on frequency bands, propagation conditions etc. LTE-MTC system and security management is enhanced compared to LTE, as numbers of devices in LTE MTC network are very large. The request defined in 3GPP TS 22.368 is

"LTE MTC optimisations shall not degrade security compared to non-MTC communications".

LTE-M eMTC (Cat M1, Cat M2) uses the technology frequency bands (400 MHz 450 MHz, 600 MHz, 700 MHz, 800 MHz, 900 MHz, 1400 MHz, 1500 MHz, 1700 MHz, 1800 MHz, 1900 MHz, 2100 MHz, 2300 MHz, 2400 MHz, 2500 MHz, 2600 MHz, 2700 MHz). The technology uses a star topology, provides a data rate of 1 Mbps for LTE-M Cat M1 and $\approx$4 Mbps DL/ $\approx$7 Mbps UL for LTE-M Cat M2 with a coverage range of 10 km that is variable and depends on frequency bands, propagation conditions etc. LTE-M technology offers SIM-based security features requiring device authentication to connect to the network. Security system and management is more complex in LTE-M (eMTC) than LTE due to massive connectivity that is supported in LTE-M (eMTC) networks.

Two innovations support LTE-M improve battery life: LTE eDRX (extended discontinuous reception) and LTE PSM (power saving mode). LTE-M delivering data rates of around 150-200kbps (1.4MHz bandwidth), with the capacity to support higher bandwidth workloads, addresses mobility credentials and provides a 50 ms/100 ms latency response.

The advantage of LTE-M over NB-IoT is its comparatively higher data rate, mobility, and voice over the network (VoLTE). LTE-M benefit from reduced complexity with devices implemented with a very simple frontend and antenna configuration.

LTE-M theoretically can be extremely power-efficient but because eDRX and PSM are only recently being deployed, their power efficiency is still evaluated.

An overview of the cellular IoT LPWAN technologies defined by 3GPP (e.g. NB-IoT, LTE-M, EC-GSM-IoT) their features and properties is presented in Table 3.5.

A comparison of the IoT personal, local, wide area networks is presented in Table 3.6.

3.8.5 Low Power Wide Area Networks (LPWAN)

3.8.5.1 LoRaWAN

LoRaWAN defines a communication protocol and network architecture for IoT low-power wide area networks (LPWANs) and is designed to address the requirements for low power consumption (i.e., long battery life), long range, and high low data rate ($<$ 2 kbit/s) while maintaining low operating and deployment costs.

Table 3.5 Overview of the cellular IoT LPWAN technologies defined by 3GPP

| | LC-LTE/MTCe | LTE-M | | | NB-IoT | | |
| | | eMTC | | | | | |
	LTE Cat 1	LTE Cat 0	LTE Cat M1	LTE Cat M2	non-BL	LTE Cat NB1	LTE Cat NB2	EC-GSM-IoT
3GPP Release	8	12	13	14	14	13	14	13
Downlink Peak Rate	10 Mbit/s	1 Mbit/s	1 Mbit/s	~4 Mbit/s	~4 Mbit/s	26 kbit/s	127 kbit/s	474 kbit/s (EDGE) 2 Mbit/s (EGPRS2B)
Uplink Peak Rate	5 Mbit/s	1 Mbit/s	1 Mbit/s	~7 Mbit/s	~7 Mbit/s	66 kbit/s (multi-tone) 16.9 kbit/s (single tone)	159 kbit/s	474 kbit/s (EDGE) 2 Mbit/s (EGPRS2B)
Latency	50-100ms	Not deployed	10-15ms			1.6-10s		700 ms-2s
Antennas	2	1	1	1	1	1	1	1–2
Duplex Mode	Full Duplex	Full or Half Duplex	Full or Half Duplex	Full or Half Duplex	Full or Half Duplex	Half Duplex	Half Duplex	Half Duplex
Device Receive Bandwidth	1.4–20 MHz	1.4–20 MHz	1.4 MHz	5 MHz	5 MHz	180 kHz	180 kHz	200 kHz
Receiver Chains	2 (MIMO)	1 (SISO)	1 (SISO)	1 (SISO)	1 (SISO)	1 (SISO)	1 (SISO)	1–2
Device Transmit Power (dBm)	23	23	20/23	20/23	20/23	20/23	14/20/23	23/33

Table 3.6 Comparison of different IoT protocols

	Personal and Local Area		Wide Area				
Network type	BLE	ZigBee	LoRaWAN	MIOTY	Sigfox	LTE-M	NB-IoT
	Mesh	Mesh	LPWA	TS-UNB	UNB	LPWA	UNB
Standard	IEEE 802.15.4	IEEE 802.15.4	Proprietary	ETSI TS 103357	Proprietary	3GPP Rel. 12-14	3GPP Rel. 13-14
Bandwidth	2 MHz	600 kHz-5 MHz	125 kHz	200 kHz	100 kHz	1.4MHz	200kHz
Cell capacity	n/a	n/a	40,000	>1,000,000	1,000,000	1,000,000	200,000
Max nodes	1,000 +	250	n/a	n/a	n/a	n/a	n/a
Max range	30 m	10-100 m	5-15 km	5-15 km	10-30 km	11 km	15 km
Throughput	2 Mbps	250 Kbps	50 Kbps	512 bps	600 bps	1 Mbps	200 Kbps
Latency	< 3 ms	15 ms	1-100 ms	n/a	>20s	10-15 ms	< 10ms
Payload	350 bytes	102 bytes	243 bytes	10 – 192 bytes	12/8 bytes (UL/DL)	–	1.6 Kbytes
Alliance	Bluetooth SIG	Zigbee Alliance	Private	MIOTY Alliance	Private	GSMA / 3GPP	GSMA / 3GPP

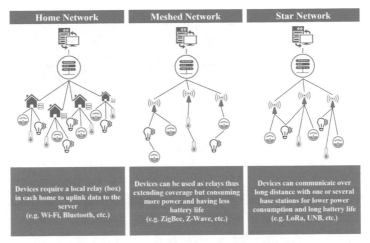

Figure 3.35   LoRa network topologies [111].

The LoRa physical layer uses chirp spread spectrum modulation a spread spectrum technique where the signal is modulated by chirp pulses (frequency varying sinusoidal pulses) hence improving resilience and robustness against interference, Doppler effect and multipath, characterised by low power usage and increased communication range allowing a single base station to cover hundreds of square kilometres. LoRa and LoRaWAN, enable long battery life for devices in the field and covers the IoT communication needs between local wireless such as Bluetooth, Wi-Fi and cellular-based wireless.

LoRaWAN features data rates of 27 kbps (50 kbps when using FSK instead of LoRa), and a single gateway can collect data from thousands of nodes deployed at 5-15 km distance. LoRaWAN networks are organised in a star of stars topology, in which gateway nodes relay messages between end devices and a central network server. End devices send data to gateways over a single wireless hop and gateways are connected to the network server through a non-LoRaWAN network (e.g. IP over Cellular or Ethernet). Communication is bi-directional, even uplink communication from end devices to the network server is preferred. LoRa network topologies are illustrated in Figure 3.35

LoRa has several key features that have made it a strong choice for organisations that are looking to achieve efficiency and cost savings with IoT. However, as use cases for IoT deployments using LoRa become more advanced, and as IoT requirements scale, a new way of approaching the problem of how to best benefit from LoRa is required. That is where Super-B comes in. Super-B is an advanced protocol that is built on top of the

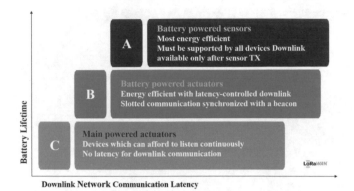

Figure 3.36   LoRaWAN classes of communications between sensors and gateways [73].

LoRaWAN protocol while utilizing the same network infrastructure as LoRa. By selecting IoT solutions using the Super-B protocol, many issues can be alleviated so that organisations can easily [73]:

- Densify IoT networks to meet their organisation's growing requirements
- Vastly improve data delivery and QoS
- Better secure IoT networks through firmware over the air (FOTA) updates
- Deliver strong return on investment (ROI) by reducing the need to add hardware to scale networks

The LoRaWAN "classes" are illustrated in Figure 3.36. LoRaWAN has three "classes" of communications between sensors and gateways [73]:

- **Class A** – The gateway is passive and in listen-only mode. Sensors utilise listen before talking and send messages whenever there is a message to send. Because of its random nature, Class A is prone to significant noise and interference, and thus packet loss, which can be exacerbated by the all-too-common practice of sending a message a few times to increase the likelihood of delivery.
- **Class B** – The network establishes a session to communicate but it's set up and torn down, forcing the sensor to search for the gateway each time it connects, creating inefficiencies. Has limited scheduled receive slots, however, lacks the efficiency needed to scale. Class B sensors frequently cause interference with Class A sensors.
- **Class C** – Has a bidirectional link that is never torn down, so is "always on" to send and receive messages.

Super-B protocol rides atop the LoRa protocol, utilizing standard LoRa but extending its structure to allow for the scheduling of messages from gateway to devices (and vice versa) while maintaining extremely low power. Super-B capitalises on the best parts of LoRa A and LoRa B to deliver scalable, secure IoT.

The LoRa technology is defined by three main parameters: spreading factor (SF), bandwidth (BW), and carrier frequency. The transfer rate varies by using orthogonal spreading factors which is a compromise between distance and emission power [108]. A pseudorandom channel hopping method is natively used in LoRaWAN to distribute transmissions over the pool of available channels, thereby reducing the collision probability.

Several companies offer LoRa-based IoT solutions for different applications. Semtech [109] has released LoRa Edge to simplify IoT deployments for indoor and outdoor asset management applications. The LoRa Edge geolocation platform integrates a LoRa transceiver with Wi-Fi and Global Navigation Satellite System (GNSS) scanning technologies. The hardware is connected to the geolocation and device management services that operate on the LoRa cloud platform. The optimisation of the capacity of the LoRaWAN network, and the possibility to perform traffic slicing for guaranteeing specific requirements in a service basis, remain as open issues. Other issues to be addressed are new channel hopping methods to meet traffic requirements when there are latency, jitter or reliability constraints (i.e. downlink ACKs for all packets), that cannot be adapted according to the noise level of each channel.

The nature of ALOHA-based access is not optimal to serve deterministic traffic and using a hybrid Time Division Multiple Access (TDMA) access on top of LoRaWAN could provide new use cases and adds more flexibility.

The location of LoRa end devices for IoT applications is necessary for different IoT use cases and GPS-based solutions too expensive and requires extra computing resources and energy consumption. TDOA-based (Time Difference Of Arrival) triangulation techniques for LoRaWAN could provide solutions for dense gateway deployments. Future developments could include the integration of cognitive radio into the LoRaWAN standard. Considering the random-based access in unlicensed bands of LoRaWAN, the performance achieved in isolated networks is cross-examined in scenarios with co-existing gateways and limited number of available channels. It is key to devise coordination mechanisms between gateways from the same or different operators to limit interference and collisions and provide co-existence mechanisms that encompass coordination and reconfiguration protocols for gateways and end-devices. [100, 110].

It should be noted that 1) the reach and autonomy of the IoT devices based LoRa connectivity will depend on the applications requirements. For instance, the reach depends not only on the transmitter performances, but also on the distance (between the end-devices and the closest gateway) and antenna gains. The battery life depends on the reading duty cycle and the number of bytes transmitted in the payload. The reach and the autonomy will be reduced when these parameters have high numbers. 2) LoRaWAN allows developing IoT private networks at cost effective for applications (e.g. in agriculture and Smart city) requiring long reach (around 10 kms) and low data rates.

3.8.5.2 Weightless – N, W, P

Weightless is a LPWAN standard offered by the Weightless Special Interest Group (SIG) in three different protocols: Weightless-N, Weightless-W, and Weightless-P. Each of these are designed to support different end-user cases and modalities.

The Weightless technology can operate in any frequency band and is currently defined for operation in license-exempt sub-GHz frequency bands (e.g. 138MHz, 433MHz, 470MHz, 780MHz, 868MHz, 915MHz, 923MHz). The approximate range is 2km (P), 5km (W, N).

Weightless-N system supports an ultra-narrowband connectivity and has many similarities with Sigfox. It is sometimes considered like a LoRa-based version of Sigfox. Wireless-N system comprises networks of partners instead of a completely end to end enclosed approach. The modulation used is BPSK and is usually intended for uplinks related to sensor data.

Weightless-N was designed to expand the range of Weightless-W and reduce the power consumption (e.g. battery lifetime up to 10 years) at the expense of data rate decrease (from up to 1 Mbps in Weightless-W to 100 kbps in Weightless-N). Weightless-N is based on the Ultra Narrow Band (UNB) technology and operates in the UHF 800-900 MHz band; it provides only uplink communication.

Weightless-W was developed as a bidirectional (uplink/downlink) solution to operate in TV whitespaces (470-790 MHz). It is based on narrowband FDMA channels with Time Division Duplex between uplink and downlink; data rate ranges from 1 kbps to 1 Mbps and battery lifetime is 3 to 5 years.

The characteristics of Weightless-P are bidirectional, fully acknowledged communication for reliability, optimised for a large number of low-complexity end devices with asynchronous uplink-dominated communication with short payload sizes (typically < 48 bytes) and standard data rates from 0.625kbps to 100kbps.

The Weightless-P systems include end devices (ED) (e.g. the leaf node in the network, low-complexity, low-cost, usually low duty cycle), base stations (BS) (e.g. the central node in each cell, with which all EDs communicate via a star topology) and base station network (BSN) (e.g interconnects all BS of a single network to manage the radio resource allocation and scheduling across the network, and handle authentication, roaming and scheduling).

Weightless-P is proposed as a high-performance two-way communication solution that can operate over 169, 433, 470, 780, 868, 915 and 923 MHz bands. Cost of the terminals and power consumption are higher than in Weightless-N, with a battery lifetime of 3 to 8 years.

In Weightless standard AES-128/256 encryption and authentication of both the terminal and the network guarantees integrity whilst temporary device identifiers offer anonymity for maximum security and privacy. OTA security key negotiation or replacement is possible whilst a future-proof cipher negotiation scheme with a minimum key length of 128 bits protects long term investment in the network integrity.

3.8.5.3 Sigfox

Working in a similar approach to Weightless-N, Sigfox is a low cost, low power, and very reliable means for connecting sensors and related devices. It is of great use in the IoT domain and is expected to remain. The Sigfox protocol focuses on attributes such as extended autonomy using very low energy, simple setup and maintenance with no configuration, quick deployment, low cost, low bandwidth, small message structure (up to 12bytes) and the possibility to be combined with Wi-Fi, BLE or cellular (e.g. GPRS, 3G, 4G, etc.).

Sigfox employs the differential binary phase-shift keying (DBPSK) and the Gaussian frequency shift keying (GFSK) that enables communication using the Industrial, Scientific and Medical ISM radio band which uses 868 MHz in Europe and 902 MHz in the US. It utilises a signal that is extremely narrowband (100 Hz bandwidth), called "Ultra Narrowband" and requires little energy, being termed "Low-power Wide-area network (LPWAN)". It is based on Random Frequency and Time Division Multiple Access (RFTDMA) and achieves a data rate around 100 bps in the uplink, with a maximum packet payload of 12 Bytes, and a number of packets per device that cannot exceed

14 packets/day. It consumes only 50 microwatts and can deliver a typical stand-by time 20 years with a 2.5Ah battery.

The network is based on one-hop star topology and requires a mobile operator to carry the generated traffic. The signal can also be used to easily cover large areas and to reach underground objects. The range is 30-50km (rural environments), 3-10km (urban environments) and the data rates varies from 10 to 1000bps.

Security is implemented in the Sigfox devices during the manufacturing process, when each Sigfox Ready device is provisioned with a symmetrical authentication key. Additional security is supported by radio technology. The Sigfox technology encryption is designed for use with short Sigfox messages. End-to-end encryption solutions are applicable to the Sigfox networks and applications.

3.8.5.4 Random-phase multiple access (RPMA) – Ingenu

Random-phase multiple access (RPMA) is a bidirectional IoT communications technology developed by Ingenu that operates in the 2.4 GHz ISM band. Ingenu claim that RPMA coverage can extend up to 70 kilometres, throughput of 19 kbps/MHz and an uplink link budget of 180 dB and 185 dB for downlink in the US. Citing security concerns, RPMA does not support IP connectivity. Instead, Ingenu choose to provide a REST API instead for data access [116]. The Ingenu proprietary LPWAN technology in the 2.4 GHz band, is based on Random Phase Multiple Access (RPMA) to provide M2M industry solutions and private networks.

The Ingenu technology provides high data rate up to 624 kbps in the uplink, and 156 kbps in the downlink. The energy consumption is higher, and the range is shorter (a range around 5-6 km) due to the high spectrum band used.

Security in RPMA wireless technology is built on 128-bit AES. It offers security features such as: mutual authentication, message integrity and replay protection, message confidentiality, device anonymity, authentic firmware upgrades and secure multicasts.

3.8.5.5 Neul

Neul technology operates in the sub-1GHz band and leverages very small slices of the TV White Space spectrum to deliver high scalability, high coverage, low power, and low-cost wireless networks.

Neul systems are based on the Iceni chip, which communicates using the white space radio to access the high-quality UHF spectrum, available due to the analogue to digital TV transition.

Data rates can be anything from a few bits per second up to 100kbps over the same single link; and devices can consume 20 to 30mA and that offer a10 to 15 years lifetime in the field.

The frequencies used by Neul are 900MHz (ISM), 458MHz (UK), 470-790MHz (White Space) providing a range of 10km.

The wireless communications links between the gateway (base station) and the network nodes are encrypted.

3.8.5.6 Wavenis

Wavenis is a wireless technology for ultra-low power and long-range Wireless Sensor Networks (WSNs) and promoted by Wavenis Open Standard Alliance.

Wavenis uses tree and star network technologies, in the 433MHz, 868MHz, 915MHz frequency spectrum providing low data rates of 9.6kbps (433 & 868MHz), 19.2kbps (915MHz) and a coverage range of approximate 1 km.

Wavenis technology is supported by 128-bit AES encryption.

3.8.5.7 WAVIoT (NB-Fi – Narrowband Fidelity)

NB-Fi (Narrowband Fidelity) is a narrow band protocol which communicates on the sub 1GHz ISM sub bands. DBPSK is used as the modulation scheme in the physical layer. WAVIoT gateways can provide -154 dBm of receiver sensitivity, and cover over 1 million nodes. On WAVIoT-developed devices, short data bursts use 50mA of current, and in idle mode – a few ʇA are used. Devices have a lifetime of up to 20 years, and a 176 dBm link budget.

WAVIoT uses a star network technology, in the 315 MHz, 433 MHz, 470 MHz, 868 MHz, 915 MHz frequency spectrum providing data rates of 10–100 bps and a coverage range of approximate 50 km.

All WAVIoT data is encrypted bidirectionally from the device to the server using an XTEA 256-bit key block cipher.

3.8.5.8 MiWi

MiWi is a wireless technology for low-power, cost-constrained networks, such as industrial monitoring and control, home and building automation, remote control, wireless sensors, lighting control, HVAC systems and automated meter reading.

MiWi uses mesh and star network technologies, in the 2.4GHz, 700MHz/800MHz/900MHz frequency spectrum providing low data rates of 20kbps and a coverage range of approximate 300m.

The MiWi protocol follows the MAC security definition specified in IEEE 802.15.4 and is based on 128-bit AES model. MiWi security mechanisms are classified as three modes: AES-CTR mode that encrypts MiWi protocol payload, AES-CBC-MAC mode that ensures the integrity of the MiWi protocol packet and AES-CCM mode that combines the previous two security modes to ensure both the integrity of the frame and encrypt the MiWi protocol payload.

3.8.5.9 MIOTY^{TM} (TS-UNB)

MIOTY^{TM} (TS-UNB) is a LPWAN, ETSI-standardised telegram-splitting ultra-narrowband (TS-UNB) technology (TS 103-357) that supports MYTHINGS. MYTHINGS is a wireless connectivity platform designed for large-scale industrial and commercial IoT networks.

Telegram-splitting is a standardised LPWAN technology in the licensed-free sub-GHz radio spectrum that feature a data rate of 512 bit/s and divides at the physical layer, an ultra-narrowband telegram into multiple equal-sised sub-packets, each of which is randomly sent at a different time and carrier frequency. Each sub-packet has a much smaller size than the original telegram, and the on-air time is reduced to 16 milliseconds. As the airtime of the sub-packets is much shorter than that of existing LPWANs, the chance of collisions with another message is very low. An algorithm in the base station permanently scans the spectrum for MIOTY sub-packets and reassembles them into a complete message. The technology has high-redundancy as up to 50% of the sub-packets can be lost without reducing the information content.

MIOTY enables energy-efficient, robust, and reliable transmission of sensor data over distances of up to several kilometres. The telegram-splitting mechanism is illustrated in Figure 3.37 [99, 103]. The mechanism allows the implementation of scalable networks for very high-density solutions. A MIOTY network is private and can have over a million devices that can transmit up to 1,5 million data packets a day to a single gateway, with no loss of information in environments with physical obstructions and poor propagation properties.

The protocol fragments data packets into numerous subpackets or telegrams and distributes them over time and frequency. MIOTY is designed to support up to 15 km range in flat terrain, up to 65,000 messages per hour, 407 bits/s, have enhanced interference-resilience features for use in shared

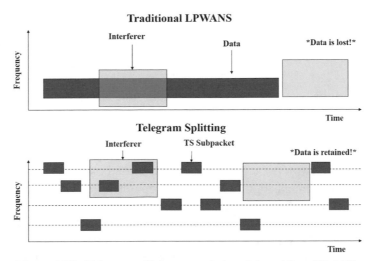

Figure 3.37 Telegram-splitting transmission. Adapted from [99, 103].

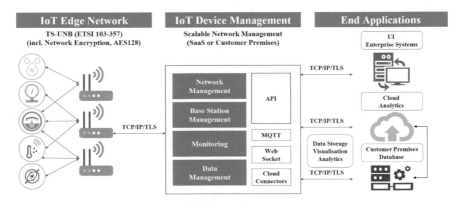

Figure 3.38 MYTHINGS network. Adapted from [99].

spectrum segments, (e.g. 868 MHz, 915 MHz ISM bands) via the time- and frequency-distribution approach, and support mobility up to approximately 80 km/h [104]. MIOTY does not support IP connectivity

An example of implementation of an IoT network based on MIOTY is illustrated in Figure 3.38 [99]. The MYTHINGS base station IoT gateway is leveraging the MIOTY wireless stack including the built-in software, the platform for device on and off-boarding, cloud/backend integration, data monitoring, network troubleshooting and indoor localisation. The MYTHINGS IoT gateway has the capacity to handle millions of messages a day from thousands of endpoints and provides a web-based user interface for management, cloud integration and MQTT interface for large-scale data collection and business analytics.

3.8.6 Satellite

During the last years, satellite service providers (e.g. Argos, Iridium, ORB-COMM, Inmarsat's Broadband Global Area Network, etc.), have integrated the IoT services into their portfolio and commercial satellite IoT initiatives (e.g. HIBER, DIAMOND, KEPLER), have started using nanosatellites for satellite IoT communication. ORBCOMM operates a satellite network dedicated to IoT providing two-way data communications in remote areas of the world via a network of low-earth orbit (LEO) satellites and ground stations. The OG2 satellites provide network redundancy, minimal line of site issues for complete global coverage, while VHF frequency furthers signal propagation.

These small satellite systems can cover oceans, rural areas, polar regions, and the integration of the nanosatellites networks with IoT networks provide new ways of monitoring the climate change, pollution, and global/regional disasters.

The miniaturisation of satellite technology for LEO (Low Earth Orbit) allows to use these satellites to serve as a backhaul for narrow-band IoT communication. The IoT applications using the satellite communication channels require data rates in the range of up to hundreds of kbps or tens of Mbps, low power consumption for battery life of the IoT nodes up to several years, compact antenna design, and coarse to even omni-directional antenna pointing.

These requirements narrow down the considerable frequency bands for Earth-to-satellite to VHF (0.03–0.3 GHz), UHF (0.3–1 GHz), L band (1–2 GHz), and S band (2–4 GHz). Frequencies, from VHF/UHF to Ka bands, are usable, with several nanosatellite using the UHF band, due to its omni-directional pattern, robustness, and low power consumption.

The IoT backbone requirements ($\sim$Mbps rate and $\sim$1000 km range) need to consider the S band, or X the X band due to the trade-off between bandwidth, directionality ($\approx 10°$), antenna (e.g., patch antenna on S/C), and transceiver size and power consumption [113].

The satellite communications offer opportunities by adding capacity on GEO (geostationary) satellites in C-, Ku- and Ka-band for direct or backhaul connectivity to deploying new LEO (low earth orbit) or HEO (highly elliptical orbit) constellations, optimised for the IoT solutions and applications.

The satellite industry responding to the IoT connectivity require solutions with low cost/low power direct to satellite connectivity and various combinations of terrestrial (cellular and LPWAN) IoT access networks and satellite backhaul.

Comparable to the cellular or Wi-Fi backhaul service, the IoT gateway backhaul over satellite emerges as a new SATCOM application segment.

The IoT market is developing around ultra-low-cost terrestrial radio transmission standards for IoT such as LoRa, Sigfox, LTE-M or NB-IoT targeting low cost per radio transmitter with dedicated gateways to concentrate larger numbers of IoT devices in range of operation. For the satellite industry connecting these gateways is leading to a new satellite application segment [114].

3.8.7 IoT Application Protocols

The IoT application protocols refer to OSI model (ISO/IEC 7498) layer 5, 6 and 7, layers that are responsible for managing communication sessions, data formatting (e.g. date translation, character encoding, data compression, encryption/decryption, presentation) and high-level APIs, resource sharing, remote file access, etc.

These layers are based on HTTP that is not suitable for resource constrained environments (e.g. verbose, requires large parsing overhead, etc.).

The requirements for many IoT applications have accelerate the developments of alternative protocols that address the resource constraint environments.

An extensive overview of the existing IoT application layer protocols is given in the following sub-sections. An illustration of the deployment of the IoT application protocols in the IoT applications is given in Figure 3.39.

3.8.7.1 The Advanced Message Queuing Protocol (AMQP)

The AMQP is largely used in the financial sector applications and applied to other types of applications. The protocol standardised by Organisation for the Advancement of Structured Information Standards (OASIS) and ISO assumes a reliable underlying transport protocol, such as TCP and is a binary message-oriented that provides message delivery guarantees for reliability, including at least once, at most once, and exactly once. This feature is very

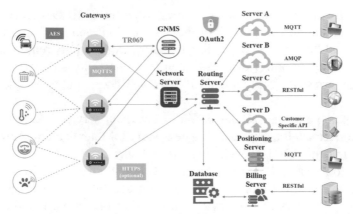

Figure 3.39 IoT applications implementation diagram. IoT application protocols deployment example.

important in the context of financial transactions (e.g., for executing credit or debit transactions). The protocol offers flow control through a token-based mechanism, to ensure that a receiving endpoint is not overburdened with more messages than it is capable of handling. AMQP. Different open-source implementations of the AMQP protocol are available.

AMQP supports both point-to-point communication and multipoint publish/subscribe interactions, defines a type system for encoding message data as well as annotating this data with additional context or metadata and the protocol can operate in simple peer-to-peer mode as well as in hierarchical architectures with intermediary nodes, e.g., messaging brokers or bridges.

The AMQP is an interoperable and cross platform messaging standard. In AMQP the messages along with a header are transmitted by the client to a broker or exchange and there is a single queue to which the message is transmitted by a producer. From the broker, the messages can be transmitted on to one or many queues. The AMQP header contains information about each byte of the message and the routing information. The broker is responsible to read headers and receive, route, and deliver messages to the client applications. The communication in AMQP protocol remains one to one between two nodes.

3.8.7.2 Constrained Application Protocol (CoAP)

CoAP is a RESTful protocol that supports the create, read, update, and delete (CRUD) verbs and in addition provides built-in support for the publish/subscribe paradigm via the new observe verb. CoAP provides a

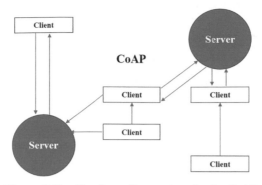

Figure 3.40 The flow of interactions for the CoAP.

mechanism where messages may be acknowledged for reliability and provides a bulk transfer mode.

Using CoAP, the client node can command another node by sending an CoAP packet that is interpreted by the CoAP server (the server may or may not acknowledge the request), which extracts the payload, and decides the action depending on its logic. The flow of interactions for the CoAP is presented in Figure 3.40.

CoAP is standardised as RFC 7252 by the IETF Constrained RESTful Environments (CORE) workgroup as a lightweight alternative to HTTP, targeted for constrained nodes in low-power and lossy networks (LLNs).

CoAP reduces the TCP overhead of seven messages required to fetch a resource by using UDP as a transport in lieu of TCP. CoAP uses short headers to reduce message sizes. IETF is further working to define mechanisms for dynamic resource discovery in CoAP via a directory service.

3.8.7.3 Distributed Data Service Real-Time Publish and Subscribe (DDS RTPS)

DDS RTPS is a data-centric application protocol using UDP as the underlying transport and standardised by Object Management Group (OMG).

The protocol supports the publish/subscribe paradigm, and which organises data into "topics" that listeners can subscribe to and receive asynchronous updates when the associated data changes. DDS RTPS supports different QoS policies for data distribution and provides mechanisms where listeners can automatically discover speakers associated with specific topics.

IP multicast or a centralised broker/server may be used to that effect. Several speakers may be associated with a single topic and priorities can be defined for different speakers, creating a redundancy mechanism for the

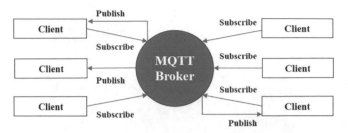

Figure 3.41   The flow of interactions for the MQTT.

architecture in case a speaker fails or loses communication with its listeners. The QoS policies include reliability, data persistence, delivery deadlines, and data freshness.

3.8.7.4 IEEE 1888

IEEE 1888, standardised by IEEE Standards Association is an application protocol for environmental monitoring, smart energy, and facility management applications, supporting reading and writing of time-series data using the Extensible Markup Language (XML) and the simple object access protocol (SOAP). The data is identified using Universal Resource Identifiers (URIs).

3.8.7.5 Message Queue Telemetry Transport (MQTT)

MQTT protocol is a binary protocol, using TCP as transport layer. The protocol was designed by IBM for enterprise telemetry and is a lightweight publish/subscribe messaging protocol standardised by OASIS. The protocol is message oriented, where messages are published to an address, referred to as a topic.

MQTT is a publish-subscribe protocol that facilitates one-to-many communication mediated by brokers with clients that can publish messages to a broker and/or subscribe to a broker to receive certain messages.

Messages are organised by topics, which essentially are "labels" that act as a system for dispatching messages to subscribers. The flow of interactions for the MQTT is presented in Figure 3.41.

Clients subscribe to one or more topics and receive updates from a client that is publishing messages for this topic. In MQTT, topics are hierarchical (like URLs), and subscriptions may use wildcards. MQTT uses a client-server architecture where clients connect to a central server (called the broker).

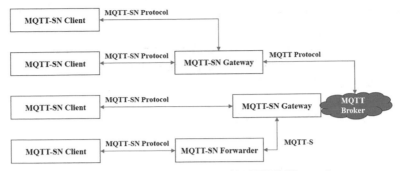

Figure 3.42   The architecture of MQTT-SN Protocol.

The protocol targets end devices where "a small code footprint" is required or where network bandwidth is limited (e.g. constrained IoT devices).

3.8.7.6 MQTT-Sensor Network (MQTT-SN)

The MQTT-SN is an open, lightweight publish/subscription protocol designed for constrained devices i.e. wireless sensor network (WSN). The protocol is based on MQTT and adapted to the specific requirements of wireless communication environments (e.g. short message length, low bandwidth, high link failures, etc.) and devices (e.g. low-cost, very low power consumption, long-battery life, limited processing and storage resources, etc.). Several changes have been introduced in MQTT-SN compared with MQTT, like the topic names are replaced by topic IDs, which reduce the overheads of transmission, topics do not need registration as they are preregistered, messages are split to send only the necessary information. The clients connect to the broker through a gateway device, which resides within the sensor network and connects to the broker and for reducing power consumption there is an offline procedure for clients who are in a sleep state so the messages can be buffered and later read by clients when they wake up.

Three components of MQTT-SN architecture are used such as MQTT-SN clients, MQTT-SN gateways (GW), and MQTT-SN forwarders as illustrated in Figure 3.42. MQTT-SN clients connect themselves to a MQTT server via a MQTT-SN GW using the MQTT-SN protocol, the MQTT-SN GW may or may not be integrated with a MQTT server. When utilising a stand-alone GW, the MQTT protocol is used between the MQTT server and the MQTT SN GW having the function to translate the messages between MQTT and MQTT-SN.

MQTT-SN clients can access a GW via a forwarder in case the GW is not directly attached to their network. The forwarder encapsulates the MQTT-SN frames it receives on the wireless side and forwards them unchanged to the GW and releases the frames it receives from the gateway and sends them unchanged to the clients.

3.8.7.7 OMA LightweightM2M (LWM2M)

The LWM2M is a device management protocol, designed to be able to extend to meet the requirements of applications by transferring service / application data. The protocol uses a simple object-based resource model with resource operations of creation/retrieval/update/deletion/configuration of attribute. The protocol provides data format support for TLV, Json, Plain Text, Opaque and transport layer support for UDP/IP and SMS. The LWM2M has M2M/IoT functionalities such as LWM2M server, access control, device, connectivity, firmware update, location, connectivity statistics and uses DTLS for security.

3.8.7.8 RESTful HTTP (REST)

The concept of Representational State Transfer (REST) uses the HTTP methods such as GET, POST, PUT, and DELETE to provide a resource oriented messaging system where the interactions can be performed simply by using the synchronous request/response HTTP commands for receiving, modifying, and sending data. REST is based on TCP/IP, uses a request/response architecture with a relative complex implementation at client side, having a larger header compared to other IoT Protocols (e.g. higher bandwidth requirements) and applying SSL/TLS for security.

3.8.7.9 Secure Message Queue Telemetry Transport (SMQTT)

The SMQTT protocol is an encryption based light weight messaging protocol based on MQTT. Compared to an MQTT session, SMQTT session has four levels: setup, encryption, publish and decryption. The protocol uses a similar MQTT broker-based architecture with the difference that both the subscriber and publisher need to register with the broker using a secret master key. The data is encrypted before being published by the publisher and then is decrypted at the subscriber end and different encryption algorithms can be used by developers.

3.8.7.10 Session Initiation Protocol (SIP)

SIP is a text-based protocol that can use different underlying transports, TCP, UDP, or SCTP and is standardised by IETF as RFC 3261. The protocol handles session establishment for voice, video, and instant messaging applications on IP networks. It also manages presence (like XMPP). The invitation messages are used to create sessions carry session descriptions that enable edge devices to agree on a set of compatible media types. The protocol uses elements called proxy servers to route requests to the user's current location, authenticate and authorise users for services, implement call-routing policies, and provide features. A registration function is defined by the protocol to enable users to update their current locations for use by proxy servers.

3.8.7.11 Streaming Text Orientated Messaging Protocol (STOMP)

The STOMP is a text-based protocol for message-oriented middleware based on TCP and uses HTTP like commands. The protocol was designed to provide interoperability among platforms, languages, and brokers. The data is communicated between a client and broker in multi-line frames containing command, header, and content. The commands used can be CONNECT, DIS-CONNECT, ACK, NACK, SUBSCRIBE, UNSUBSCRIBE, SEND, BEGIN, COMMIT or ABORT.

3.8.7.12 Very Simple Control Protocol (VSCP)

VSCP is an open source standard protocol for M2M, IoT that enables low-cost devices to be networked together with computers and/or to operate as autonomous system, whatever the communication channels are used. The protocol utilises an event-based architecture, provides mechanisms for device discovery, identification, configuration and has support for secure device firmware updates. VSCP is an application level protocol that uses CAN, RS-232, Ethernet, TCP/IP, MQTT, 6LowPan or other protocols as it's transport mechanism and work over cable and over the air.

3.8.7.13 Extensible Messaging and Presence Protocol (XMPP)

XMPP is a message-centric protocol based on the Extensible Markup Language (XML) that use TCP as underlying transport with an option to run XMPP over HTTP. The protocol was designed for instant messaging, contact list, and presence information maintenance and extended to several other applications, including network management, video, voice-over IP, file sharing, social networks, and online gaming, etc. XMPP is used for many IoT smart grid applications. The XMPP Standards Foundation (XSF) actively develops open extensions to the protocol.

An overview of selected IoT application protocols is presented in Table 3.7.

3.9 IoT Trustworthiness

The trustworthiness of IoT technologies and applications is critical to the acceptance and adoption of the technology, and many ethical issues must be addressed along the way to developing these technologies. The IoT applications are advancing and interact, cooperate, and collaborate with humans and animals. From the point of view of system design, the trustworthiness of IoT technologies are directly connected to the concept of dependability. Assuring dependability is ensuring the basis for trust in IoT technologies. The concept integrates the elements of availability, reliability, safety, security resilience, privacy and it embraces "privacy and security by design" as a model for an implementable IoT application.

The trustworthiness of IoT technologies and applications needs to consider trust semantics, metrics, models, IoT platforms, trusted IoT network computing, operating systems, software, and applications, while addressing the trust in mobile, wireless communications and risk and reputation management. Figure 3.43 illustrates the many channels for IoT applications security breaches. In this context, IoT applications need to embed mechanisms to continuously monitor security and stay ahead of the threats posed by interactions with other IoT applications and environments.

Trust is based on the ability to maintain the security of the IoT system and the ability to protect application/customer information, as well as being able to respond to unintended security or privacy breaches. In the IoT, it is important to drive security, privacy, data protection and trust across the whole IoT ecosystem.

3.9.1 Trust and Privacy in IoT Through Distributed Identity Management

IoT upcoming technologies rapidly thrust digitalisation into many application domains, such as smart cities, supply chain, industrial control, and healthcare

Table 3.7 Overview IoT application protocols

IoT Protocol	Functions	Transport	Format	Applications	SDO
AMQP	Message orientation, queuing, and pub/sub Data transfer with delivery guarantees (at least once, at most once, exactly once)	TCP	Binary	Financial services	OASIS
CoAP	REST resource manipulation via CRUD Resource tagging with attributes Resource discovery through RD	UDP	Binary	Low power and lossy networks	IETF
DDS (RTPS)	Pub/sub messaging with well-defined data types Data discovery Elaborate QoS	UDP	Binary	Real-time distributed systems (military, industrial, etc.)	OMG
IEEE 1888	Read/write data into URI Handling time-series data	SOAP/ HTTP	XML	Energy and facility management	IEEE
MQTT	Lightweight pub/sub messaging Message queuing for future subscribers	TCP	Binary	Enterprise telemetry	OASIS
SIP	Manage presence Session establishment Data transfer (voice, video, text)	TCP, UDP, SCTP	XML	IP telephony	IETF
XMPP	Manage presence Session establishment Data transfer (text or binary)	TCP HTTP	XML	Instant messaging	IETF XSF

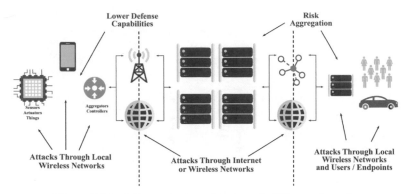

Figure 3.43   IoT security breaches through different channels.

systems. One of the most critical issues and challenges in IoT is to secure communication between IoT devices and internet.

The proliferation of connected IoT devices has led to an increased attack surface that leads to new cybersecurity risks that can compromise device security and data security. Smart healthcare medical systems, machine-to-machine communications (M2M), intelligent transportation systems (ITS), Industry 4.0 have heightened needs for trustworthy components with guaranteed authenticity, integrity, and confidentiality.

To address such security and privacy challenges, different authentication and privacy-preserving solutions are emerging to empower trusted IoT communications and enable anonymous mechanisms to protect sensitive and private data [191–193].

IoT is a service-providing infrastructure, where devices (things) can operate autonomously and need to be uniquely identifiable. Albeit cryptography is applied to protect communicated data when third parties are involved, this prerequisite mutual authentication between the communicating entities.

Maintaining a permanent unique identifier is a privacy risk, as the device can potentially be cloned, and IoT infrastructure maybe misused or wrong information can be provided to hide alerts or to get secret user information.

To guarantee security and privacy in IoT authentication protocols it is fundamentally required to guarantee mutual authentication, identity secrecy, device anonymity, non-traceability, forward security, availability [194].

To deal with these requirements, authentication protocols for IoT environments involve two or three phases and use various cryptographic mechanisms, including physical unclonable functions (PUF), digital signature, private-key cryptography, and public-key cryptography.

Generally, mutual authentication consists of two stages: the enrolment stage, where every node should be identified within the system, and the authentication stage, where a number of handshake messages are exchanged between the end node device and the server, which result is a session key to be used for upcoming communication [195, 196], and with provision of hardware crypto-cores [203].

By using three phases, registration, login and authentication phase, and password change phase, Chung et al. [197] can provide anonymity, hop-by-hop authentication, and non-traceability. Multiple trusted authority with role separation [205] have been proposed to authenticate communication in secure vehicular ad-hoc networks (VANETs) and through using identity-based aggregate signatures [206]. Recently, focusing on IoT setting, proposals involve Near Field Communications (NFC) secure element (SE)-based mutual authentication and attestation for IoT access with a user device, such as a smartphone [204].

Improved public-key-based mechanisms have been investigated, with most existing authentication protocols to be based on elliptic curve cryptography (ECC) to deal with the capabilities of constrained-resources devices compared to RSA algorithm.

Example proposals feature mutual authentication, non-traceability, and session key agreement [198]. Considering M2M communications in IIoT environment, lightweight authentication schemes are needed. Proposals involve a device equipped with a Secure Element (SE), which is authenticated by a network element equipped with a Trusted Platform Module (TPM) in two phases, registration and authentication [199].

Isolated execution environments through a proxy are a prominent solution in interconnected Cyber-Physical systems (CPSs), such as vehicles, given the penetration of the IoT paradigm in vehicles have raised the collection of a huge amount of data [207].

In IIoT, privacy-preserving biometric-based provable secure authentication protocol with ECC has also been proposed, in which a user authenticates himself to a gateway to agree on a session key for all future communications to be made secure and then accesses sensory data of a node [199].

In IIoT with a fog layer Attribute Credential-based Public Key Cryptography (AC-PKC) scheme [201] meets authentication and privacy-preserving access control requirements through employing a two level verification scheme which requires a fog node to generate a signature for the command it issues for an actuator.

The actuator on receiving this, must perform the two-level verification to authenticate that the command was indeed issued by a trusted fog node.

As IoT deployments mature, standardisation efforts have recognised and developed OAuth 2.0 authorisation framework [211] through using cryptographic tokens and authorisation and resource servers. The authorisation consent by the resource owner is provided after the owner is authenticated by an authorisation server; however, the authentication procedure is not part of OAuth 2.0. Authorisation is provided for different levels of access, such as read and write/modify, which are termed scopes, and for a specific time interval.

3.9.2 Decentralised Identification in IoT

The term decentralisation is recently often used to describe a security approach where data are spread over many network nodes in an approach to reduce chances of one point being vulnerable to introducing network threats or risks and anyhow limit or modify the network's or devices intended purpose.

Decentralisation includes aspects such as control level, access and ownership spread around the whole network nodes/actors that comprise it. Recently, this decentralisation refers to a shifting from centralised to distributed modes of network configurations. In mind of IoT devices identification in a network, decentralisation can play a predominant role as through this many issues of IoT ecosystems being brokered in a server/client approach. Up to now, this configuration seemed adequate and 'enough' but as the IoT devices' number and complexity increase, this is not considered as suitable anymore.

In the framework of IoT devices and solutions, core components of the IoT infrastructure and networks were cloud servers raising single point of failures largely in the IoT networks. In mind of critical applications (health, automotive, production etc.), decentralised approaches have a lot to offer regarding shared approaches.

Blockchain and other technologies recently support this decentralisation including concepts of trusting credentials in the network and validation of key modifications, self-sovereignty in identities involving self-trust on control of own identity and zero-knowledge proofs that actually proof identities and data without revealing any secret information. These are being discussed later in this chapter.

Always thinking of the IoT domain, the adoption of decentralised approaches for devices' identification is expected to strongly reduce installation costs as well as costs for large (centralised) data centres or related infrastructures. This is also expected to support distribution of both storage and computational requirements while support single point of failures risks.

IoT devices to be used and managed when deployed in different applications require a unique identification. From simple device description to more sophisticated appliances as USB serial naming, security keys, cryptography keys need device identification.

Today device identification is done using the device's network address (IP, LPWAN, etc.) , hardware identifier as in the case of RFID, simple hardcoding it into the firmware, separately flashing info into the FLASH memory, or random generating by the first run of the IoT device..

Some microcontrollers such as STM32 provided a unique 96-bit ID encoded identifier that is generated during the manufacturing process. This unique Identifier consists of 3 parts: X and Y coordinates on the wafer expressed in BCD format, the lot number and wafer number. All these solutions are not 100% secure in all situations where it is important to prove the real identity of a device.

Several researches and trials are proposing to shift from a centralised server-client paradigm connecting to cloud via internet to architectures where industrial processes exchange information directly in a Peer-to-Peer (P2P) fashion (i.e., machine-to-machine (M2M) connections), hence promoting the decentralisation of computations across participating entities. Until today in security frameworks, digital identity is supported on centralised, commonly third-party, information systems which raise single point of failure concerns for an infrastructure.

Additionally, centralised identity management has privacy challenges with the possibility of digital identity exposure if the central authority is compromised. Furthermore, IoT devices with a hardware identifier can present privacy problems due to the identifier facilitating tracking and correlation attacks.

With the advent of blockchain, Identity Management (IdM) systems are switching from traditional web-centric approach or identity federation approaches, towards the self-sovereign identity (SSI) paradigm. Blockchain technology allows transferring digital assets (such as IoT data) in a decentralised fashion using the ledger, i.e., distributed ledger technology (DLT) [202], without intermediary central third-parties, while enabling public verifiability as well as provenance of the digital transactions and data. Meanwhile, cryptographic mechanisms (such as asymmetric encryption algorithms, digital signature, and hash functions) guarantee the integrity of data blocks in the blockchains.

Therefore, a blockchain ensures non-repudiation of transactions, while each transaction in the blockchain is traceable to every user with the attached historic timestamp.

The ledger is immutable, meaning that past transactions cannot be modified by any entity registering transactions in the blockchain, and is shared and synchronised across all participating nodes. This way, the blockchain guarantees that the ledger cannot be tampered with, and that all the data held by the blockchain is trustworthy.

In IoT environment, millions of constrained smart entities with scarce capabilities to enforce proper security mechanisms, strive to cope with cyber-attacks that may leak their communicated data, and ultimately, sensitive, and private information of their owners/users.

Besides, in IoT, user privacy controls are difficult to apply, as the smart objects usually act on behalf of the user without user control and consent, undermining the adoption of the minimal personal disclosure principle. Blockchain brings a fully decentralised root of trust avoiding central authorities, thus creating trust across initially non-trusted or even unknown users and things [202, 208].

Blockchain-based identity schemes encrypt a user's identity, hash it and add its attestations to the blockchain ledger. These attestations are later used to prove the user's identity. Three important schemes and concepts determine the broader landscape of DLT-based IdMs:

- Decentralised Identity: This identity solution is like the conventional identity management solutions where credentials from a trusted service are used. The key modification is the storage of validated attestations on a distributed ledger for later validation by a third party.
- Self-sovereign identity: A user or device is the entity who owns and controls his/her identity without heavily relying on central authorities. It provides a framework to enable exchange of information and propagation of trust between peers.
- Zero knowledge proofs: Zero-knowledge protocols provide that an entity can prove the knowledge of its secret associated to public data without revealing any information about the secret. Essentially, one party (the prover) can prove to another party (the verifier) that they know a value x, without conveying any information apart from the fact that they know the value x. It is used in blockchain to perform authentication without giving the secret to other party.

A distributed identifier (DID) is a new type of identifier without a central issuing and controlling agency that creates and controls the identifier. Instead, DIDs are created and managed by the identity owner, an approach known as self-sovereign identity [210].

DID is acting as a permanent identifier which never changes and is resolvable. It acts as a persistent verifiable identifier through cryptography and can be used to encrypt communication channels for safe and secure messaging. A typical DID contains one or more public keys that can be used to authenticate the DID.

One or more services that can be used to for interaction via protocols supported by those services. Finally contains metadata such as digital signatures, timestamps, and other cryptographic proofs. An entity can have any different number of DIDs for many different purposes.

Current strengths and challenges of applying DLT to identity management together with the evaluation of three proposals (i.e., uPort, ShoCard, and Sovrin) are analysed in [202].

Currently, there are multiple competing DID technologies, in which the Decentralised Identity Foundation [209] is promoting interoperability among implementations. Today DIDs are an elegant solution in IoT to solve identity challenges, such as generated data insecurity, fraudulent identities, or third-party controlled IDs. Since DID are globally resolvable several IoT services can be enabled in a secure manner.

Hyperledger Indy is the only developed ledger to be permissioned. Even though anyone can read the blockchain's contents, only allowed entities can write to it. Since it is permissioned, it does not need proof-of-work to eliminate spam, making the transaction delay in the order of a few seconds (duration of network consensus).

To deploy DIDs and verifiable credentials in IoT, the IoT device should have:

- Sufficient performance for cryptographic operations,
- Enough energy to perform the required operations,
- Non-volatile storage space to store the code and cryptographic keys,
- Sufficient entropy source to generate random cryptographic keys.

Storing the keys on the device can present an unacceptable security risk of key leakage unless the device utilises a secure element, e.g., a trusted platform module (TPM), or embrace a proxy solution to act as a guardian for the keys.

In the scope of Self-Sovereign Identity model as shown in Figure 3.44, the blockchain acts as distributed and reliable identity verifier, providing

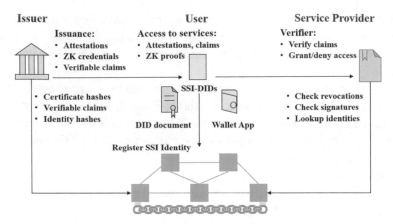

Figure 3.44   Blockchain-based self-sovereign identity management.

provenance and verifiability of identities. Thus, the ledger provides a crypto-graphic root of trust, which facilitates identity management without external authorities.

The major technical challenge that the SSI technology currently faces is its infancy and lack of widespread usage. It must be noted that privacy enhancing by using decentralised identifiers for the IoT devices is necessary for protecting the privacy of the users and owners of the devices, but similar care has to be taken with all other elements of the system to truly protect the privacy of users.

Additionally, in the direction of W3C community group [212], standard mechanisms must exist to define transparent interfaces and common data models that can be used in the exchange of information among parties to ensure that different entities (agents, devices, etc.) can interact seamlessly.

IoT in conjunction with blockchain, can assist to track, process and exchange transactions among connected devices. However, the development of new privacy-preserving approaches should manage the cost of cryptographic operations, to foster the adoption of blockchain technologies.

Today, lightweight scalable blockchain solutions can address the challenges of traditional security and privacy methods in IoT environment, centralisation, lack of privacy or safety threats. Nevertheless, careful infrastructure tuning should consider control of periodic integrity and authentication verification in the blockchain, to effectively prevent malicious nodes from intruding, to resist DDoS attacks and to prevent tampering with the device firmware.

3.10 Discussion

The rise of the IoT has brought countless applications and numerous opportunities, from the personal sensing applications, all-time connected sensors, smartphone, and wearable computers, to cloud and IoT and intelligent devices for robotic applications. Next wave of Internet of Intelligent Things is yet to come and has qualitatively augmented capacities to collect, manage and process data towards the personal-centric applications that are in high demand today and in the years to come.

Next-generation digital technologies fuelled by varieties of IoT, Industrial IoT, Tactile IoT, Internet of Robotic Things, Intelligent Internet of Things, Artificial Intelligence of Things, Internet of Things Senses, are developing to enable personalisation of services and interactions between things, humans and environments by unifying the physical, digital, virtual and cyberspaces into a continuum of experiences.

By bringing more intelligence to the IoT, a crucial characteristic will be provided to the future systems, which is the capacity for an awareness of the surrounding information through the integration of multiple pieces of information. We are yet to see the seamless integration of the Internet, intelligent things, and AI.

Acknowledgements

The IoT European Research Cluster – European Research Cluster on the Internet of Things (IERC) maintains its Strategic Research and Innovation Agenda (SRIA), considering its experiences and the results from the on-going exchange among European and international experts.

The present document builds on 2010, 2011, 2012, 2013, 2014, 2015, 2016, 2017 and 2018 Strategic Research and Innovation Agendas.

The IoT European Research Cluster SRIA is part of a continuous IoT community dialogue supported by the EC DG Connect – Communications Networks, Content and Technology, E4 – Internet of Things Unit for the European and international IoT stakeholders. The result is a lively document that is updated every year with expert feedback from on-going and future projects financed by the EC. Many colleagues have assisted over the last few years with their views on the IoT Strategic Research and Innovation agenda document. Their contributions are gratefully acknowledged.

Contributing Projects and Initiatives

SmartAgriFood, EAR-IT, ALMANAC, CITYPULSE, COSMOS, CLOUT, RERUM, SMARTIE, SMART-ACTION, SOCIOTAL, VITAL, BIG IoT, VICINITY, INTER-IoT, symbIoTe, TAGITSMART, bIoTope, AGILE, Be-IoT, UNIFY-IoT, ARMOUR, FIESTA, ACTIVAGE, AUTOPILOT, CREATE-IoT, IoF2020, MONICA, SYNCHRONICITY, U4IoT, BRAIN-IoT, ENACT, IoTCrawler, SecureIoT, SOFIE, CHARIOT, SEMIoTICS, SerIoT.

References

[1] OpenFog Reference Architecture for Fog Computing, OPFRA001. 020817, 2017.

[2] L. M. Vaquero and L. Rodero-Merino, "Finding your way in the fog: Towards a comprehensive definition of fog computing," SIGCOMM Comput. Commun. Rev., vol. 44, no. 5, pp. 27–32, Oct. 2014. Online at: http://doi.acm.org/10.1145/2677046.2677052

[3] R. van der Meulen, "Edge computing promises near real-time insights and facilitates localized actions", Web article, 2018. Online at: https://www.gartner.com/smarterwithgartner/what-edge-computing-means-for-infrastructure-and-operations-leaders/

[4] J. Morrish, M. Hatton and M. Arnott, "Global IoT Forecast Insight Report 2020", Transforma Insights, 2020. Online at: https://transforma insights.com/research/reports/global-iot-forecast-insight-report-2020

[5] ETSI White Paper, "Mobile Edge Computing A key technology towards 5G", #11, (09/2015), online at: https://www.etsi.org/image s/files/ETSIWhitePapers/etsi_wp11_mec_a_key_technology_towards _5g.pdf

[6] ETSI, "Mobile Edge Computing (MEC); Framework and Reference Architecture", GS MEC 003 V1.1.1 (03/2016).

[7] Ahmed, E. Ahmed, "A Survey on Mobile Edge Computing", IEEE, Int'l Conf. on Intelligent System and Control ISCO 2016. Online at: https://www.researchgate.net/publication/285765997

[8] ETSI, "Multi-access Edge Computing (MEC); Framework and Reference Architecture", GS MEC 003, V2.1.1 (01/2019).

[9] ETSI White Paper, "MEC in 5G networks". #28, (06/2018).

[10] ETSI White Paper, "Network Transformation; (Orchestration, Network and Service Management Framework)", #32, (10/2019). Online

at: https://www.etsi.org/deliver/etsi_gs/MEC/001_099/003/02.01.0
1_60/gs_MEC003v020101p.pdf

[11] S. Ibrahim, H. Jin, B. Cheng, H. Cao, S. Wu, and L. Qi, "CLOUDLET: towards mapreduce implementation on virtual machines," in Proceedings of the 18th ACM International Symposium on High Performance Distributed Computing, HPDC 2009, Garching, Germany, June 11-13, 2009, 2009, pp. 65–66. Online at: https://dl.acm.org/doi/10.1145/155 1609.1551624

[12] UN Environment Programme, "Emissions Gap Report 2019". Online at: https://wedocs.unep.org/bitstream/handle/20.500.11822/30797/EG R2019.pdf?sequence=1&isAllowed=y

[13] Sangwon Suh, et. al., United Nations Environment Programme, "Green Technology Choices: The Environmental and Resource Implications of Low-Carbon Technologies", 2017. Online at: https://www. resourcepanel.org/reports/green-technology-choices

[14] M. Satyanarayanan, P. Bahl, R. Caceres and N. Davies, "The Case for VM-Based Cloudlets in Mobile Computing," in IEEE Pervasive Computing, vol. 8, no. 4, pp. 14–23, Oct.–Dec. 2009. Online at: https://doi.org/10.1109/MPRV.2009.82

[15] European Commission, " The European Green Deal", COM(2019) 640 final, 2019. Online at: https://ec.europa.eu/info/sites/info/files/europ ean-green-deal-communication_en.pdf

[16] United Nations, "Transforming our World: The 2030 Agenda for Sustainable Development" 2015. Online at: https://sustainabledevelopme nt.un.org/content/documents/21252030%20Agenda%20for%20Sustai nable%20Development%20web.pdf

[17] European Commission, "Circular Economy Action Plan – For a cleaner and more competitive Europe", 2020. Online at: https://ec.e uropa.eu/jrc/communities/en/community/city-science-initiative/docu ment/circular-economy-action-plan-cleaner-and-more-competitive0

[18] R. Arshad, S. Zahoor, M. A. Shah, A. Wahid and H. Yu, "Green IoT: An Investigation on Energy Saving Practices for 2020 and Beyond," in IEEE Access, vol. 5, pp. 15667–15681, 2017. Online at: https://doi. org/10.1109/ACCESS.2017.2686092

[19] N. Kaur and S. K. Sood, "An Energy-Efficient Architecture for the Internet of Things (IoT)," in IEEE Systems Journal, vol. 11, no. 2, pp. 796–805, June 2017, doi: 10.1109/JSYST.2015.2469676

[20] T. Qiu, A. Zhao, R. Ma, V. Chang, F. Liu, and Z. Fu. A task-efficient sink node based on embedded multi-core SoC for Internet of Things.

Future Generation Computer Systems, 82, pp. 656–666, 2018. https://doi.org/10.1016/j.future.2016.12.024

[21] S. Murugesan, "Harnessing Green IT: Principles and Practices," in IT Professional, vol. 10, no. 1, pp. 24–33, Jan.-Feb. 2008, doi: 10.1109/MITP.2008.10

[22] Elijah – Cloudlet-based Edge Computing. Online at: http://elijah.cs.cmu.edu/

[23] Open Edge Computing Initiative. Online at: https://www.openedgecomputing.org/

[24] OpenStack. Online at: https://www.openstack.org/

[25] M. Satyanarayanan, W. Gao and B, M Lucia, "The Computing Landscape of the 21st Century", HotMobile '19: Proceedings of the 20th International Workshop on Mobile Computing Systems and Applications, February 2019, pp. 45–50, Online at: https://dl.acm.org/doi/10.1145/3301293.3302357

[26] Y. Wang, "Definition and categorization of dew computing," Open Journal of Cloud Computing (OJCC), vol. 3, no. 1, pp. 1–7, 2016.

[27] M. Gushev, "Dew Computing Architecture for Cyber-Physical Systems and IoT", Internet of Things, Volume 11, September 2020. Online at: https://doi.org/10.1016/j.iot.2020.100186

[28] P. P. Ray, "An Introduction to Dew Computing: Definition, Concept and Implications," in IEEE Access, vol. 6, pp. 723–737, 2018. Online at: https://doi.org/10.1109/ACCESS.2017.2775042

[29] Tractica Report, "Artificial Intelligence for Edge Devices". Online at: https://www.tractica.com/research/artificial-intelligence-for-edge-devices/

[30] D. Schatsky, N. Kumar and S. Bumb, "Intelligent IoT – Bringing the power of AI to the Internet of Things", December 2017. Online at: https://www2.deloitte.com/insights/us/en/focus/signals-for-strategists/intelligent-iot-internet-of-things-artificial-intelligence.html#endnote-sup-1

[31] J. Fu, Y. Liu, H. Chao, B. K. Bhargava and Z. Zhang, "Secure Data Storage and Searching for Industrial IoT by Integrating Fog Computing and Cloud Computing," in IEEE Transactions on Industrial Informatics, vol. 14, no. 10, pp. 4519–4528, Oct. 2018. Online at: https://doi.org/10.1109/TII.2018.2793350

[32] M. Jbair, B. Ahmad, M. H. Ahmad and R. Harrison, "Industrial cyber physical systems: A survey for control-engineering tools," 2018 IEEE Industrial Cyber-Physical Systems (ICPS), St. Petersburg, 2018,

pp. 270–276. Online at: https://doi.org/10.1109/ICPHYS.2018.8387 671

[33] M. Frey et al., "Security for the Industrial IoT: The Case for Information-Centric Networking," 2019 IEEE 5th World Forum on Internet of Things (WF-IoT), Limerick, Ireland, 2019, pp. 424–429. Online at: https://doi.org/10.1109/WF-IoT.2019.8767183

[34] A. H. Sodhro, S. Pirbhulal and V. H. C. de Albuquerque, "Artificial Intelligence-Driven Mechanism for Edge Computing-Based Industrial Applications," in IEEE Transactions on Industrial Informatics, vol. 15, no. 7, pp. 4235–4243, July 2019. Online at: https://doi.org/10.1109/TII.2019.2902878

[35] R. Minerva, G. M. Lee and N. Crespi, "Digital Twin in the IoT Context: A Survey on Technical Features, Scenarios, and Architectural Models," in Proceedings of the IEEE. Online at: https://doi.org/10.1109/JPROC.2020.2998530

[36] E. Y. Song, M. Burns, A. Pandey and T. Roth, "IEEE 1451 Smart Sensor Digital Twin Federation for IoT/CPS Research," 2019 IEEE Sensors Applications Symposium (SAS), Sophia Antipolis, France, 2019, pp. 1–6. Online at: https://doi.org/10.1109/SAS.2019.8706111

[37] T. Catarci, D. Firmani, F. Leotta, F. Mandreoli, M. Mecella and F. Sapio, "A Conceptual Architecture and Model for Smart Manufacturing Relying on Service-Based Digital Twins," 2019 IEEE International Conference on Web Services (ICWS), Milan, Italy, 2019, pp. 229–236. Online at: https://doi.org/10.1109/ICWS.2019.00047

[38] Y. Lu, X. Huang, K. Zhang, S. Maharjan and Y. Zhang, "Communication-Efficient Federated Learning for Digital Twin Edge Networks in Industrial IoT," in IEEE Transactions on Industrial Informatics. Online at: https://doi.org/10.1109/TII.2020.3010798

[39] Z. Tsai, " The Emerging Role of AI in Edge Computing", November 2018. Online at: https://www.rtinsights.com/the-emerging-role-of-ai-in-edge-computing/

[40] M. Dsouza, "Intelligent Edge Analytics: 7 ways machine learning is driving edge computing adoption in 2018", August 2018 Online at: https://hub.packtpub.com/intelligent-edge-analytics-7-ways-machine-learning-is-driving-edge-computing-adoption-in-2018/

[41] STMicroelectronics, "Neural Networks on STM32", online at: https://www.st.com/content/st_com/en/about/innovation---technology/artificial-intelligence.html

[42] J. Lin, W. Yu, N. Zhang, X. Yang, H. Zhang and W. Zhao, "A Survey on Internet of Things: Architecture, Enabling Technologies, Security and Privacy, and Applications," in IEEE Internet of Things Journal, vol. 4, no. 5, pp. 1125–1142, Oct. 2017. Online at: https://doi.org/10.1 109/JIOT.2017.2683200

[43] J. Ni, K. Zhang, X. Lin, X.S. Shen, "Securing fog computing for internet of things applications: challenges and solutions", IEEE Commun. Surv. Tutor. 20 (1) (2018) pp. 601–628. Online at: https://doi.org/10.1 109/COMST.2017.2762345.

[44] A.C. Baktir, A. Ozgovde, C. Ersoy, "How can edge computing benefit from software-defined networking: a survey, use cases, and future directions", IEEE Commun. Surv. Tutor. 19 (4) (2017) pp. 2359–2391. Online at: https://doi.org/10.1109/COMST.2017.2717482

[45] M. Marjanović, A. Antonić and I. P. Žarko, "Edge Computing Architecture for Mobile Crowdsensing," in IEEE Access, vol. 6, pp. 10662–10674, 2018. Online at: https://doi.org/10.1109/ACCESS.2018.2799 707

[46] A List of All Human Senses. Online at: https://www.scribd.com/doc ument/251594575/A-List-of-All-Human-Senses

[47] J. Sachs et al., "Adaptive 5G Low-Latency Communication for Tactile Internet Services," in Proceedings of the IEEE, vol. 107, no. 2, pp. 325–349, Feb. 2019. https://doi.org/10.1109/JPROC.2018.28645 87

[48] D. Van Den Berg et al., "Challenges in Haptic Communications Over the Tactile Internet," in IEEE Access, vol. 5, pp. 23502–23518, 2017. Online at: https://doi.org/10.1109/ACCESS.2017.2764181

[49] A. Aijaz, M. Dohler, A. H. Aghvami, V. Friderikos and M. Frodigh, "Realizing the Tactile Internet: Haptic Communications over Next Generation 5G Cellular Networks," in IEEE Wireless Communications, vol. 24, no. 2, pp. 82–89, April 2017. Online at: https://doi. org/10.1109/MWC.2016.1500157RP

[50] K. Antonakoglou, X. Xu, E. Steinbach, T. Mahmoodi and M. Dohler, "Toward Haptic Communications Over the 5G Tactile Internet," in IEEE Communications Surveys & Tutorials, vol. 20, no. 4, pp. 3034–3059, Fourthquarter 2018. Online at: https://doi.org/10.1109/COMST. 2018.2851452

[51] S. K. Sharma, I. Woungang, A. Anpalagan and S. Chatzinotas, "Toward Tactile Internet in Beyond 5G Era: Recent Advances, Current

Issues, and Future Directions," in IEEE Access, vol. 8, pp. 56948–56991, 2020. Online at: https://doi.org/10.1109/ACCESS.2020.2980369

[52] S. Knab, and R.Rohrbeck. "Why intended business model innovation fails to deliver: insights from a longitudinal study in the German smart energy market." Proceedings of the R&D Management Conference, Stuttgart, Germany, June 3–6, 2014.

[53] O. Vermesan and J. Bacquet (Eds.). Next Generation Internet of Things. Distributed Intelligence at the Edge and Human Machine-to-Machine Cooperation, ISBN: 978-87-7022-008-8, River Publishers, Gistrup, 2018. Online at: https://european-iot-pilots.eu/wp-content/uploads/2018/11/Next_Generation_Internet_of_Things_Distributed_Intelligence_at_the_Edge_IERC_2018_Cluster_eBook_978-87-7022-007-1_P_Web.pdf

[54] O. Vermesan and J. Bacquet (Eds.). Cognitive Hyperconnected Digital Transformation. Internet of Things Intelligence Evolution, ISBN: 978-87-93609-11-2, River Publishers, Gistrup, 2017. Online at: https://www.riverpublishers.com/research_details.php?book_id=456

[55] A. Gluhak, O. Vermesan, R. Bahr, F. Clari, T. Macchia, M. T. Delgado, A. Hoeer, F. Boesenberg, M. Senigalliesi and V. Barchetti, "Report on IoT platform activities", 2016, online at http://www.internet-of-things-research.eu/pdf/D03_01_WP03_H2020_UNIFY-IoT_Final.pdf.

[56] IoT Analytics GmbH, IoT Platforms Company Landscape 2020. Online at: https://iot-analytics.com/product/iot-platforms-landscape-database-2020/

[57] M. Heller, "How to choose a cloud IoT platform". Online at: https://sg.channelasia.tech/article/print/679602/how-choose-cloud-iot-platform/

[58] O. Vermesan and P. Friess (Eds.). Digitising the Industry Internet of Things Connecting the Physical, Digital and Virtual Worlds, ISBN: 978-87-93379-81-7, River Publishers, Gistrup, 2016. Online at: https://www.riverpublishers.com/research_details.php?book_id=396

[59] O. Vermesan and P. Friess (Eds.). Building the Hyperconnected Society – IoT Research and Innovation Value Chains, Ecosystems and Markets, ISBN: 978-87-93237-99-5, River Publishers, Gistrup, 2015. Online at: https://www.riverpublishers.com/research_details.php?book_id=307

[60] S. Nakamoto. Bitcoin: A Peer-to-Peer Electronic Cash System. Online at: https://bitcoin.org/bitcoin.pdf

[61] Opportunities and Use Cases for Distributed Ledgers in IoT, GSMA 2018, online at: https://www.gsma.com/iot/wp-content/uploads/2018/09/Opportunities-and-Use-Cases-for-Distributed-Ledgers-in-IoT-f.pdf

[62] O. Vermesan, et. al., "The Next Generation Internet of Things – Hyperconnectivity and Embedded Intelligence at the Edge", in O. Vermesan and J. Bacquet (Eds.). Next Generation Internet of Things – Distributed Intelligence at the Edge and Human Machine-to-Machine Cooperation, ISBN: 978-87-7022-008-8, River Publishers, Gistrup, 2018, pp. 19–91.

[63] O. Vermesan, et. al., "Internet of robotic things: converging sensing/actuating, hypoconnectivity, artificial intelligence and IoT Platforms", in O. Vermesan and J. Bacquet (Eds.). Cognitive Hyperconnected Digital Transformation Internet of Things Intelligence Evolution, ISBN: 978-87-93609-10-5, River Publishers, Gistrup, 2017, pp. 97–155.

[64] What is Blockchain? Blockchainhub Berlin, updated July 2019. Online at: https://blockchainhub.net/blockchain-intro/

[65] K. Loupos, B. Caglayan, A. Papageorgiou, B. Starynkevitch, F. Vedrine, C. Skoufis, S. Christofi, B. Karakostas, A. Mygiakis, G. Theofilis, A. Chiappetta, H. Avgoustidis, George Boulougouris – Cognition Enabled IoT Platform for Industrial IoT Safety, Security and Privacy – The CHARIOT Project, IEEE International Workshop on Computer Aided Modeling and Design of Communication Links and Networks (CAMAD), 11–13 September 2019, Limassol, Cyprus, DOI: 10.1109/CAMAD.2019.8858488.

[66] Papageorgiou, T. Krousarlis, K. Loupos, A. Mygiakis, DPKI: A Blockchain-Based Decentralized Public Key Infrastructure System, Global IoT Summit 2020, 3rd Workshop on Internet of Things Security and Privacy (WISP), 3–5 June 2020, Dublin.

[67] M. Swan, Blockchain: Blueprint for a New Economy, O'Reilly Media, Inc., 2015.

[68] K. Christidis, M. Devetsikiotis, Blockchains and smart contracts for the internet of things. IEEE Access 4, pp. 2292–2303, 2016.

[69] What is a blockchain operating system and what are the benefits? Introducing Overledger from Quant Network. Online at: https://medium.com/@CryptoSeq/what-is-a-blockchain-operating-system-and-what-are-the-benefits-c561d8275de6

[70] D. Yaga, P. Mell, N. Roby, and K. Scarfone. Blockchain Technology Overview. Technical Report. NISTIR. 2018. Online at: https://doi.org/ 02

[71] Outlier Ventures Research, Blockchain-Enabled Convergence – Understanding The Web 3.0 Economy, online at https://gallery.ma ilchimp.com/65ae955d98e06dbd6fc737bf7/files/Blockchain_Enabled _Convergence.01.pdf

[72] What is a blockchain? https://www2.deloitte.com/content/dam/Deloit te/ch/Documents/innovation/ch-en-innovation-deloitte-what-is-block chain-2016.pdf

[73] LongView Whitepaper, Super-B for IoT: Improving QoS in LoRa Networks, 2019. Online at: https://www.longviewiot.com/resource s/assets/white-paper-super-b-protocol-pdf/

[74] M. Wright, 5G – A brave new future. Telestra. Presentation at Mobile World Congress, Barcelona, 2018.

[75] S. Lin, H.F. Cheng, W. Li, Z. Huang, P. Hui, C. Peylo, Ubii: physical world interaction through augmented reality, IEEE Trans. Mob. Comput. 16 (2017) 872–885.

[76] X. Sun, N. Ansari, EdgeIoT: mobile edge computing for the Internet of Things, IEEE Commun. Mag. 54 (2016) 22–29.

[77] Edge Computing Market, MarketsandMarkets Report 2017, online at: https://www.marketsandmarkets.com/Market-Reports/edge-computin g-market-133384090.html

[78] Ericsson Mobility Report, June 2018, online at: https://www.ericsson .com/assets/local/mobility-report/documents/2018/ericsson-mobility -report-june-2018.pdf

[79] Ericsson Mobility Report, June 2019, online at: https://www.ericsson .com/49d1d9/assets/local/mobility-report/documents/2019/ericsson -mobility-report-june-2019.pdf

[80] Ericsson Mobility Report, November 2019, online at: https://www.er icsson.com/4acd7e/assets/local/mobility-report/documents/2019/emr-november-2019.pdf?_ga=2.258613962.1969153473.1594892941--1 856930902.1594892941&_gac=1.212672800.1594892967.EAIaIQob ChMIkb6xoL_R6gIVS9OyCh3EWA8tEAAYASAAEgJNYPD_BwE

[81] Ericsson ConsumerLab, "10 Hot Consumer Trends 2030 – Internet of the senses", December 2019, Online at:

[82] Ericsson Mobility Report, June 2020, online at: https://www.ericsson .com/49da93/assets/local/mobility-report/documents/2020/june2020 -ericsson-mobility-report.pdf

[83] Cloud IoT Edge, online at: https://cloud.google.com/iot-edge/

[84] Hyperledger, online at: https://www.hyperledger.org/

[85] Enterprise Ethereum Alliance (EEA), online at: https://entethalliance .org/

[86] Hyperledger Fabric, online at: https://www.ibm.com/blockchain/hyp erledger/fabric-support

[87] G. Herr, J. Lyon, and S. Gillen, "Industrial intelligence: Cognitive analytics in action," presentation at EMEA Users Conference, Berlin, 2016.

[88] Research and Markets, online at: https://www.researchandmarkets.c om/

[89] State of the IoT 2018: Number of IoT devices now at 7B – Market accelerating, IOT ANALYTICS, online at: https://iot-analytics.com/st ate-of-the-iot-update-q1-q2--2018-number-of-iot-devices-now-7b/

[90] IDTechEx, Comparison of Low Power Wide Area Networks (LPWAN) for IoT 2018–2019, online at https://www.idtechex.com/research/arti cles/comparison-of-low-power-wide-area-networks-lpwan-for-iot-20 18--2019-00014777.asp

[91] Wi-Fi HaLow, online at: https://www.wi-fi.org/discover-wi-fi/wi-fi-halow

[92] H. Haas, "LiFi is a paradigm-shifting 5G technology", Reviews in Physics 3 (2018), pp. 26–31.

[93] Wi-SUN FAN Overview, IETF LPWAN working group, online at: ht tps://tools.ietf.org/id/draft-heile-lpwan-wisun-overview-00.html

[94] SEMTECH, online at: https://www.semtech.com/

[95] T. Ryberg, BERG INSIGHT, NB-IoT networks are here, now it's time to make business, online at: https://www.iot-now.com/2018/07/04/851 56-nb-iot-networks-now-time-make-business/

[96] NB-IoT, CAT-M, SIGFOX and LoRa Battle for Dominance Drives Global LPWA Network Connections to Pass 1 Billion By 2023, ABIre- search, 2018, online at: https://www.abiresearch.com/press/nb-iot-cat -m-sigfox-and-lora-battle-dominance-drives-global-lpwa-network-c onnections-pass-1-billion-2023/

[97] Wi-SUN FAN Overview, online at: https://tools.ietf.org/id/draft-heile-lpwan-wisun-overview-00.html

[98] Opportunities and Use Cases for Distributed Ledgers in IoT, GSMA 2018, online at: https://www.gsma.com/iot/wp-content/uploads/2018 /09/Opportunities-and-Use-Cases-for-Distributed-Ledgers-in-IoT-f.p df

[99] BEHRTECH, "The Ultimate Guide io Wireless Connectivity for Massive Scale IoT Deployments", online at: www.behrtech.com

[100] Lavric, "LoRa High-Density Sensors for Internet of Things", Journal of Sensors, 2019.

[101] ETSI – European Telecommunications Standards Institute, "Short range devices; Low Throughput Networks (LTN); Protocols for radio interface A", ETSI TS 103 357, 2018.

[102] T. Lauterbach, "MYTHINGS vs LoRa: A comparative study of Quality-of-Service under external interference", 2019, online at: https://behrtech.com/resources/lora-vs-mythings/

[103] MIOTY, Radiocrafts AS, online at: https://radiocrafts.com/products/mioty-network/

[104] mioty® – The Wireless IoT Technology; Fraunhofer IIS, online: http://mioty.de/

[105] ANT/ANT+ Defined, online at: https://www.thisisant.com/developer/ant-plus/ant-antplus-defined/

[106] "ANT message protocol and usage: Application notes," Dynastream Innovations Inc., 2007, online at: https://www.thisisant.com/resources/ant-message-protocol-and-usage/

[107] S. Khssibi, H. Idoudi, A. V. D. Bossche, T. Val, and L. A. Saidane, "Presentation and analysis of a new technology for low-power wireless sensor network," International Journal of Digital Information and Wireless Communications (IJDIWC), vol. 3, no. 1, pp. 75–86, 2013, online at: https://oatao.univ-toulouse.fr/12426/1/Khssibi_12426.pdf

[108] "AN1200.22 LoRa^{TM} Modulation Basics", online at: https://semtech.my.salesforce.com/sfc/p/#E0000000JelG/a/2R00000010Ju/xvKUc5w9yjG1q5Pb2IIkpolW54YYqGb.frOZ7HQBcRc

[109] P. Pachuca, "LoRa Edge^{TM} Explained: How LR1110 Drives Smarter Geolocation", Semtech's Corporate Blog, 2020, online at: https://blog.semtech.com/lora-edge-explained

[110] F. Adelantado, X. Vilajosana, P. Tuset-Peiro, B. Martinez, J. Melia-Segui and T. Watteyne, "Understanding the Limits of LoRaWAN," in IEEE Communications Magazine, vol. 55, no. 9, pp. 34–40, Sept. 2017, online at: doi: 10.1109/MCOM.2017.1600613.

[111] N. Ducrot, D. Ray, A. Saadani, O. Hersent, G. Pop, and G. Remond, "LoRa Device Developer Guide," Orange, Connected Objects and Partnership. Technical Document., Apr. 2016, online at: https://developer.orange.com/wp-content/uploads/LoRa-Device-Developer-Guide-Orange.pdf

[112] Minaburo, A. Pelov, and L. Toutain, LP-WAN Gap Analysis, IETF Std., Feb. 2016, online at: https://tools.ietf.org/pdf/draft-minaburo-lp-wan-gap-analysis-00.pdf

[113] Z. Yoon, W. Frese, and K. Briess. "Design and Implementation of a Narrow-Band Intersatellite Network with Limited Onboard Resources for IoT". Sensors (Basel). 2019 Sep;19(19), online at: doi:10.3390/s19194212. PMID: 31569831; PMCID: PMC6806246.

[114] H. Urlings, "Satellite IoT: A Game Changer for the Industry?", 2019, online at: http://satellitemarkets.com/satellite-iot-game-changer-industry

[115] W. Webb, "Weightless: A Bespoke Technology for the IoT", online: http://www.weightless.org/news/weightless-a-bespoke-technology-for-the-iot

[116] Ingenu. How RPMA Works: The Making of RPMA, online at: https://www.ingenu.com/portfolio/how-rpma-works-the-making-of-rpma/

[117] Sigfox Technology, online at: https://www.sigfox.com/en/what-sigfox/technology

[118] Weightless Specification, online at: http://www.weightless.org/about/weightless-specification

[119] Gupta, R.K. Jha, A survey of 5G network: architecture and emerging technologies, IEEE Access 3 (July) (2015) 1206–1232.

[120] M. Iwamura, "NGMN View on 5G architecture", Proceedings of the IEEE Vehicular Technology Conference, 2015, pp. 1–5.

[121] NGMN 5G Initiative White Paper, February 2015, online at: https://www.ngmn.org/wp-content/uploads/NGMN_5G_White_Paper_V1_0.pdf

[122] Afolabi, A. Ksentini , M. Bagaa , T. Taleb , M. Corici , A. Nakao ,Towards 5G network slicing over multiple-Domains, in IEICE Trans. Commun. (11) (2017) 1992–2006.

[123] ITU-T Y.3011, Framework of Network Virtualization for Future Networks, January 2012, online at: https://www.itu.int/rec/T-REC-Y.3011-201201-ITRGPP23.799, Study on Architecture for Next Generation System, Rel.14 (December 2016), online at: https://portal.3gpp.org/desktopmodules/Specifications/SpecificationDetails.aspx?specificationId=3008

[124] X. de Foy, A. Rahman , "Network slicing-3GPP Use case", InterDigital Communications, LLC, 2017, online at: https://tools.ietf.org/pdf/draft-defoy-netslices-3gpp-network-slicing-02.pdf

[125] J. Crawshaw, "Network Slicing: OSS/BSS Key to Commercial Success", Tech Mahindra, 2019 online at: https://cache.techmahindra.com/static/img/pdf/oss-bss-key-to-commercial-success.pdf

[126] Al-Fuqaha, M. Guizani, M. Mohammadi, M. Aledhari, M. Ayyash, Internet of Things: a survey on enabling technologies, protocols, and applications, IEEE Commun. Surv. Tutorials 17 (June (4)) (2015) 2347–2376.

[127] M.R. Palattella, M. Dohler, A. Grieco, G. Rizzo, J. Torsner , T. Engel, L. Ladid , Internet of Things in the 5G era: enablers, architecture, and business models, IEEE J. Sel. Areas Commun. 34 (February (3)) (2016) 510–527.

[128] M. Hasan, E. Hossain, D. Niyato, Random access for machine-to-machine communication in LTE-advanced networks: issues and approaches, IEEE Commun. Mag. 51 (June (6)) (2013) 86–93.

[129] M. Maier, M. Chowdhury, B.P. Rimal, D.P. Van, The tactile internet: vision, recent progress, and open challenges, IEEE Commun. Mag. 54 (May (5)) (2016) 138–145.

[130] *Progress on 3GPP IoT*, February 2016, [Online]. Available: http://www.3gpp.org/news-events/3gpp-news/1766-iot_progress

[131] *Release 14*, February 2016, [Online]. Available: http://www.3gpp.org/release-14.

[132] S. Nakamoto, Bitcoin: A Peer-to-Peer Electronic Cash System, online at: https://bitcoin.org/bitcoin.pdf

[133] S. Popov, The Tangle, 2018, online at: https://assets.ctfassets.net/r1dr6vzfxhev/2t4uxvsIqk0EUau6g2sw0g/45eae33637ca92f85dd9f4a3a218e1ec/iota1_4_3.pdf

[134] IOTA, online at: https://iota.org

[135] Ripple, online at: https://ripple.com

[136] Sovrin, online at: https://sovrin.org

[137] BigchainDB, online at: https://www.bigchaindb.com

[138] Digital Twin consortium. Online at: https://www.digitaltwinconsortium.org/index.htm

[139] R., Stark, and T., Damerau, Digital Twin. Springer Berlin Heidelberg, Berlin, Heidelberg, 2019, pp. 1–8. Online at: http://dx.doi.org/10.1007/978-3-642-35950-7_16870-1.

[140] Digital Twin towards a meaningful framework-Arup. Online at: https://www.arup.com/perspectives/publications/research/section/digital-twin-towards-a-meaningful-framework

[141] Industrial Data. Online at: https://www.iso.org/committee/54158.html

[142] Digital Twin Manufacturing Framework. Online at: https://www.iso.or g/standard/75066.html

[143] JETI. Online at: https://jtc1info.org/technology/jeti/

[144] ISO/TC 184 – Automation systems and integration. Online at: https: //www.iso.org/committee/54110.html

[145] P2806 – System Architecture of Digital Representation for Physical Objects in Factory Environments. Online at: https://standards.ieee.org /project/2806.html#Standard

[146] Requirements and capabilities of a digital twin system for smart cities. ITU-T work programme. Online at: https://www.itu.int/ITU-T/workpr og/wp_item.aspx?isn=16396

[147] buildingSMART International, Position Paper, "Enabling an Ecosystem of Digital Twins". Online at: https://www.buildingsmart.org/wp-c ontent/uploads/2020/05/Enabling-Digital-Twins-Positioning-Paper-Final.pdf

[148] IoTwins H2020 project. Online at: https://www.iotwins.eu/

[149] Virtual Singapore. National Research Foundation (NRF), Singapore. Online at: https://www.nrf.gov.sg/programmes/virtual-singapore

[150] University of Cambridge. UK. National Digital Twin Programme. Online at: https://www.cdbb.cam.ac.uk/what-we-do/national-dig ital-twin-programme

[151] J Konečný, H.B., McMahan, F.X., Yu, P., Richtárik, A.T., Suresh, D., Bacon. "Federated learning: Strategies for improving communication efficiency". Online at: arXiv preprint arXiv:1610.05492

[152] OpenMined – open-source community. Online at: https://www.open mined.org/

[153] P., Kairouz, H. B., McMahan, et. al., "Advances and Open Problems in Federated Learning," Dec. 2019. Online at: https://arxiv.org/abs/1912 .0497

[154] K. Panetta, 5 Trends Emerge in the Gartner Hype Cycle for Emerging Technologies, 2018, online at: https://www.gartner.com/smarterwithg artner/5-trends-emerge-in-gartner-hype-cycle-for-emerging-technolo gies-2018/

[155] N. J. Nilsson, The Quest for Artificial Intelligence: A History of Ideas and Achievements, Cambridge, UK Cambridge University Press, 2010.

[156] IEEE, Tactile Internet Emerging Technologies Subcommittee, online at: http://ti.committees.comsoc.org/

[157] The Tactile Internet ITU-T Technology Watch Report ITU-T, 2014, online at: https://www.itu.int/dms_pub/itu-t/oth/23/01/T2301000023 0001PDFE.pdf

[158] G. Batra, A. Queirolo, and N. Santhanam, McKinsey & Company, 2018, Artificial intelligence: The time to act is now, online at: https: //www.mckinsey.com/industries/advanced-electronics/our-insights/ar tificial-intelligence-the-time-to-act-is-now

[159] 5G Network Architecture A High-Level Perspective, White Paper, HUAWEI Technologies CO., LTD., 2016, online at: https://www-file. huawei.com/-/media/CORPORATE/PDF/mbb/5g_nework_architect ure_whitepaper_en.pdf?la=en&source=corp_comm

[160] 5G Security Architecture White Paper, HUAWEI Technologies CO., LTD., 2017, online at: https://www-file.huawei.com/-/media/CORPO RATE/PDF/white%20paper/5g_security_architecture_white_paper_e n-v2.pdf?la=en&source=corp_comm

[161] Network Slicing Use Case Requirements, GSMA, April 2018, online at: https://www.gsma.com/futurenetworks/wp-content/uploads/2018/ 07/Network-Slicing-Use-Case-Requirements-fixed.pdf

[162] O. Holland et al., "The IEEE 1918.1 "Tactile Internet" Standards Working Group and its Standards," in Proceedings of the IEEE, vol. 107, no. 2, pp. 256–279, Feb. 2019. Online at: https://doi.org/10.1109/ JPROC.2018.2885541

[163] Z. S. Bojkovic, B. M. Bakmaz and M. R. Bakmaz, "Vision and enabling technologies of tactile internet realization," 2017 13th International Conference on Advanced Technologies, Systems and Services in Telecommunications (TELSIKS), Nis, 2017, pp. 113–118. Online at: https://doi.org/10.1109/TELSKS.2017.8246242

[164] L. Zou, *et al.,* "Novel tactile sensor technology and smart tactile sensing systems: A review," *MDPI Sensors*, vol. 17, 2017.

[165] F. Dressler, "Towards the Tactile Internet: Low Latency Communication for Connected Cars," online at: http://conferences.sigcomm.org/si gcomm/2017/files/tutorial-c2c.pdf

[166] W. E. Forum, "Industrial internet of things: Unleashing the potential of connected products and services," Jan. 2015, World Economic Forum, Geneva, Switzerland, White Paper REF 020315.

[167] T. H. Szymanski, "Securing the Industrial-Tactile Internet of Things with Deterministic Silicon Photonics Switches," in IEEE Access, vol. 4, pp. 8236–8249, 2016. Online at: https://doi.org/10.1109/ACCESS .2016.2613512

[168] M. Maier, M. Chowdhury, B. P. Rimal and D. P. Van, "The tactile internet: vision, recent progress, and open challenges," in IEEE Communications Magazine, vol. 54, no. 5, pp. 138–145, May 2016. Online at: https://doi.org/10.1109/MCOM.2016.7470948

[169] M. Maier, "FiWi access networks: Future research challenges and moonshot perspectives," 2014 IEEE International Conference on Communications Workshops (ICC), Sydney, NSW, 2014, pp. 371–375. Online at: https://doi.org/10.1109/ICCW.2014.6881225

[170] M. Chowdhury and M. Maier, "Collaborative Computing for Advanced Tactile Internet Human-to-Robot (H2R) Communications in Integrated FiWi Multirobot Infrastructures," in IEEE Internet of Things Journal, vol. 4, no. 6, pp. 2142–2158, Dec. 2017. Online at: https://doi.org/10.1109/JIOT.2017.2761599

[171] Spectrum of Seven Outcomes for AI, online at: https://www.constellat ionr.com/

[172] ITU-T, Internet of Things Global Standards Initiative, http://www.itu. int/en/ITU-T/gsi/iot/Pages/default.aspx

[173] International Telecommunication Union – ITU-T Y.2060 – (06/2012) – Next Generation Networks – Frameworks and functional architecture models – Overview of the Internet of things

[174] O. Vermesan, P. Friess, P. Guillemin, H. Sundmaeker, et al., "Internet of Things Strategic Research and Innovation Agenda", Chapter 2 in Internet of Things – Converging Technologies for Smart Environments and Integrated Ecosystems, River Publishers, 2013, ISBN 978-87-92982-73-5

[175] Parks Associates, Monthly Wi-Fi usage increased by 40% in U.S. smartphone households, online at https://www.parksassociates.co m/blog/article/pr-06192017

[176] Gluhak, O. Vermesan, R. Bahr, F. Clari, T. Macchia, M. T. Delgado, A. Hoeer, F. Boesenberg, M. Senigalliesi and V. Barchetti, "Report on IoT platform activities", 2016, online at http://www.internet-of-things -research.eu/pdf/D03_01_WP03_H2020_UNIFY-IoT_Final.pdf.

[177] IoT Platforms Initiative, online at https://www.iot-epi.eu/

[178] IoT European Large-Scale Pilots Programme, online at https://europe an-iot-pilots.eu/

[179] S. Moore, (2016, December 7) Gartner Survey Shows Wearable Devices Need to Be More Useful, online at http://www.gartner.co m/newsroom/id/3537117

[180] Digital Economy Collaboration Group (ODEC). Online at http://arch ive.oii.ox.ac.uk/odec/

[181] Connect building systems to the IoT, online at http://www.electronics-know-how.com/article/1985/connect-building-systems-to-the-iot

[182] S. Kejriwal and S. Mahajan, Smart buildings: How IoT technology aims to add value for real estate companies The Internet of Things in the CRE industry, Deloitte University Press, 2016, online at https://www2.deloitte.com/content/dam/Deloitte/nl/Documents/real-estate/deloitte-nl-fsi-real-estate-smart-buildings-how-iot-technology-aims-to-add-value-for-real-estate-companies.pdf

[183] ORGALIME Position Paper, 2016, online at http://www.orgalime.org/sites/default/files/position-papers/Orgalime%20Comments_EED_EPBD_Review%20Policy%20Options_4%20May%202016.pdf

[184] The EU General Data Protection Regulation (GDPR) (Regulation (EU) 2016/679).

[185] RERUM, EU FP7 project, www.ict-rerum.eu

[186] Digital Agenda for Europe, European Commission, Digital Agenda 2010–2020 for Europe. Online at http://ec.europa.eu/information_society/digital-agenda/index_en.html

[187] O. Vermesan, P. Friess, G. Woysch, P. Guillemin, S. Gusmeroli, et al., " Europe's IoT Strategic Research Agenda 2012", Chapter 2 in The Internet of Things 2012 New Horizons, Halifax, UK, 2012, ISBN 978-0-9553707-9-3

[188] O. Vermesan, et al., "Internet of Energy – Connecting Energy Anywhere Anytime" in Advanced Microsystems for Automotive Applications 2011: Smart Systems for Electric, Safe and Networked Mobility, Springer, Berlin, 2011, ISBN 978-36-42213-80-9

[189] M. Yuriyama and T. Kushida, "Sensor-Cloud Infrastructure – Physical Sensor Management with Virtualized Sensors on Cloud Computing", NBiS 2010: 1–8

[190] Test Considerations for 5G New Radio, White Paper, Keysight Technologies, April 2018, online at: http://literature.cdn.keysight.com/litweb/pdf/5992--2921EN.pdf

[191] M. A. Ferrag, L. A. Maglaras, H. Janicke, J. Jiang, and L. Shu, "Authentication Protocols for Internet of Things: A Comprehensive Survey", in Security and Communication Networks, Hindawi, 2017, doi: 10.1155/2017/6562953

[192] V. Hassija, V. Chamola, V. Saxena, D. Jain, P. Goyal and B. Sikdar, "A Survey on IoT Security: Application Areas, Security Threats, and

Solution Architectures," in *IEEE Access*, vol. 7, pp. 82721–82743, 2019.

[193] M. Banerjee, J. Lee, K.-K. R. Choo, "A blockchain future for internet of things security: a position paper", in Digital Communications and Networks, Vol. 4, Issue 3, 2018, pp. 149–160, doi:10.1016/j.dcan.2017.10.006

[194] N. Chikouche, P.-L. Cayrel, El H. M. Mboup, B. O. Boidje, "A privacy-preserving code-based authentication protocol for Internet of Things", in Journal of Supercomputing, Springer Verlag, 2019, doi:10.1007/s11227-019-03003-4

[195] Schmitt, C.; Noack, M.; Stiller, B. TinyTO: Two-way authentication for constrained devices in the Internet of Things. In Internet of Things; Elsevier: Amsterdam, The Netherlands, 2016; pp. 239–258.

[196] M. Jan, P. Nanda, M. Usman, and X. He, "PAWN: A payload-based mutual authentication scheme for wireless sensor networks," Concurrency Computation, 2016.

[197] Y. Chung, S. Choi, Y. S. Lee, N. Park, and D. Won, "An enhanced lightweight anonymous authentication scheme for a scalable localization roaming service in wireless sensor networks," Sensors, vol. 16, no. 10, article no. 1653, 2016.

[198] Maarof, M. Senhadji, Z. Labbi, M. Belkasmi, "Authentication protocol for securing internet of things", In Proceedings of the Fourth International Conference on Engineering & MIS 2018. ACM, pp 29:1–29:7

[199] A. Esfahani et al., "A Lightweight Authentication Mechanism for M2M Communications in Industrial IoT Environment," in IEEE Internet of Things Journal, vol. 6, no. 1, pp. 288–296, Feb. 2019. Online at: https://doi.org/10.1109/JIOT.2017.2737630

[200] X. Li, J. Niu, M. Z. A. Bhuiyan, F. Wu, M. Karuppiah and S. Kumari, "A Robust ECC-Based Provable Secure Authentication Protocol With Privacy Preserving for Industrial Internet of Things," in IEEE Transactions on Industrial Informatics, vol. 14, no. 8, pp. 3599–3609, Aug. 2018. Online at: https://doi.org/10.1109/TII.2017.2773666

[201] X. Yao, H. Kong, H. Liu, T. Qiu and H. Ning, "An Attribute Credential Based Public Key Scheme for Fog Computing in Digital Manufacturing," in IEEE Transactions on Industrial Informatics, vol. 15, no. 4, pp. 2297–2307, April 2019. Online at: https://doi.org/10.1109/TII.2019.2 891079

[202] P. Dunphy and F. A. P. Petitcolas, "A First Look at Identity Management Schemes on the Blockchain," in IEEE Security & Privacy, vol. 16, no. 4, pp. 20–29, July/August 2018. Online at: https://doi.org/10.1109/MSP.2018.3111247

[203] G. Kornaros, O. Tomoutzoglou and M. Coppola, "Hardware-Assisted Security in Electronic Control Units: Secure Automotive Communications by Utilizing One-Time-Programmable Network on Chip and Firewalls," in IEEE Micro, vol. 38, no. 5, pp. 63–74, Sep./Oct. 2018. Online at: https://doi.org/10.1109/MM.2018.053631143

[204] D. Sethia, D. Gupta and H. Saran, "NFC Secure Element-Based Mutual Authentication and Attestation for IoT Access," in IEEE Transactions on Consumer Electronics, vol. 64, no. 4, pp. 470–479, Nov. 2018. Online at: https://doi.org/10.1109/TCE.2018.2873181

[205] D. Mbakoyiannis, O. Tomoutzoglou, and G. Kornaros, "Secure over-the-air firmware updating for automotive electronic control units", In Proceedings of the 34th ACM/SIGAPP Symposium on Applied Computing (SAC '19), pp. 174–181, 2019. https://doi.org/10.1145/3297280.3297299

[206] L. Zhang, C. Hu, Q. Wu, J. Domingo-Ferrer and B. Qin, "Privacy-Preserving Vehicular Communication Authentication with Hierarchical Aggregation and Fast Response," in IEEE Transactions on Computers, vol. 65, no. 8, pp. 2562–2574, 1 Aug. 2016. Online at: https://doi.org/10.1109/TC.2015.2485225

[207] G. Kornaros, et al., "Towards Holistic Secure Networking in Connected Vehicles through Securing CAN-bus Communication and Firmware-over-the-Air Updating", Journal of Systems Architecture, vol. 109, pp. 101761, 2020. https://doi.org/10.1016/j.sysarc.2020.101761

[208] J. Bernal Bernabe, J. L. Canovas, J. L. Hernandez-Ramos, R. Torres Moreno and A. Skarmeta, "Privacy-Preserving Solutions for Blockchain: Review and Challenges," in IEEE Access, vol. 7, pp. 164908–164940, 2019. Online at: https://doi.org/10.1109/ACCESS.2019.2950872

[209] Decentralized identity foundation," 2018, https://identity.foundation/.

[210] C. Allen, "The path to self-sovereign identity," 2016.

[211] IETF, https://datatracker.ietf.org/doc/draft-ietf-ace-oauth-authz/?include_text=1

[212] W3C, A Primer for Decentralized Identifiers", Draft Community Group Report 19 Jan. 2019, https://w3c-ccg.github.io/did-primer/

[213] Wi-Fi Alliance, https://www.wi-fi.org/

[214] S. Behrens. "What's the Status of Wi-Fi 6?". Paessler, 3 May 2019.

[215] M. Kapko. "5G, WiFi 6 Set to Battle for Control of Enterprise Private Networks. SDxCentral, 29 January 2020.

[216] The 3rd Generation Partnership Project (3GPP), http://www.3gpp.org/

[217] Narrow Band IoT & M2M – IoT Global Ecosystem. GSA, April 2020, https://gsacom.com/paper/iot-global-ecosystem-april-2020-summary/

[218] R. Merritt. IoT Nets in Two-Horse LPWAN Race. EETimes, 7 May 2019, https://www.eetimes.com/iot-nets-in-two-horse-lpwan-race/#

4

Real-time Management of Energy Consumption in Water Resource Recovery Facilities Using IoT Technologies

Rita Alves[1], Mário Nunes[2], Augusto Casaca[2], Pedro Póvoa[1] and José Botelho[1]

[1]Águas do Tejo Atlântico, Portugal
[2]INOV Inesc Inovação/Inesc-ID, Portugal

Abstract

The real-time management of energy consumption in Water Resource Recovery Facilities is key for the intended implementation of demand-side management in those Facilities, with the objective of optimizing energy consumption. For this objective, special purpose energy meters have been designed, implemented and deployed. The energy meters measure several electrical parameters for the distinct processes and equipment in the Water Resource Recovery Facilities, transmitting those measurements to a central platform where they are stored and treated. The communication architecture chosen for this purpose is based on the Internet of Things paradigm, using Wi-Fi or LoRa protocols for communication within the Facilities, a private network for external access to the water company servers and MQTT as the higher layer protocol for data transfer. At the moment, thirty-seven energy meters are deployed in two Water Resource Recovery Facilities in Lisbon, Portugal, and some of the measurement results obtained are shown in the article.

4.1 Introduction

Águas do Tejo Atlântico (AdTA) is a leading company operating in the environmental sector in Portugal, and its mission is to contribute to the pursuit

185

of national objectives in the wastewater collection and treatment within a framework of economic, financial, technical, social and environmental sustainability.

AdTA has the responsibility to manage and operate the wastewater multi-municipality system of the Great Lisbon and West of Great Lisbon areas, guaranteeing the quality, continuity and efficiency of the water public services, in order to protect the public health, population welfare, the accessibility to the public services, the environmental protection and economic and financial sustainability of the sector, in a framework of equity and tariff stability, contributing also to the regional development and planning.

AdTA manages 103 Water Resource Recovery Facilities (WRRF), 268 pumping stations and 1093.40 km of main sewage system, and treats around $194 \, \text{Mm}^3/\text{year}$, serving a population of 2.4 million inhabitants and a covered area of $4,145 \, \text{km}^2$.

A WRRF refers to a facility designed to treat municipal wastewater and recover sub-products, such as bio-solids, treated wastewater and energy. The level of treatment at a plant will vary based on the water bodies classification and the specific processes involved. Treatment processes may include biological, chemical, and physical treatment.

Energy is often the second highest operating cost at a WRRF after the labor cost; it can be about 50% of a utility's total operating costs [1]. The reason for this is that most of the processes that occur in a WRRF require energy for their operation and are intensive energy consumers, such as aeration and elevation. In 2018, the energy consumption at AdTA was about 92,206,887 kWh and represented 49% (€ 8,378,240) of the operating costs.

In the WRRF, the most significant energy consumption is related with biological treatment (aeration) and pumping. However, it is expected to have an increase of energy consumption in the wastewater industry due to: more demanding legal discharge limits that in some cases may imply the use of energy intensive technologies; sludge advanced treatment such as drying, incineration and pyrolysis; and natural ageing of sewage systems resulting in increase of infiltrations, with the consequent pumping and transportation costs [2].

Nowadays, also new challenges appear due to the limited water and nutrient resources, to the existence of the circular economy framework, and to climate changes concerns associated with the fossil fuel consumption. In this context, a new paradigm for the use of the domestic wastewater was created where domestic wastewater is being looked more as a resource than a waste and the wastewater treatment plants now are known as WRRF, where it is possible to recover nutrients and water to achieve a more sustainable use of the wastewater energy potential, and become a driver for the circular economy [3].

AdTA is a consortium member of the running European Union H2020 research project "Intelligent Grid Technologies for Renewables Integration and Interactive Consumer Participation Enabling Interoperable Market Solutions and Interconnected Stakeholders", which has the acronym InteGrid. The main objectives of InteGrid are to test the flexibility of electrical energy consumption for domestic and industrial consumers, to test energy storage systems and to make forecasts of renewable energy production and consumption. The main role of AdTA in this project is to manage the flexibility of its internal processes in order to minimize the energy costs according to the market and to the requirements of the electrical grid operators, leading to an optimization of the AdTA internal processes, through a demand side management paradigm.

The AdTA role focusses into the flexibility of energy consumption and into the new challenges of turning a wastewater treatment plant into a WRRF. This is an important path towards its objective of achieving operational optimization and flexibility of processes. To optimize and make processes more flexible it is essential to know the performance of the processes. Thus, one of the variables to monitor is the energy consumption of the different processes and equipment. This article is related with the specific work done for real-time monitoring of the electrical energy consumption in WRRF by using Internet of Things (IoT) technologies. This work is being done by AdTA, within the InteGrid project, in collaboration with INOV, a Portuguese research institute.

The use of IoT technologies, at the communication and platform levels, means a state-of-the-art solution for the management of energy consumption in WRRF, which may become an important vertical domain in the vast area of IoT applications.

A first article has already described the basic ideas of the project and the plans for its implementation [4]. The present article, besides introducing the main architectural concepts, has a strong focus into the deployment done and some of the obtained results. Section 2 reports on the related work. Section 3 is dedicated to the IoT based communication architecture adopted in the project and Section 4 deals with the implementation of the energy meter. Sections 5 and 6 describe the deployment and the obtained results, respectively. We conclude in Section 7.

4.2 Related Work

Currently, more than optimizing energy generation and distribution, it is the demand side management that is receiving increasing attention by research and industry [5]. Energy systems are sensitive to energy consumption spikes and, therefore, measurements have to be taken, either to optimize energy generation and distribution or to reduce and shift peak power demands.

Energy operating costs significantly depend on the energy tariff structure applied, but in most energy studies, the energy consumption is multiplied by an average energy price. Different time-of-use and/or peak penalty charges may change the cost efficiency of a control solution completely [6]. There still exists a gap between energy consumption and costs since there is no generalized cost model describing current energy tariff structures to evaluate operating costs at WRRF [7].

An application of a real energy pricing structure was applied to a calibrated model for evaluating operational strategies in two large WRRF, in the context of the Portuguese "SmartWater4Energy" project [8]. The importance and need of mathematical modelling for energy optimization of specific energy costs at real WRRF processes was assessed. Time periods with potential for further optimization were identified, supporting a smart grid basis in terms of water and energy markets that respond to the demands. This work was developed in the AQUASAFE platform [9, 10], where the different models and the data from different sensors were included for calibrating and evaluating the models. However, most of the energy consumption measurements in the scope of this project are done in an indirect way, through the Supervisory Control and Data Acquisition (SCADA) system of AdTA. The AQUASAFE platform is a structure that allows managing information from different sources (e.g., SCADA systems, SQL, FTP servers) necessary to obtain an adequate response in the context of the management. The AQUASAFE platform was developed at AdTA, integrating data already existing in the company, in order to produce answers to the specific needs of the operation and management (e.g., overflows, energy management, emergency response). The measured data, mainly from water flows, is imported in real-time and the models run periodically in the forecast mode in simulation scenarios chosen by the user. The AQUASAFE platform is in use at AdTA nowadays.

Thus, access to real-time total energy consumption of WRRF is important, but there is also a need to have accurate measurements on the energy consumption and other electrical parameters in real-time for the WRRF and their equipment/processes. This is done having in view the objective of checking and monitoring the implementation of efficiency/optimization measures and the flexibility of electrical energy consumption at a WRRF, since it is not always possible to evaluate the effectiveness of the measures implemented by accessing only the total energy consumption of a WRRF. Those detailed measurements need to be communicated to the AQUASAFE platform, which has analytical capability to extract information from the data being measured and then it will be possible to design a strategy for shifting peak loads and reducing energy consumption. These are the main reasons that originated the developments described in this article.

On what concerns the development of dedicated energy meters to measure several electrical parameters like current, voltage and power in real-time, and adapted for wireless sensor network communications, there is previous work already done for the smart grid environment [11, 12]. In the present case, the energy meters will be different from those ones as, now, they must have the following distinctive characteristics: (i) to be adapted to the constraints of a WRRF deployment; (ii) to measure the parameters required in WRRF environment; (iii) to comply with the IoT communications paradigm and platform architecture; (iv) to be low cost, as a large number of meters is required to cover the universe of WRRF.

4.3 Communication Architecture

The IoT based communication architecture adopted for this case considers each meter and the respective equipment to which is connected as a "thing". The data generated from all the things is stored in the database of a central AdTA server. The communication between the things and the server is done through the private AdTA wide area network for security reasons.

From the universe of the IoT communication protocols [13], we have considered that Wi-Fi [14] and LoRa [15] would be two appropriate standard communication technologies to employ in this use case.

Wi-Fi is a well-known technology, having low cost communication modules for the meters, a low-cost Access Point (AP), and high data rates (Mbps). As low cost is a key objective, the choice of Wi-Fi as one of the selected technologies looked appropriate.

LoRa was the second technology selected for the use case. LoRa is the physical layer containing the wireless modulation utilized to create a long-range communication link. The complete stack of protocols used over LoRa is known as LoRaWAN (LoRa Wide Area Network). Compared to Wi-Fi, LoRa has lower data rates (Kbps) and the cost of the LoRa gateway is higher than the Wi-Fi AP. However, its installation might be simpler than Wi-Fi in more complex networks and it enables a longer communication distance than Wi-Fi, which is very useful for the communication between some remote equipment in the WRRF and the LoRa gateway.

The decision taken for the InteGrid pilots was to test Wi-Fi in one WRRF (located at Chelas) and LoRa in another WRRF (located at Beirolas), both in the Lisbon urban area. As it is required to transmit the data from the energy meters to the Control Centre, we had to establish a communication architecture that would allow a seamless and secure transmission. The chosen architecture is shown in Figure 4.1, both for the Wi-Fi and the LoRa access cases.

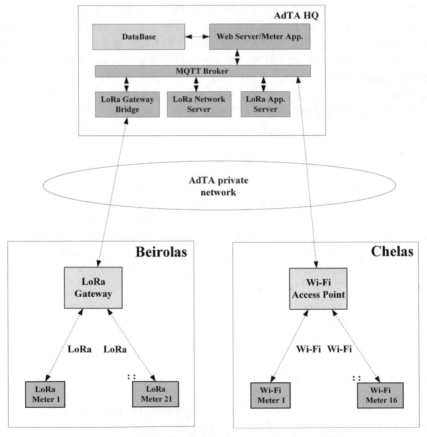

Figure 4.1 Wi-Fi and LoRa based communication architecture.

There might be several Wi-Fi APs installed in a WRRF. The WRRF energy meters are deployed in different WRRF equipment, e.g., recirculating pumps, equalization pumps, ventilators, etc. The meters communicate to the nearest Wi-Fi AP, sending a message containing the meter identification, followed by the electrical measurements. The data is forwarded from the Wi-Fi AP to an AdTA Server located at the Control Centre of AdTA Headquarters (HQ) via the AdTA private communication network. The data is uploaded into the database via Java Script Object Notation (JSON) format commands. The users can access the data either via a direct connection to the database or indirectly through a Web server.

Figure 4.2 shows a simplified protocol stack of the Wi-Fi based access network. A conventional TCP/IP stack of protocols is used over the Wi-Fi Medium Access Control (MAC) and Physical (PHY) layers. The Wi-Fi AP converts Wi-Fi into Ethernet and communicates with the AdTA server via the AdTA private wide area network.

For the communication system with LoRa at the physical layer, the communication architecture is similar to the one indicated for the Wi-Fi case, having as a main difference the use of a Gateway instead of the Wi-Fi AP.

Figure 4.3 shows the protocol stack of LoRaWAN, which comprises four layers: Radio Frequency layer, Physical layer, Media Access Control (MAC) layer and Application layer.

The Radio Frequency layer defines the radio frequency bands that can be used in LoRa. We adopted the 868 MHz band available for Industry, Scientific and Medical (ISM) applications in Europe. The LoRa physical layer implements a derivative of the Chirp Spread Spectrum (CSS) scheme. CSS aims to offer the same efficiency in range, resolution and speed of acquisition, but without the high peak power of the traditional short pulse mechanism. The MAC layer defines three classes of end nodes, respectively Class A (baseline), B (beacon) and C (continuous). In this project we use

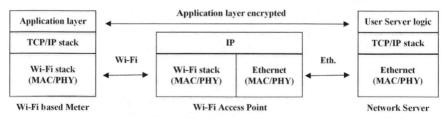

Figure 4.2 Protocol stack of the Wi-Fi access network.

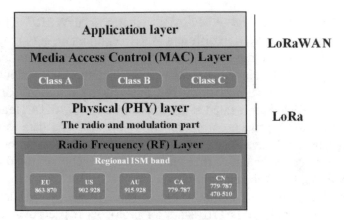

Figure 4.3 LoRaWAN communication layers.

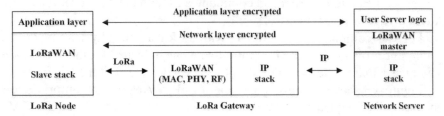

Figure 4.4 Protocol stack of the LoRa access network.

only Class A, since it is the most energy efficient and the only one that is mandatory. To achieve this high energy efficiency, the nodes in this class are 99% of the time inactive and are only ready to receive after transmitting a message. The Application layer is related with the user application.

The basic components of the LoRaWAN architecture are the following: nodes, gateways and network server. The nodes are the simplest elements of the LoRa network, where the sensing or control is undertaken. The gateway receives the data from the LoRa nodes and transfers them into the backhaul system. The gateways are connected to the network server using standard IP connections. On this way, the data communication uses a standard protocol, but any other communication network, either public or private, can be used. The LoRa network server manages the network, acting to eliminate duplicate packets, scheduling acknowledgements, and adapting data rates. Figure 4.4 shows a simplified protocol stack of the interconnected components.

The high-level communication between an energy meter and the AdTA server uses the Message Queuing Telemetry Transport (MQTT) protocol.

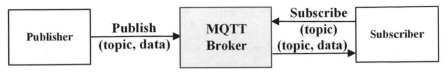

Figure 4.5 MQTT communication architecture.

MQTT is an IoT connectivity protocol. It was designed as a lightweight broker-based publish/subscribe messaging protocol, which is open, simple, lightweight and easy to implement. These characteristics make it ideal for use in constrained environments, for example, where the network has low bandwidth or is unreliable, as is the case of wireless sensor communications, or when run on an embedded device with limited processor or memory resources, as is the case of the energy meters.

In the MQTT architecture the elements that generate information are called Publishers and the elements that receive information are called Subscribers. The Publishers and Subscribers are interconnected through the MQTT broker, as shown in Figure 4.5.

The energy meter contains a MQTT client and the server a MQTT broker. Periodically, e.g., every minute, the meter sends a MQTT Publish message to the MQTT broker, located in the server, with the following structure: Meter ID, Voltage, Current, Power, Power Factor, Energy, Service Time, Timestamp. For configuration of the different parameters in the energy meter the MQTT broker uses Subscribe messages with different topics, namely: Change of the measurement period, Change of the communication parameters (specific of Wi-Fi or LoRa), Set date/time, Set the initial value of the energy counter, Set current transformer ratio, Set Power Threshold (define the power threshold to consider that the equipment is in service), Set Meter mode (tri-phasic or 3 x mono-phasic).

4.4 The Energy Meter

The energy meter was designed to allow the monitoring, not only from energy consumption, but also of other electrical parameters like current, voltage, power and power factor.

The energy meter is housed in a 6U DIN-rail ABS polymer enclosure (69 mm × 87 mm × 66 mm) using 4 pins for the three voltage phases and neutral connection at the bottom (V1, V2, V3, N) and 6 pins for the connection of 3 current transformers at the top, one for each phase (I1+, I1−, I2+, I2−, I3+, I3−).

At the top there are two additional indicators, on the left side the L1 red led blinks when there is a message being transmitted. On the right side, the L2 green led blinks when there is energy consumption.

There are two versions of the energy meter, one with Wi-Fi communication and another with LoRa communication.

The energy meter block diagram is shown in Figure 4.6 and comprises four main modules: Measurement module, Processing & Communication module, Galvanic Isolation module and AC/DC dual power supply module.

The Measurement module and the Processing & Communication module are galvanic isolated for user protection, namely for the antenna connector and antenna cable handling. This requires a dual power supply with one of them connected to the Measurement module and the other to the Processing & Communication module. The Galvanic Isolation module is required to be connected to both.

The power, power factor, energy consumption and service time are calculated internally in the energy meter from the voltage and current readings. They are transmitted to the AdTA server, via MQTT protocol messages, using the following units for the different parameters: Voltage: 0.1 Volt, Current: 0.1 Ampere, Power: Watt, Power Factor: 0–100 (100 corresponds to Power

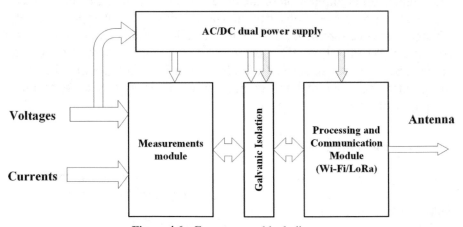

Figure 4.6　Energy meter block diagram.

Factor $= 1$), Energy: 0.1 kWh (accumulated value), Service Time: minutes (accumulated value), Time Stamp: seconds.

4.5 Deployment

The meters were deployed in two large WRRF in the Lisbon urban area, located at Chelas and Beirolas, respectively. Taking into account their relatively large size, these WRRF were considered as very suitable for the use case demonstrations.

In 2018, the Chelas subsystem (including WRRF and pumping stations on drainage system) treated $12,950,357$ m^3 and consumed 6,488,170 kWh, which is about 6.7% of the treated wastewater and 7% of the energy consumption in the AdTA universe (6% of the energy costs). On the other hand, in 2018, the Beirolas subsystem (including WRRF and pumping stations on drainage system) treated $14,573,955$ m^3 and consumed 6,703,237 kWh, which is about 7.5% of the treated wastewater and 7.3% of the energy consumption in the AdTA universe (8.3% of the energy costs).

In the AdTA universe, these two subsystems represent 14.3% of energy costs and also produce energy that is consumed in their internal processes. In Beirolas, 14.1% of energy consumption is from co-generation energy production and in Chelas, it is about 24.5%.

Sixteen energy meters with the Wi-Fi module were installed in Chelas, while twenty-one meters with the LoRa module were installed in Beirolas. The first objective was to test both communication technologies in order to make an evaluation of their strong and weak aspects, from the technical and economic points of view. The second objective is that AdTA is able to perform demand side management operations in both of them in the context of the InteGrid project. By having the knowledge on the real-time energy consumption and on the values of other electrical parameters, in a demand side management situation AdTA will be able to shift loads in a controlled way so that the impact is minimized in the WRRF.

The Chelas WRRF, shown in Figure 4.7, covers an area of around $37,500$ m^2 (250 m $\times$ 150 m) in a central area of Lisbon.

Figure 4.8 shows the plant of the Chelas WRRF, where Pi signals the points where the meters equipped with Wi-Fi are located and AP indicates the location of a Wi-Fi Access Point. The number of meters in each AP

Figure 4.7 Aerial view of the Chelas WRRF.

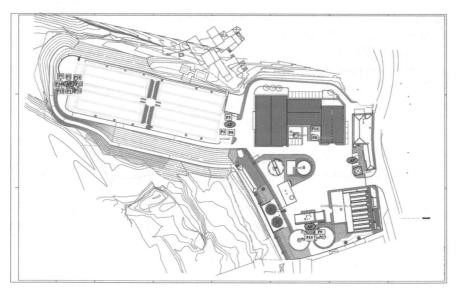

Figure 4.8 Wi-Fi based deployment at Chelas WRRF.

is variable, depending on the topology of the Wi-Fi network. The meters are connected to different WRRF equipment, such as pumps and ventilators. There are 4 Wi-Fi APs deployed to interact directly with the meters. However, due to the large area size and to the propagation characteristics in

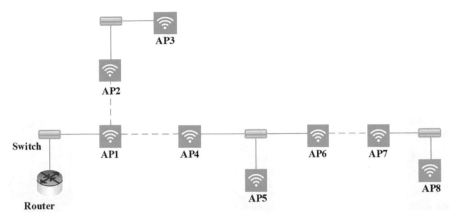

Figure 4.9 Wi-Fi network at the Chelas WRRF.

the WRRF environment (e.g., meters installed in metallic cupboards), four more additional APs were installed in a wireless mesh configuration. The deployed Wi-Fi network is shown in Figure 4.9.

As shown in Figure 4.9, the Wi-Fi network requires the deployment of three radio links, namely between AP1 - AP2, AP1 - AP4 and AP6 - AP7. To avoid interferences, these radio links operate in the 5 GHz ISM band, while the communication with the meters is done in the 2.4 GHz ISM band. AP1, AP3, AP5 and AP8 are the APs that directly contact with the meters.

The Beirolas WRRF, shown in Figure 4.10, covers a larger area of around 100,000 m^2 (400 m × 250 m) in Lisbon.

Figure 4.11 shows the plant of the Beirolas WRRF, where Pi are the locations where the LoRa meters are placed. G is the location of the LoRa gateway and antenna. As the range of the LoRa technology is higher than the Wi-Fi technology, only a single Gateway is needed to cover all the area of the Beirolas WRRF, with an external antenna located on the roof of the same building where is the gateway, on the left of the figure. The location of the LoRa antenna is not ideal (it should be located on a more central point of the WRRF), but it was deployed there to simplify the connection with the AdTA private network.

Figure 4.10   Aerial view of the Beirolas WRRF.

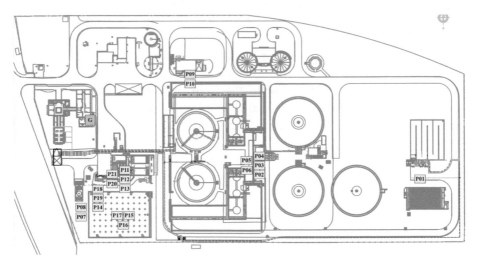

Figure 4.11   LoRa based deployment at Beirolas WRRF.

4.6 Measurement Results

To monitor the energy meters deployed in the different WRRF a web platform was developed, allowing to view the main electric parameters that have been measured.

As seen in Figure 4.12, the web platform screen shows on the left panel the meters that are online and on the right the different electrical parameters of the selected meter (MED19 in the example), namely the Service Time (T), Current (I), Voltage (V), Energy (E), Power (P) and Power Factor.

For each parameter, it is shown the value in each of the three phases (first three columns) and the average or total value in the fourth column. In this case, it is shown the average for service time, current, voltage and power factor, and the total value, resulting from the sum in the three phases, for energy and power.

The service time and energy are accumulated values, the others are instantaneous values captured when the message is sent, typically every minute.

Associated with each measurement, at the bottom of each box, we can see a timestamp showing the instant in which the measurement was made. This is especially important for the LoRa based meters where due to bandwidth limitations it is not possible to send all measurements in a single message.

In the database, only a subset of the electrical measurements is stored, namely the total power of the three phases, the total accumulated energy of the three phases and the service time, together with the meter id and time stamp.

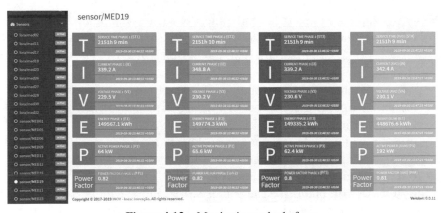

Figure 4.12 Monitoring web platform.

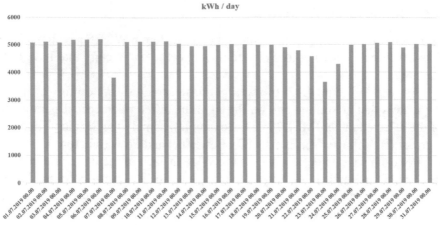

Figure 4.13 Consumption of meter 19 in July 2019.

Based on those values, we can create different graphics to show the consumption of the different meters. In Figure 4.13 we present an example of the consumption (in kWh) of the meter MED19 in July 2019.

In addition to the meter web platform, an application was also developed to configure the different meters, namely the measurement period, the energy threshold to activate the service time counter, the current transformer ratio, and the electrical parameters or internal registers that can be read on demand.

Besides the referred functionality, this application allows also to display, in graphical form, the relation between the voltage and current of the three phases, which is very useful in the meter installation, as it allows an easy detection of a wrong configuration and correct it on the spot. Figure 4.14 shows an example of this type of graphic. As seen in the figure, the voltages of the 3 phases are shifted by 120° as expected, and the currents are shifted around 32° to the right of the corresponding voltages, meaning that an inductive load is present.

Another interesting functionality of the developed application is to allow the calculation of the Discrete Fourier Transform of the voltage and current of a specific meter in order to calculate the respective harmonics. In the example presented in Figure 4.15, we can see a low harmonic content of both the voltage and current, still with significant higher values for the current.

At a next step, the data provided from the energy meters will be integrated in the AQUASAFE platform. The data is stored in the database of the AdTA server in real time as described and, automatically, the AQUASAFE imports all the energy meter information from the data files to the AQUASAFE

Angles Phase 1,2,3 – Voltage/Current

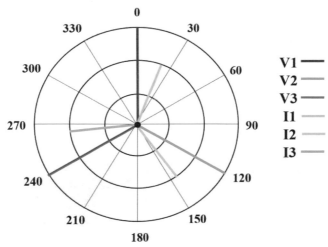

Figure 4.14 Polar Graphic of the voltage and current angles.

database recurring to an importer already installed in the AQUASAFE platform.

AdTA has also been working on the so-called flexibility matrixes and on their integration into the AQUASAFE platform. For the provision of the energy flexibility available in the WRRF, the solution developed in the InteGrid project is a Virtual Power Plant (VPP). A VPP is a cloud-based distributed power plant that aggregates the capacities of distributed energy resources for the purposes of enhancing power generation, as well as trading power on the electricity market. It was installed a File Transfer Protocol (FTP) server that connects the flexibility requests from the Distribution System Operator, the real time energy consumption of the WRRF and the WRRF available flexibility (flexibility matrix). Periodically, AQUASAFE will provide the positive and negative flexibility available in the WRRF to the VPP FTP Server. The FTP Server will then send an activation request with the flexibility demand needed and a Short Message Service (SMS) to the operator in the control room. The flexibility provision is activated manually by the operator, through the SCADA system, by turning on/off equipment or increasing/decreasing the power of the equipment according to the activation request.

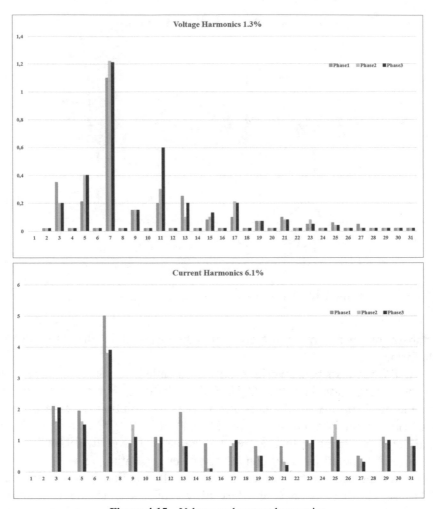

Figure 4.15   Voltage and current harmonics.

4.7 Conclusion

An IoT based platform for real-time management of energy consumption in WRRF was presented. The developed work included the design and implementation of energy meters for measuring different electrical parameters, the deployment of those meters in two WRRF in the Lisbon urban area, the deployment of two IoT communication protocols (Wi-Fi and LoRa) for the access networks in the WRRF and the transmission of the data from either Wi-Fi APs or LoRa Gateway to a central server and database, via MQTT, where data analytics can be performed.

The objective of these two pilots was, in first place, to evaluate the chosen IoT communication technologies in the WRRF environment. The second objective was the enabling of demand side management operations, having in view the shifting of loads from peak load situations so that a better balance of the energy consumption can be achieved.

Concerning the evaluation of the Wi-Fi and LoRa technologies, both have proved satisfactorily for the WRRF environment, although due to the higher data rate, Wi-Fi allows having shorter intervals to make the measurements. On the other hand, the LoRa network is more straightforward to deploy in a WRRF than the Wi-Fi. For future deployments in other WRRF, any of those two protocols is technically adequate for real-time energy measurement and respective energy management operations, and the respective cost will be an important factor for decision.

From the web server where the data is stored and through a special purpose application developed, we are already able to perform some data analytics on the collected data. Examples were given of showing the energy consumption of a meter along a period of time, determination of a polar graphic of the voltage and current angles, and display of the current and voltage harmonics.

For future work, the implementation of the automatic data transfer to the AQUASAFE platform will permit integrating data analytics in that platform, which is already part of the working environment in AdTA. Also demand side management operations, based on the use of flexibility matrixes, will be shown for the WRRF in Chelas and Beirolas.

Finally, this work has also to do with the so-called smart wastewater management, which is an important component of the smart city concept. It is worthwhile noticing that by using IoT technologies, this solution is up to date with the status of communications and platforms in smart cities, which will enable the integration of this solution in a smart city environment.

Acknowledgements

The research leading to this work is being carried out as a part of the InteGrid project (Intelligent Grid Technologies for Renewables Integration and Inter-active Consumer Participation Enabling Interoperable Market Solutions and Interconnected Stakeholders), which received funding from the European Union's Horizon 2020 Framework Programme for Research and Innovation under grant agreement No. 731218. The sole responsibility for the content of this publication lies with the authors.

List of Abbreviations

AdTA	Águas do Tejo Atlântico
AP	Access Point
DIN	German Institute for Standardization
FTP	File Transfer Protocol
IoT	Internet of Things
IP	Internet Protocol
JSON	Java Script Object Notation
LoRa	Long Range
LoRaWAN	LoRa Wide Area Network
MAC	Media Access Control
MQTT	Message Queuing Telemetry Transport
PHY	Physical
SCADA	Supervisory Control and Data Acquisition
SMS	Short Message Service
TCP	Transmission Control Protocol
VPP	Virtual Power Plant
WRRF	Water Resource Recovery Facilities

References

[1] United States Environmental Protection Agency, Energy efficiency for water and wastewater utilities. 2015, online at: https://www.epa.gov/sustainable-water-infrastructure/water-and-energy-efficiency-utilities-and-home

[2] P. Póvoa, "Contribution for modelling strategies and operational management of WRRF with combined sewer systems", Ph.D. thesis, Instituto Superior Técnico, University of Lisboa, 2017.

[3] International Water Association, Water Utility Pathways in a Circular Economy, 2016, online at: http://www.iwa-network.org/wp-content/uploads/2016/07/IWA_Circular_Economy_screen.pdf

[4] M. Nunes, R. Alves, A. Casaca, P. Póvoa and J. Botelho, "An Internet of Things based Platform for Real-Time Management of Energy Consumption in Water Resource Recovery Facilities", First IFIP

International Cross-Domain Conference, IFIP IoT 2018, pp. 121–132, Springer Open, Poznan, Poland, September 2018.

[5] P. Palensky and D. Dietrich, Demand side management: demand response, intelligent energy systems, and smart loads, Ind. Inform. IEEE Trans. 7(3), pp. 381–388, 2011.

[6] P. Póvoa, A. Oehmen, P. Inocêncio, J. S. Matos and A. Frazão, "Modelling energy costs for different operational strategies of a large Water Resource Recovery Facility", Water Science Technology, May 2017; doi: 10.2166/wst.2017.089

[7] I. Aymerich, L. Rieger, D. Sobhani, D. Rosso and L. Corominas, "The difference between energy consumption and energy cost: Modelling energy tariff structures for water resource recovery facilities", Elsevier, Water Research, Volume 81, pp. 113–123, September 2015, online at: https://www.sciencedirect.com/science/article/pii/S0043135415002705

[8] SmartWater4Energy project, 2015, online at: http://smartwater4energy.hidromod.pt/

[9] AQUASAFE: A R&D complement to Bonn Network tools to support Water Safety Plans implementation, exploitation and training, International Water Association Newsletter, Vol. 1, Issue 3, 2009.

[10] AQUASAFE Newsletter, September 2012, online at: http://www.aquasafeonline.net/en/principal.asp

[11] A. Grilo, A. Casaca, M. Nunes, A. Bernardo, P. Rodrigues and J. Almeida, "A Management System for Low Voltage Grids", Proceedings of the 12th IEEE PES PowerTech Conference, (PowerTech 2017), Manchester, United Kingdom, June 2017.

[12] N. Silva, F. Basadre, P. Rodrigues, M. Nunes, A. Grilo, A. Casaca, F. Melo and L. Gaspar, "Fault detection and location in low voltage grids based on distributed monitoring", Proceedings of the IEEE International Energy Conference (Energycon 2016) Conference, Leuven, Belgium, April 2016.

[13] Keysight Technologies, The Internet of Things: Enabling technologies and solutions for design and test, 2017, online at: https://literature.cdn.keysight.com/litweb/pdf/5992-1175EN.pdf?id=2666018

[14] IEEE 802.11 b/g/n standard, IEEE, 2009.

[15] LoRa Alliance, LoRaWAN 1.1 specification, 2017.

5

IoT Solutions for Large Open-Air Events

Francesco Sottile[1], Jacopo Foglietti[1], Claudio Pastrone[1],
Maurizio A. Spirito[1], Antonio Defina[1], Markus Eisenhauer[2],
Shreekantha Devasya[2], Arjen Schoneveld[3],
Nathalie Frey[3], Hors Pierre-Yves[4], Paolo Remagnino[5],
Mahdi Maktabdar Oghaz[5], Karim Haddad[6],
Charalampos S. Kouzinopoulos[7], Georgios Stavropoulos[7],
Patricio Munoz[8], Sébastien Carra[8], Peeter Kool[9]
and Peter Rosengren[9]

[1]LINKS Foundation, Italy
[2]Fraunhofer-Institut für Angewandte Informationstechnik, Germany
[3]DEXELS, the Netherlands
[4]OPTINVENT, France
[5]Kingston University, United Kingdom
[6]Brüel & Kjaer Sound & Vibration Measurement, Denmark
[7]CERTH/ITI, Greece
[8]Acoucité, France
[9]CNet Svenska AB, Sweden

Abstract

This chapter presents the main results of the MONICA project, one of the five
large-scale pilot projects funded by the European Commission. MONICA
focuses on the adoption of wearable IoT solutions for the management of
safety and security in large open-air events as well as on the reduction of
noise level for neighbours. The project addresses several challenges in eleven
pilots of six major European cities using a large number of IoT wearables and
sensors. The chapter first introduces all MONICA challenges in the context of
large open-air events and then presents the corresponding adopted technical
solutions, the defined IoT architecture and the perspective in integrating a
wide range of heterogeneous sensors. On one side, the focus is on the solutions that have been adopted to improve the crowd management, crowd safety

and emergency responses by using wearables for both visitors and the security staff at the events, including also the adoption of video processing and data fusion algorithms to estimate the number of visitors and its distribution in the event area and to detect suspicious activity patterns. On the other hand, it describes how innovative Sound Level Meters (SLMs) can be deployed to monitor the sound propagation within the event area while reducing the noise impact on the neighbourhood.

5.1 Introduction

Sustainable urban development is recognised as a key challenge at the global level. The "Mapping Smart Cities in EU" report issued by ITRE (Industry, Research and Energy Committee) of the European Parliament points out that more than half of all European cities with more than 100,000 residents have implemented or planned measures to have "smarter" cities. The report defines six elements of Smart City characteristics: Smart Governance, Smart Economy, Smart Mobility, Smart Environment, Smart People, and Smart Living. Smart Living, the subject of MONICA, is defined as ICT that enables lifestyles, behaviour and consumption. It implies also healthy and safe living in a culturally vibrant city with diverse cultural facilities and incorporates good quality housing and accommodation.

On this background, the main focus of the MONICA project, one of the five large-scale IoT pilots funded by the European Commission, is on how to leverage on the use of multiple existing and novel IoT technologies to address real-world challenges in open-air event scenarios. More specifically, MON-ICA demonstrates a large-scale ecosystem that uses innovative wearables and IoT devices with closed-loop back-end services integrated into an interoperable, cloud-based platform capable of offering a multitude of simultaneous and targeted applications. In order to demonstrate the ability of deploying IoT solutions to address real challenges, three Smarter Living ecosystems with high societal relevance have been chosen: Security, which tackles the problem of managing public security and safety during big events; Acoustic, concerning the impact of noise propagation in neighbourhood areas while carrying out open-air festivals; IoT Platform, to demonstrate to developers and service providers the ability of the MONICA platform to cooperate with other Smart City IoT platforms, Open Data portals, and generic enablers. All the ecosystems will be demonstrated in the scope of large-scale city events; however, they have general applicability for any dynamic or fixed large-scale application domain in smart cities such as open markets, fairs, cultural venues, sports events etc.

There are eleven main events targeted by MONICA and organised in six major European cities. The following provides a brief overview of the involved pilot cities, explaining their nature, characteristics and challenges.

Tivoli Gardens in Copenhagen. Tivoli Gardens in Copenhagen, Denmark, is a famous amusement park and pleasure garden that attracts around 4.5 million visitors annually. Tivoli organises several outdoor concerts during Friday evenings of the summer months. The number of complaints by the residents for noise pollution has increased over the years and the flow control in the venue is difficult and requires a lot of security personnel.

Kappa FuturFestival and Movida in Turin. Kappa FuturFestival in Turin, Italy, takes place once a year for two days. Various national and international artists for electronic dance music attract more than 45,000 participants for the occasion. The second pilot in Turin is the Movida in the San Salvario district: lots of bars, restaurants and liquor stores with thousands of visitors moving around the district. The city of Turin wants to address the noise and security situations in these events.

Dom and Port Anniversary in Hamburg. The city of Hamburg, Germany, attracts thousands of visitors by hosting several events every year. The focus of MONICA is both on the Hamburger DOM, the biggest public festival in Northern Germany held in the centre of Hamburg city three times a year, and on the Hamburg Port Anniversary, which comprises maritimes parades, historic sailboats and ships, music, fireworks and food. Both events attract a vast number of visitors and the major concern is crowd and crew management.

Fête des Lumières and Nuits Sonores in Lyon. In the city of Lyon, France, urban festivals are part of the local culture. For four nights in December, during Fête des Lumières a variety of different artists light up buildings, streets, squares and parks all over the city. Concerning music instead, Nuits Sonores festival has been known as one of the great European meetings in the realm of innovative music and creativity, taking over different locations of the city for six days and nights. City of Lyon has identified that the security and safety of participants are the most critical areas to be improved for these events, moreover, reducing complaints from residents about noise pollution is also an objective.

Pützchen's Markt and Rhein in Flammen in Bonn. The federal city of Bonn, Germany, hosts two large-scale events: Pützchen's Markt and Rhein in Flammen. Pützchen's Markt is one of the oldest fairs in Germany and is held every year on the second weekend of September. Huge number of visitors gathering in the narrow streets of the residential area makes the event more challenging for managing security and safety deployments. Rhein in Flammen is an open-air festival which takes place in the public park Rheinaue with three performing stages and fireworks on the banks of Rhein which makes the venue highly crowded. The main challenge that the city of Bonn is interested to handle is the staff management for both pilots. In addition, noise control is one of the important aspects to be addressed for Rhein in Flammen.

Headingley Stadium in Leeds. Headingley Stadium outside Leeds, England, is the home of Yorkshire County Cricket Club, Leeds Rhinos rugby league team and Yorkshire Carnegie rugby union team. It can accommodate 21,000 people and already has security arrangements with statically installed cameras across the stadium and body worn cameras by the security staff. The stadium authorities wish to increase the security by automated video analysis of the streamed images. In addition, they want to add staff monitoring and emergency evacuation planning.

Related MONICA use cases are described in detail in the next subsection.

5.1.1 Main MONICA Use Cases

Crowd and Capacity Monitoring

Crowd and capacity monitoring applications are useful to predict and handle emerging incidents in large open-air events. These applications are primarily based on data from crowd wristbands and CCTV cameras. Some features are reported as follows:

- Crowd count: to know the number of people being in a specific area.
- Crowd density: estimating crowd size and density shown as a heat map.
- Crowd flow: detecting anomalies such as fights, explosions or high-risk queues.
- Redirection: redirect visitors to safer areas using large screens or APPs.

Health, Security and Safety Incidents

Other applications support not only the detection of incidents but also the reporting and handling of them in order to provide a timely response. Relevant applied devices include smart-glasses, staff wristbands, activity recognition wearables, CCTV cameras, mobile phones and environmental sensors (e.g. for the detection of strong wind). Some features are reported as follows:

- Notify staff/guards about type of incident and related location, for instance, through smart-glasses and staff wristbands.
- Report back to control room, for instance, through smart-glasses, staff wristbands.
- Detect person falling or fights, for instance, through CCTV, activity recognition wearables.
- Report incident using a mobile phone APP.
- Monitor wind speed, for instance, through environmental sensors, to detect high risk for tall amusement rides.

Missing Person

Knowing the exact position of staff members all the time is a high priority when incidents occur as they need assessment or assistance. Moreover, quickly finding a child (who is missing) or a friend is also relevant at some events. For the location of people, staff and crowd wristbands, tracker GPS devices, and mobile phones could be used. Some features are reported as follows:

- High-precision location (staff wristbands).
- Approximate location (crowd wristbands, trackers GPS devices, mobile phones).

Sound Monitoring and Control

Organisers of concerts want to give their performers and audiences the best music experience, but they also wish to comply with local regulations on environmental sound exposure. This produces a dilemma when you talk about outdoor concerts in residential areas. Since high sound pressure levels are necessary for optimal concert sound, there is a risk of the audience becoming disappointed and artists turning down invitations to perform if regulations say that volumes have to be turned down. And even if regulations are met, event organisers still have to deal with residents living next to the venue who complain about the noise coming from the concerts, and that it is affecting their quality of life.

To help solve these challenges, MONICA has developed and deployed an acoustic system providing sound zones at the venue. The system consists of novel sound field control schemes which provide an optimised sound field in the audience area (called bright zone) while minimising the exposure to noise in neighbouring areas (called dark zones). Thus, at the front of the stage, the music can be louder for a better concert experience, whereas the sound level is reduced outside the concert area and in a selection of quiet zones. The sound levels are automatically controlled by the system, adjusting for changes in weather or size of audiences. Additionally, information about the sound levels is displayed and used for different purposes, i.e. moving to a quieter zone, checking that regulations are kept, studying health aspects etc.

The acoustic system can be divided into three main sub use cases (namely Sound Field Control System, Quiet Zones and Monitoring) that are presented in the following subsections.

Sound Field Control System. Based on an array of loudspeakers and integration with the existing sound system at the venue, MONICA has developed an Adaptive Sound Field Control System (ASFCS), providing an optimised sound field in front of the stage whereas reducing the sound level outside the zone. To accommodate for the variable weather conditions i.e. wind, temperature and humidity, which add complexity to the propagation of sound, the ASFCS employs an adaptive model that adjusts for changes in climate and audience configuration.

The data used for updating the propagation model come from various IoT-enabled devices at the venue such as SLMs, wind and temperature sensors. The IoT devices are integrated with the ASFCS using the MONICA cloud platform. As a result, the ASFCS can continually update the sound propagation model and the loudspeaker signals, using the data from the devices.

Silent Zone System. In addition to the ASFCS, a Silent Zone System provides local silent zones within and close to the loud event area. These quiet zones can be used by security and/or medical staff as well as by the public for any need.

Sound Monitoring. For awareness purposes and to comply with acoustic regulations in the city, IoT SLMs are used to measure sound levels at strategic locations and send the information to organisers, public services, the audience and the neighbourhood. Other IoT enabled devices such as smartphones and wristbands can also be used to measure non-calibrated sound levels.

5.2 MONICA IoT Platform and Data Modelling

5.2.1 MONICA Functional Architecture

The MONICA functional architecture has been iteratively defined considering the use cases and related functional and non-functional requirements. The MONICA functional architecture has been inspired by the reference High Level Architecture (HLA) developed by the Working Group 3 (WG3) of the AIOTI [1]. Thus, following the HLA, the MONICA architecture has been subdivided in different layers as depicted in Figure 5.1. The main difference with respect to the HLA is the definition of two additional layers namely *Device Layer* and *Edge Layer*. In particular, the *Device Layer* includes all the IoT wearables (e.g., crowd wristbands, staff wristbands and smartglasses) and IoT sensors, which can be either fixed (e.g., SLMs, loudspeakers, cameras, environmental sensors) or mobile (e.g., wireless SLMs, cameras installed in a blimp). The MONICA architecture uses a distributed approach where different gateways (GWs) are deployed at the edge of the environment in which they operate to minimise the delay that both network and cloud modules may introduce. In fact, these GWs have to process in real-time big amount of raw data generated by a large number of IoT wearables and sensors. Thus, the MONICA GWs belong to the *Edge Layer* that incorporates Artificial Intelligence (AI) methods and techniques. In particular, the *Edge Layer* includes: Wristband-GW running localisation algorithms, Security Fusion Node (SFN) executing video analytic algorithms, SLM-GW

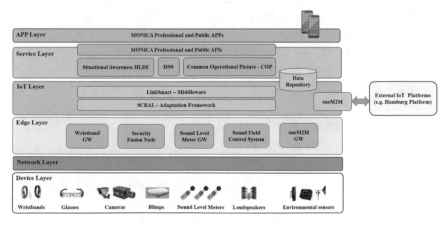

Figure 5.1 MONICA functional architecture.

processing sound data from SLMs, and oneM2M-GW collecting data from environmental sensors. All the functionalities offered by the GWs along with the states of IoT devices are semantically represented by the *IoT Layer* according to the OGC SensorThings standard [2]. In particular, the *IoT Layer* is composed of the following three sub-components:

- The *Adaptation Layer*, here represented by SCRAL, provides technology-independent management of IoT physical resources and uniform mapping of data into a standard representation that can be easily handled by the upper platform modules;
- *Middleware*, here represented by LinkSmart, offers data storage and directory services for resources registered in the IoT platform;
- *External IoT Platform Connectors* that handle the communication with external IoT platforms and the integration of data coming from outside (e.g. from the Hamburg Smart City platform) integrated using the oneM2M standard APIs.

Going upward, there is the *Services Layer* that implements the intelligence of the platform and integrates specific high-level processing modules providing technical solutions to meet the application requirements. The service modules are combined with knowledge base components and a Decision Support System (DSS), whose aim is to propose a set of intervention strategies to assist human operators in gathering context-sensitive information and decision making. Before the *APP Layer*, the MONICA system exposes specific Application Programming Interfaces (APIs) that provide service access points for MONICA application developers and external application developers that want to access MONICA functionalities and information streaming from the platform. Additionally, MONICA provides efficient deployment and monitoring tools that ease the platform deployment process as well as the operational monitoring of GWs, communication networks and system services.

5.2.2 IoT Middleware Platform

The IoT layer of the MONICA platform consists of an adaptation layer and a middleware. "Adaptation" means the chance of having transparent access to physical information without technology-specific knowledge, while a middleware offers storage and directory services. The Smart City Resource Adaptation Layer (SCRAL) is the adaption layer adopted in MONICA, that provides technology independent management of physical resources and uniform mapping of data into standard representation. On the other hand, the LinkSmart middleware, together with an OGC SensorThings compliant solution, is deployed as back-end infrastructure, giving support for storage utilities as well as event forwarding solutions and monitoring tools.

5.2.2.1 SCRAL

SCRAL is an interface layer that supports uniform and transparent access to physical resources for sensing and monitoring purposes. Its main function is to ease the integration of devices providing the platform with a standard way to communicate with them. The integration process is implemented by the Field Access Level, a SCRAL sublayer in charge of the physical communication with resources and external platforms. In particular, this sublayer includes specific adapters for each type of protocol and network configuration that can be used either as a web server or client, depending on the type of API available for a given resource. Data gathered from the access level are afterwards delivered to a Core module, a virtualisation layer adopting an Open Geospatial Consortium (OGC)-based Semantic Framework, and finally stored into GOST, a Go implementation of the OGC SensorThings API (see Section 5.2.3 for more details about the standard and data model). The API is only accessible by administrators having authenticated access via a VPN connection. However GOST can also be configured to read Observations from an event broker by subscribing to a pre-defined topic. Further details about Event Broker and LinkSmart services are presented in the following section.

5.2.2.2 LinkSmart

LinkSmart is an open source IoT platform providing support for data storage and management, service and resource directory, stream mining and machine learning. LinkSmart services follow a microservice pattern. Hence, they can be easily integrated and orchestrated based on a concrete use case. In the context of the MONICA project, LinkSmart provides:

- Service orchestration with Service Catalog.
- Brokering solutions with Message Queue Telemetry Transport (MQTT) protocol.
- Authorization proxy service with Border Gateway.

LinkSmart Service Catalog[1] provides a directory of all the services running in the MONICA ecosystem. The Service Catalog contains entries of everything

---

[1] https://docs.linksmart.eu/display/SC

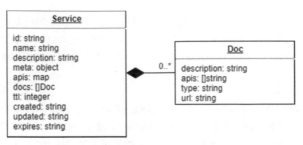

Figure 5.2 Service catalog data model.

that is meant to be discovered by other applications and services. A discoverable entry "service" has the data model as shown in Figure 5.2. Each service has a unique "id" and a user defined "name". The detailed description about the service is added in the optional "description" field. Each registered service has an expiry time beyond which the entry in the registry shall be removed for the service. This is called Time to Live (TTL) and service entry has to be updated before TTL seconds. "Docs" links to external resources such as OpenAPI specifications or wiki pages containing the documentation about the service. This enables direct service to service interaction if valid semantics are followed. "APIs" represent service endpoints, subscription or publish topics and the protocols to be used for communication such as MQTT, HTTP or SOAP. Service entry also contains the details about when it is "created", when the entry is recently "updated" and when the entry shall be "expired". The expiration value is calculated based on TTL. Service Catalog provides both HTTP and MQTT based interfaces. HTTP API exposed by the Service Catalog provides the CRUD functions for editing and managing the catalog. Service Catalog also uses MQTT for service registration and de-registration by interacting through a pre-defined MQTT topic.

Event broker provides a message bus for efficient asynchronous communication of sensor data streams implementing the publish/subscribe communication pattern. MQTT is indeed considered as the ideal protocol for low bandwidth, high latency and non-reliable connections which exactly describes a typical IoT scenario. MQTT broker is responsible for maintaining the clients and forwarding the messages across the clients. In the MONICA context different services and resources communicate to each other through MQTT, for example, to get notified when a new device is registered to the platform or also to generate new statistics feeding the application layer. In addition, being the broker a service itself, it can be discovered with the help of the Service Catalog. Concerning security, a MQTT broker can be vulnerable at different levels, such as network, transport or protocol level. In the network level, using a Virtual Private Network (VPN) for communication between the clients could save a broker from external attacks. In other cases, the LinkSmart Border Gateway is used.

LinkSmart Border Gateway[2] provides a single point of entry for accessing all the services running in IoT autonomous systems (IoT-AS) such as the MONICA cloud. It provides the functionality to intercept all the MQTT and HTTP REST API requests and performs authentication and authorisation for these requests with the help of identity providers conforming to the OpenID Connection Protocol. In the context of MONICA, Keycloak[3] is used as the Identity Provider. Authorisation rules are maintained as user or group attributes in Keycloak and can be defined by administrators based on path/topic and methods, e.g. to limit permissions for certain users to perform only HTTP GET requests on a certain REST API or to allow only MQTT subscriptions on certain topics. The rules are provided by the OpenID Connect provider as a custom claim in the access token to the Border Gateway. To improve performance, Border Gateway caches the access tokens during the span of their lifetime in a Redis database.

5.2.3 Semantic Framework

A Semantic Framework is a semantic data model that implements a distributed knowledge management infrastructure. More specifically, the MONICA Semantic Framework is a conceptual data model focused on the semantic virtualisation of both IoT resources and services employed in the project. This model supports various information domains and adopts the SensorThings API standard [2]: an OGC standard providing an open and unified framework to interconnect IoT sensing devices, data and applications over the web. It is an open standard addressing both the syntactic and semantic interoperability of the IoT; thus, it supports the exchange of information in a shared meaning among entities of a different nature belonging to the same system. It follows REST principles, the JSON encoding, the OASIS OData protocol, the URL conventions and also provides a MQTT extension allowing users and devices to publish and subscribe updates from devices. The UML of the OGC data model is reported in Figure 5.3.

---

[2]https://docs.linksmart.eu/display/BGW
[3]https://www.keycloak.org/

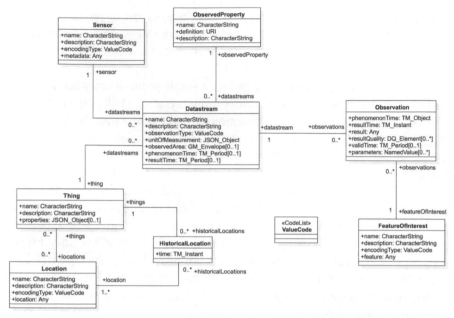

Figure 5.3   OGC SensorThings data model.

The metadata generated by the SCRAL concerning "Things", "Sensors" etc. as well as the concrete "Observations" (i.e. the sensor data) are permanently stored in the backend using an open source PostgreSQL database. In addition, the OGC server API provides endpoints for maintaining the metadata and writing and retrieving the data over the web in JSON format. The OGC server can also subscribe to different MQTT topics in order to fetch observations from sensors and to store them. Most of the sensor data collected during the MONICA pilot events are stored in the OGC server. Some of the exceptions include raw camera measurements which are processed locally. In case of multimedia data, only aggregated information is stored in the OGC server. The OGC server also acts as the intermediate storage for all the transient data created by different high-level services. Finally, the collected data can be analysed and visualised (live or post-event) using the open source platform Grafana [2, 3] in combination with the Grafana OGC Sensor-Things Plugin [4]. This offers powerful capabilities for data analysis and data presentation.

5.3 Security and Safety Solutions for Large Open-Air Events

5.3.1 Introduction

This section focuses on security and safety solutions based on wearables as well as video analytics. Different types of wearables have been developed within MONICA. One is a low-cost wristband providing coarse localisation based on a sub-GHz radio and used for crowd monitoring. Another type of wristband has been developed for the security staff. This is a more expensive device than the crowd wristband as it is equipped with a display supporting staff communication with the Command Centre (CC) and adopts UWB technology providing a more accurate localisation. A third type of wearable has been developed by MONICA proving tracking capabilities based on GNSS and LoRa communication; thus, getting rid of a cumbersome anchors deployment so it is able to cover very large areas with reduced cost. Finally, a smart glasses application has been developed to support the security staff to quickly share information about situations during an event.

In addition, MONICA has developed solutions for crowd management and monitoring using existing camera infrastructure, where Artificial Intelligence (AI) modules implementing deep learning algorithms have been used to model complex crowd behaviours and characteristics. Several crowd analysis algorithms have been developed including crowd counting and density estimation, crowd flow analysis, gate counting and crowd anomaly detection.

5.3.2 Crowd Wristband

The crowd wristband is a low-cost wearable device based on a bi-directional sub-GHz radio technology (compliant with the standard ETSI EN 300 220 V3.2.1) able to achieve a radio link larger than 100 m. Within the MONICA project, the crowd wristbands have been used for monitoring the density and location of a very large number of people at open-air events. The collected crowd monitoring data can be used to create heat maps of the crowd; thus, showing crowd size and densities. Moreover, the crowd wristband integrates an RFID interface to support access control and cashless payments. Two LEDs are also available on the crowd wristband, which can be used for entertainment purposes and for crowd control.

The first large-scale implementation of the crowd wristband was deployed by DEXELS in 2014 during two weekends of the Tomorrowland festival in

Figure 5.4 LED show with crowd wristbands in front of the main stage at Tomorrowland 2014.

Boom, Belgium. Each weekend a total number of 125,000 crowd wristbands were active for three days. The system supported "scanning" all 125,000 wristbands in 8 minutes. Within the MONICA project, the wristbands infrastructure has been integrated with the IoT platform and it was demonstrated first in the Tivoli Garden (in Copenhagen, the 26th of April 2019) and then in Lyon in occasion of the Woodstower festival (30th–31st August 2019), where more than 6,200 wristbands were distributed to the audience.

In total, 60 base stations (BSs) were used in Tomorrowland (see image in Figure 5.4), 15 BSs in Tivoli and 15 BSs in Woodstower, spread over the festival area, connected using Ethernet cables and powered by Power over Ethernet (PoE).

5.3.2.1 Technology overview

The crowd wristband can appear in multiple bracelet incarnations. It could be a leather bracelet, a textile wristband or a silicon wearable as in MONICA (see Figure 5.5). The current version of the crowd wristband contains the following components: radio chip and MCU (SoC solution), CR2032 battery, two bright RGB LEDs, RFID/NFC chip, one button, clock, and two antennas (HF RFID and UHF).

Together these parts cooperate to form the crowd wristband solution. The software running on the MCU manages the operation of the wristband. It controls the radio communication, the LEDs and the behaviour of the button press. The MCU can wake up from its deep sleep mode in several ways. On wakeup, the MCU starts its normal operation by listening to radio messages

Figure 5.5 Crowd wristband.

transmitted from a BS. The BS messages synchronise the wristband clocks and send commands either to a particular wristband or to all wristbands. The commands instruct the MCU to light up the LEDs. Each wristband can be addressed separately by means of a unique ID. Depending on the application, this unique ID could be associated with personal details of the visitor wearing the wristband. However, for crowd monitoring purposes, the wristbands in MONICA were anonymously distributed to the audience of the festivals. The button can be used for several user inputs. One example is "Friend-Connect" or "Contact Sharing" as demonstrated during IoT-Week 2019 in Aarhus. In particular, two people can connect and exchange personal information. The only thing that needs to be done to connect is holding two wristbands in close proximity of each other and pressing the button until the LEDs flash green in both wristbands. Upon this action, the two wristbands' identifiers are included in a "*connect*" message and sent to the MONICA platform that in turn sends it to an external cloud platform related to the IoT-Week APP. It should be mentioned that this "Contact Sharing" functionality was supported by two preliminary steps where the user first registers himself/herself using the APP and then associates this with his/her own MONICA wristband. Therefore, all the personal information was managed by the IoT-Week APP and the related cloud platform.

5.3.2.2 Crowd wristband TDMA protocol

The crowd wristband protocol uses parallel time slots to allow for higher throughput of wristband messages. Four radio channels (TRX, RX1, RX2, RX3) are used simultaneously as shown in Figure 5.6, where TRX is called "Pilot Channel". Wristbands messages are sent using a TDMA protocol. Moreover, a CSMA phase in the TDMA scheme is used for sending "urgent" messages like "wristband connects" and button pushes.

The wristband protocol relies on tight clock-synchronisation to support the TDMA protocol. The clocks on the wristband are synchronised by pilot messages sent by each BS. Depending on a unique ID every BS sends its pilot

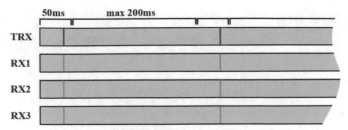

Figure 5.6   Crowd wristband TDMA protocol.

message in a predefined time slot. The length of a time slot, for both pilots and wristband messages, is set to 3 ms. A maximum number of 16 BSs time slots initiate a new "communication window" as shown in Figure 5.6 (see red interval). The wristband protocol supports a maximum of 16 BSs per "Pilot Channel" (TRX). Hence, the pilot-phase of the messaging windows always takes 50 ms. A wristband uses the pilot message to synchronise its local clock. Besides clock synchronisation, a pilot message contains the "wristband ID range" parameters and optional LED commands. The "wristband ID range" is used to define the logical range of wristband IDs (WUIDs) that need to be polled. The range is defined by a start- and an end-value. The WUID of a wristband is first masked to fall within the polling range before its reporting time slot/channel is determined.

The remainder of the messaging window is used to send a wristband "reporting message". Each wristband is assigned its unique timeslot and channel (1 out of 4) to send this message. The reporting window size can be adjusted from 0 ms to 200 ms, resulting in a total communication window of 50–250 ms.

A total number of 6 Pilot Channels can be defined in an event area, resulting in a maximum infrastructure of 96 BSs. The default operation mode is 250 ms, resulting in a maximum reporting throughput of almost 6,400 wristbands per second in an event area (thanks to the 4-fold parallelism and 6 different Pilot Channels). A wristband determines the "strongest" Pilot Channel by scanning other channels every 30 s. By keeping a list of strongest channels, it decides whether it is time to switch to another Pilot Channel. This effectively implements a channel handover procedure for the wristbands allowing for larger areas that can be supported by the protocol. This still limits the maximum area size that can be covered by the wristbands; exploring ways to bypass this limitation is part of future work.

The 50 ms mode is used for low-latency LED operation, allowing a new LED command to be sent every 50 ms. This comes at the expense of not being able to report wristband messages during 50 ms of operation. Moreover, urgent message mode for "wristband connect" and "button push" messages can be sent using a CSMA protocol providing a responsive and correct implementation of the so called "Friend-Connect" feature.

5.3.2.3 Infrastructure for crowd wristbands

The deployment scheme of the crowd wristband infrastructure is depicted in Figure 5.7. As it can be observed, the crowd wristbands need a dedicated infrastructure of BSs that communicate with each other and with the wristbands. The maximum safe range between a wristband and a BS is 75 m. This implies that a wristband must always be at maximum 75 m away from a BS in order to have coverage. This characteristic can be used to design and setup the BS infrastructure for a specific venue. Since there is a limit to the number of BSs (96) there is a limit to the area that can be covered. Hence, currently the spatial scalability is limited. The number of wristbands that is currently supported is limited by the 3 bytes that are used to identify a wristband. There is no inherent limitation to the number of wristbands in the protocol itself. The BS radio is controlled by an ARM based PC board running Linux and the BS software. The BSs themselves are joined together in a software cluster. A unique redundant-communication protocol has been developed that enables the use of multiple physical communication layers between the BSs.

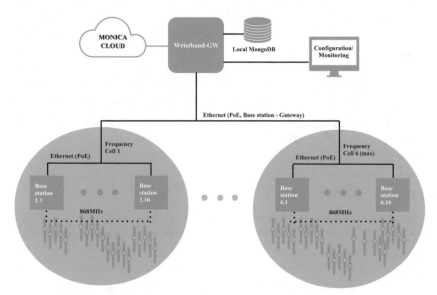

Figure 5.7 Crowd wristbands infrastructure deployment.

TCP/IP based communication, both Ethernet and Wi-Fi, as well as several low-bandwidth wireless communication technologies are supported between BSs and the GW. Altogether this creates a highly fault-tolerant communication channel. If for example the Ethernet or Wi-Fi infrastructure fails, the messages are still sent using the alternative available wireless infrastructure, making the system independent of the festival's infrastructure. A typical communication use case is a message that originates from a wristband, being received by one or more BSs and further transported to the GW.

The GW is deployed locally on the festival premises and it sends wristbands messages to the MONICA cloud. This setup enables the mobile APPs that are running on the visitor's smartphones to interact with the cloud system. The messages that are received from the wristband are collected by the GW that performs a real-time triangulation to drive the crowd monitoring system, heat map visualisation and individual wristband tracking. Since Received Signal Strength Indicator (RSSI) is used as range measurement, the typical localisation accuracy of crowd wristbands is 15–20 m. A management console is available for operators to control the entire system.

5.3.2.4 Services enabled by crowd wristbands

The location of the visitors can be used to calculate a crowd density. The resolution of this discrete density field is typically 5 m × 5 m. This is a useful feature for crowd monitoring, i.e. knowing the number of visitors in various event areas at any instant. This could also be used to detect high-risk queues (or at least high-risk densities) based on the maximum capacity of these areas. The locations collected by the crowd wristbands are used to create a current overview of the crowd distribution in the event area. These crowd services run in the MONICA cloud where also the DSS implements algorithms to detect over capacity or high-risk queues. The crowd density can be visualised in a dashboard application running in the CC of the event. In addition to the DSS, this information can be used by CC staff to detect hot spots in crowd densities.

5.3.3 Staff Wristband

The staff wristband (see image in Figure 5.8) comes with more features than the crowd wristband. The staff wristband provides a more accurate tracking capability. While the typical accuracy of crowd wristbands is 10–20 m, the accuracy of the staff wristband is less than 50 cm thanks to the adoption of Ultra-Wide Band (UWB) radio technology, compliant to the standard ETSI EN 302 065-2 (UWB Location Tracking).

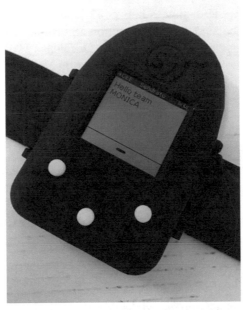

Figure 5.8 UWB-based staff wristband.

In addition, the staff wristband has an integrated BLE radio that can be used to communicate with a smartphone or smart glasses and also features a LCD screen that can be used to notify or instruct the user.

The fundamental building blocks of the staff wristband system are anchors and tags. Anchors are fixed location UWB nodes, containing at least one so called *master anchor* that is responsible for collecting all the data (wirelessly or wired) from the other anchors. Anchors send/receive messages to/from mobile tags. These messages are used in the localisation process as well as for communicating so called user payloads. These payloads can include e.g. data from sensors integrated into the tags.

The system uses UWB-based geometrical localisation. In particular, the ranging measurements use a Two Way Ranging (TWR) Time of Arrival (TWR ToA) method. The TWR does not require any synchronisation of the clocks at all, however this comes at the expense of having to communicate at least three messages between tag/anchor before the range can be determined. This means that with TWR, less tags can be tracked in a certain amount of time compared to a Time Differential of Arrival TDoA method. Using TWR, the system is able to support 1,200 location updates per second. Hence, 1,200 tags can be ranged running at an update rate of 1 Hz.

The next step in the process is localisation that calculates the position of the tag based on the distances between the tag and the (visible) anchors. The position is calculated using a lateration algorithm. It is worth remarking that the ranging accuracy achieved by the Decawave UWB chip is ± 10 cm in Line-of-Sight (LoS) conditions and ± 30 cm in Non-Line-of-Sight (NLoS) conditions; thus, there is always additive (white) noise present in the measured distances. As a consequence, an exact (closed form) solution of the lateration problem is not possible. One has to rely on an optimisation procedure to calculate the location. Typically, a Non-Linear Least Square (NLLS) method is used. In case of tracking a moving object, additional methods are used. Jitter in the calculated track is usually mitigated using some smoothing or filtering method. In MONICA it was used either an Extended Kalman Filter (EKF) or an Extended Finite Impulse Response (EFIR) in combination with an NLoS detection and mitigation method. These methods result in low jitter while still having acceptable latencies (<500 ms in case of a 20 Hz update rate).

Figure 5.9 shows the deployment scheme of the UWB staff wristband, where a cluster of UWB anchors are managed by a master anchor that in turn sends the ranging measurements to the GW that runs the lateration algorithm.

5.3.3.1 Technology overview

The UWB staff wristband integrates the following of components: $1.3''176 \times 176$ display, Bluetooth LE, USB and wireless charging, DecaWave DW1000 UWB radio, ARM Cortex M4, and a 400 mAh battery. Furthermore, several sensors and actuators are available: light sensor, IR proximity, pressure sensor, temperature sensor, humidity sensor, microphone, 9 axis IMU (accelerometer, gyroscope, magnetometer), 2 buttons and haptic feedback.

5.3.3.2 Services enabled by staff wristbands

Security staff localisation. The staff wristbands allow for the localisation of staff members. As such, they help to implement several use cases. Besides staff localisation, the wristbands can be used to notify staff members by sending text messages that are displayed on the LCD screen of the wristband. Moreover, the buttons can be used to send notifications to the CC.

Health/security incidents. By leveraging the IMU of the staff wristband, certain health incidents can be detected. A wrist-worn accelerometer is used for recognition of abnormal activities related to stewards and security staff visitors in crowded environments.

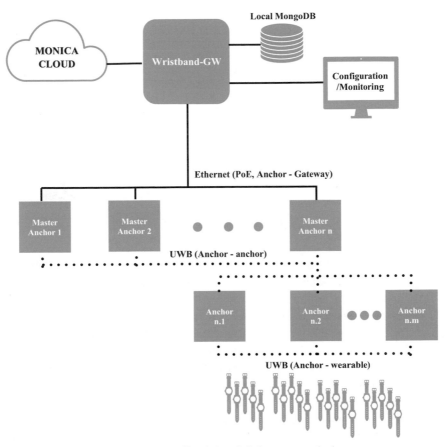

Figure 5.9   UWB staff wristbands infrastructure deployment.

5.3.4 Staff Tracking Device

Tracking devices based on the Global Navigation Satellite System (GNSS) have also been developed in MONICA for staff members to be used in large open-air events. Contrary to UWB staff wristbands (presented in 5.3.3), these tracking devices do not need the installation of UWB anchors for localisation purposes as they rely on signals from satellites. The devices record latitude and longitude data from the GNSS module and periodically send them to BSs using Long Range (LoRa) technology [5]. The LoRa BSs or gateways are integrated with LoRa signal transceivers and forward these data to the MONICA IoT platform. The rest of this section presents both the GNSS tracking devices and the LoRa Gateways.

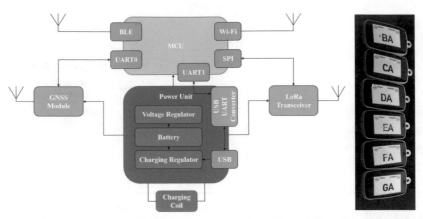

Figure 5.10 Modular diagram of HW components in trackers (left) and enclosed tracking devices (right).

The trackers are encapsulated in a small car key sized form factor. The hardware components are described in Figure 5.10. They contain an 1100 mAh battery which is wirelessly chargeable and lasts for more than 22 hours of active usage. An MCU is responsible for executing the tracker logic for collecting and sending the location information. The MCU is connected to Bluetooth Low Energy (BLE) and Wi-Fi transceivers. But these modules are deactivated in normal operations and used only for special purposes such as software update. The MCU is connected via a Serial Peripheral Interface (SPI) to the LoRa transceiver.

The GNSS module is connected to the MCU via a Universal Asynchronous Receiver Transmitter (UART). The GNSS module calculates the position every second with an accuracy of up to two meters. The localisation accuracy depends on many factors such as presence of high-rise buildings around the tracker. As soon as the MCU is powered on, it sends a registration request to the IoT Platform. The MCU is usually in deep sleep mode in order to save battery. It wakes up at a fixed interval to send the last recorded geo location coordinates. The trackers also have a programmable LED to show the status of the trackers so that the users will know whenever the status of the tracker (e.g. turned off, sending data, sending emergency beacons). The tracker has a button which is used for turning the device on or off and to initiate emergency beacons.

The LoRa gateway is a Raspberry PI device enclosed in a waterproof box. The gateway is integrated with a LoRa transceiver. One or more gateways are strategically installed in the venue in such a way that cover as much area as possible. One gateway can cover an area spanning thousands of meters. The

Figure 5.11 Signal coverage of two LoRa gateways (Red and Green) in the Rhein in Flammen event.

gateways are PoE. The data transferred from the tracker to the LoRa gateway include the device name, latitude, longitude, battery level, Signal to Noise Ratio (SNR) and accuracy of the geo-location. The gateway reads this data and forwards the data received from the trackers to the IoT Platform over HTTP.

Around 45 GNSS trackers have been successfully used by volunteers of the fire brigade, police, public order office and emergency response in pilot events at Bonn in 2019 (Rhein in Flammen and Pützchens markt). There were no concrete failure scenarios except minor signal breaks. The signal coverage was sufficient to contain the entire event venue just with two gateways as shown in Figure 5.11. Message duplication was one of the problems while reading LoRa channels. The reason behind duplication was the spreading of the messages across different time slots. This was filtered in the DSS before sending to the Common Operational Picture (COP) dashboard so that duplicate events are not generated.

Future work with respect to the GNSS tracker is an addition of an energy efficient user interface (e.g. an LCD) showing the messages from the CC. As of now, a simple LED is used blinking different patterns to give the user feedback. This needs to be improved with a descriptive display showing the messages and status.

Figure 5.12   ORA-2 smart glasses.

5.3.5 Smart Glasses

The ORA-2 smart glasses enable hands free mobile computing and Augmented Reality (AR) applications such as remote maintenance, logistics, remote training, situation awareness, and much more. It can run applications as a standalone wearable computer and can connect to a network via Wi-Fi and to any smart device via Bluetooth.

The ORA-2, image shown in Figure 5.12, features a disruptive transparent retinal projection technology. The virtual screen of the ORA-2 has two configurations allowing both "augmented reality" and "glance" modes. This "Flip-Vu" feature allows the image to be either directly in the wearer's field of view or just below.

The ORA-2 is equipped with a dual core processor GPU, camera, microphone, sound, inertial sensors, Wi-Fi, Bluetooth, GPS, ambient light sensor, and a high capacity rechargeable battery, and for a better ergonomic, a small Bluetooth remote controller has been added. The patented ORA-2 smart glasses hardware platform comes complete with its own flexible Android SDK in order to develop apps and fine tune the user experience.

An application, called MonicOra, has been developed by the Optinvent team for the MONICA project, which is based on a simple architecture so that the security staff can use it after few learning instructions. The goal of this equipment is to quickly share information about situations during an event. The security staff using the smart glasses can send predefined messages, audios, photos and videos to the CC, which can be visualised in the COP dashboard. In turn the CC (through the COP dashboard) can respond using the same way of communication, to one, or some/all smart glasses. The MonicOra APP fits many security use cases involving policemen, security staff, firemen, stadium stewards by changing few configurations with a quick update. Controlled by a little Bluetooth remote controller in hand, the smart glasses can easily manage all the applications using only 3 buttons.

One important aspect of this device is the battery lifetime. Since display, LED, Wi-Fi, GPS and a lots of other components are working together, it has been learned how to increase the activity in two ways. First, a power bank can be provided to be connected when the APP is running. Second, the users have been instructed to follow the correct usage. This equipment has to be in a sleep mode most of the time and be active very quickly to communicate with the COP. The most important thing is that the security staff work is supported by the smart glasses but are aware of the environment as much as possible at the same time.

Due to the technology, the AR glasses, which have the littlest optical engine size, provide a good experience in middle and low luminosity environments. This system has to be enhanced a lot if the amount of luminosity is high. In fact, all the MONICA demonstrations have been performed during low luminosity ambient so that the camera could be more sensitive and efficient.

5.3.5.1 Services enabled by smart glasses

The security staff equipped with ORA-2 glasses is completely aware of his environment because of the transparent lightguide (in sleep mode, no light disturbs the eye) and could easily report any situation by sending predefined messages, recording audio, picture or video (in active mode). Since the smart glasses are constantly tracked by GPS, the COP dashboard can send back the right information to the closest wearer(s) to find, for instance, a criminal sending an ID (for police officers), a plan with security exits (for stadium stewards), or audio and written translation for searching a lost child etc. With this equipment, combined with the other devices involved in the MONICA project, either a city or an event organiser can better manage all the deployed security forces in a more efficient way to prevent, take care, respond and quickly solve many situations.

5.3.6 Crowd Heat Map Module

5.3.6.1 High level data fusion anomaly detection module overview

A High-Level Data Fusion and Anomaly Detection (HLDF-AD) module represents the first stage of heterogeneous data aggregation and intelligent algorithm aiming to provide more complex and refined information as

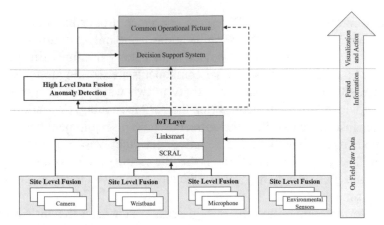

Figure 5.13 Simplified version of the MONICA architecture.

Table 5.1 Raw data available for HLDF-AD module

On Field Raw Data	Input Source	Main Fields
Crowd density global	Cameras	Geo Spatial density map that identifies people distribution
Gate people counting	Cameras	Entrance-Exit people events
People flow	Cameras	People movement trend
Wristband positions	Wristbands	Single person localisation
Sound level	SLMs	Geo spatial Sound Heat Map

requested by the DSS and COP modules. More specific, such a module acquires raw data broadcasted by MONICA IoT wearables and sensors deployed in the field and tries to fuse them in order to generate added value information and feed the upper level's components with real status information of the monitored area and eventually notifying anomaly conditions. Figure 5.13 shows a simplified version of the MONICA architecture just to show the data flow and highlight the role of the HLDF-AD module. From a general perspective, the MONICA ecosystem foresees a first stage of onsite elaboration (if possible) for each kind of sensors involved. The HLDF-AD module can combine them in order to complete or certify different input sources data and extend meaning towards higher-level modules.

Table 5.1 lists the most important raw data available for the HLDF-AD module. It shall be taken into account that these table reflects a complete set of MONICA sensors that might not be available for all monitored sites as it happened during some pilot executions.

Based on the inputs reported in Table 5.1, the HLDF-AD module should be able to produce both detection in terms of instantaneous on-site observation and prediction information as an extension of historical data collection and trend analysis on that. For the second stage, a huge amount of historical information is required.

5.3.6.2 Crowd Heat Map Algorithm overview

Among different types of output information, the most significant example is represented by the Crowd Heat Map estimation, which is an estimation of 2D East-North people geographic distribution with respect to a reference position. The HLDF-AD module combines raw data coming from cameras and wristbands in order to calculate people distribution in a monitored geographic area and estimate the total amount of people attending a specific event. In particular, the input information data exploited by such algorithm are Crowd Density Global, Gate People Counting, and Wristband Positions. In principle, Crowd Heat Map and Crowd Density Global could be the same information. However, both information can be combined into a unique geospatial matrix that describes a more refined instantaneous people distribution. It is worth remarking that Crowd Density Global might be affected by issues depending on camera configuration. For instance, the number of cameras or the related coverage could be too low with respect to the event area, and a wrong setup of them (e.g. too high, too low, very low environment visibility) might have impact on the results. Moreover, the possibility for the HLDF-AD module to perform cross-check between camera results and other data sources allows to enhance the confidence level of the output result.

The Crowd Density Global is the most important input from HLDF-AD perspective in view of Crowd Heat Map provisioning. The HLDF-AD module acquires them and, after a clean-up procedure, follows some logic steps considering also historical information and Gate People Counting. This helps to eventually correct the total amount of people based on the principle that the total number of people at time T_1 equals the total number of people at time T_0 plus the number of people entered/exited to/from the monitored area. It is assumed that the monitored area is fenced, where people can enter or exit just passing through the gates monitored by the MONICA cameras. For instance, if at time T_0 there are 100 people in the monitored area and it has been estimated that 10 people have entered between T_1 and T_0, then at T_1 the HLDF-AD module assumes that there are 110 people. Potential differences raised by the Crowd Density Global have been cleaned by the HLDF-AD algorithm that estimates a new refined Crowd Heat Map. The next step considers Wristband Positions data. The HLDF-AD module calculates another Crowd Heat Map considering the Wristband Positions, i.e. counting the number of positions inside each geospatial cell of the matrix as 2D East-North distance with respect to a reference position. The new computed Crowd Heat Map can be used as the first output in case of missing cameras or in order to refine the confidence level of the Crowd Density Global computed at the first stage. This data process shall be carried out by means of a rough estimation of the percentage of people who take and use the wristbands with respect to the total amount of people attending the event. Moreover, it should be assumed that the distribution of the people using the wristband is equal to the distribution of the people not wearing the wristbands. Furthermore, it must be taken into account that the percentage of people with a wristband should be significant with respect to the real total amount of people in order to obtain reliable results (at least 20%).

5.3.6.3 Crowd Heat Map at the Woodstower event

During the Woodstower festival (in Lyon, 30th–31st August 2019) the Crowd Heat Map service performed by the HLDF-AD was tested. Since this pilot did not have any cameras installed, the Crowd Heat Map was estimated taking as input only the wristbands' positions from the Wristband-GW. In this pilot, the MONICA partners anonymously distributed up to 6,200 wristbands to the audience of the musical event. The position of each wristband was estimated by the Wristband-GW and transmitted every 4 minutes to the MONICA cloud. The HLDF-AD module stored temporarily each wristbands' position, and every 4 minutes calculated the Crowd Heat Map. This service was registered into the Service Catalogue of the MONICA platform. The size of the monitored area was 300 m by 200 m. The representation was performed on a 10 m × 10 m cell size.

Figure 5.14 shows an example of the estimated Crowd Heat Map based on crowd wristband positions. It can be seen that, the crowd was mainly concentrated in front of the four stages.

5.3.7 Crowd Management and Monitoring Using CCTV Cameras

With the help of one of the most prevalent IoT technologies such as IP surveillance cameras, the MONICA project offers intelligent and autonomous solutions for crowd management problems based on cameras. These solutions require minimum human intervention and can be scaled up indefinitely with no compromise on reliability. MONICA solutions for crowd management and

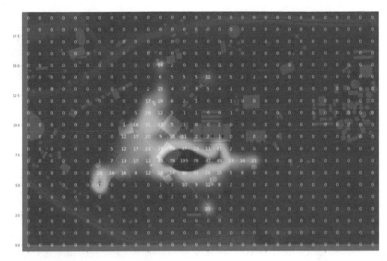

Figure 5.14 Example of a crowd heat map estimated during the Woodstower event.

monitoring can be adopted to the majority of the existing surveillance infrastructure which significantly reduces the costs yet improves the performance and productivity. Unlike classical labour-intensive monitoring systems which suffer from issues like scalability, reliability and cost-effectiveness, intelligent surveillance systems can be scaled up indefinitely with no compromise on reliability. In the last decade, the computer vision community has pushed on crowd behaviour analysis and has made a lot of progress in this field. The emergence of deep learning and Convolutional Neural Network (CNN) in the last decade has boosted the performance of image classification techniques and has started having a positive impact on crowd behaviour analysis. CNNs have gained ground in crowd monitoring and behaviour analysis. In MONICA, deep learning has been used to model complex crowd behaviours and characteristics. Several crowd analysis algorithms including crowd counting, localisation, density estimation, crowd flow analysis, gate counting, crowd anomaly detection, fight detection, and object detection have been developed in MONICA and deployed in several pilot plans across Europe. The following subsections describe some of these algorithms in more detail.

5.3.7.1 Crowd counting and density estimation

Crowd counting and density estimation are of great importance in computer vision due to its essential role in a wide range of surveillance applications including crowd management and public security. However, drastic scale

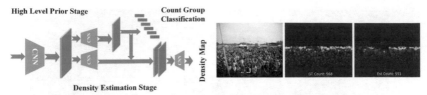

Figure 5.15　Left: CNN for crowd counting and density estimation. Right: the original image, the Ground truth and the predicted density map.

variations, the clutter background scene, and severe occlusions make it challenging to generate high-quality crowd density maps. MONICA offers crowd counting and density estimation algorithms based on CNN. The proposed solution aimed to address a wide variety of crowd density levels by incorporating a high-level prior into the deep convolutional neural network. The high-level prior learns to classify the count into various groups whose class labels are based on the discretised number of people present in the scene. The count class label allows us to estimate coarse count of people in the given regardless of scale variations thereby enabling the network to learn more discriminative global features. The high-level prior is jointly learned along with density map estimation using a cascade of CNN networks as shown in the following Figure 5.15. The two tasks (crowd count classification and density estimation) share an initial set of convolutional layers which is followed by two parallel sets of networks that learn high dimensional feature maps relevant to high-level prior and density estimation, respectively. The global features learned by the high-level prior are concatenated with the feature maps obtained from the second set of convolutional layers and further processed by a set of fractionally strided convolutional layers to produce high resolution density maps.

5.3.7.2 Crowd flow analysis

Crowd flow is another informative metric in crowd management and behaviour analysis. There is a critical capacity where flow begins to decrease as the crowd's density increases. A dense crowd with high flow magnetite poses a serious safety threat and might lead to a human stampede. In visual surveillance, optical flow algorithms have become an important component of crowded scene analysis. The application of optical flow allows crowd motion dynamics of hundreds of individuals to be measured without the need to detect and track them explicitly, which is an unsolved problem for dense crowds. MONICA uses Flownet CNN which first produces representations of

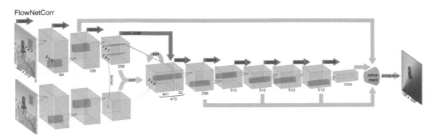

Figure 5.16 CNN for crowd flow estimation.

the two temporally consecutive images separately, and then combines them together in the 'correlation layer' and learn the higher representation together. The representations then will be up-sampled to reproduce the flow map in original image scale. The correlation layer is used to perform multiplicative patch comparisons between two feature maps. Figure 5.16 shows the Flownet architecture used to measure crowd flow and dynamic.

5.3.7.3 Fight detection

Fight and public violence are prevalent phenomena in large public crowds. Aside from the possible injuries and fatalities, violence can cause damage to public assets and disturb public peace and tranquillity. An automated fight detection model could be a great addition to the existing surveillance systems deployed in public large events such as concerts, stadiums, and amusement parks. The proposed fight detector in MONICA uses a modular architecture, with distinguished components, which were responsible for the fundamental operations, such as video capturing and event detection. Particularly, a computer vision algorithm constitutes the core of the system and it captures mid-term motion patterns by isolating regions of interest in each frame of a video stream and forming point trajectories through the optical flow. After that, the enclosed trajectories over the last frames extract a motion histogram for each region.

5.3.7.4 Pipelining in the security fusion node

The Security Fusion Node (SFN) sits as a separate component to the Video Processing Pipeline and the video algorithms; however, in practice they often run on the same processing node. The SFN acts as an interface between the outputs of the algorithms and the higher-level MONICA services. By having awareness of all the cameras (via a camera registration process), the

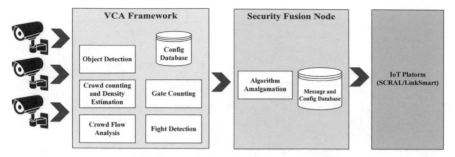

Figure 5.17   High-level overview of where the SFN is located within the MONICA architecture.

SFN is broadly speaking a lightweight multithreaded REST API responsible for forwarding messages from the edge layer up to the cloud and adding additional information where required. The SFN is a kind of central gateway concept that facilitates the platform for information fusion. As such, this node takes as input the output from the algorithms running on the local cameras, whether they be IP cameras, smart glasses or camera equipment mounted on a blimp. Figure 5.17 shows the SFN architecture relative to MONICA.

5.4 Sound Monitoring Solutions for Large Open-Air Events

5.4.1 Introduction

The MONICA project aims at applying IoT technologies for the management of large-scale outdoor cultural events. These types of events come often with various disturbances to the neighbourhood. One of the most noticeable is noise. Moreover, among the different pilot tests in the MONICA project, many of them concern music, where the acoustic experience is prominent for the participants. For both reasons, noise impact on the neighbourhood and quality of sound for the audience, sound monitoring is an important piece of the management of such outdoor events. Additionally, the noise level is often regulated by local legislations, specifying generally a threshold that should not be exceeded otherwise exposing possibly the event organisers to penalties. It is then important to be able to set thresholds and alerts in the sound management system.

A sound monitoring system is made of IoT SLMs that are deployed at different places, in the neighbourhood and in the venue. These units send

Figure 5.18   Sound level meter box.

different acoustic data to a cloud platform, where they are recorded, possibly further processed and sent to an interface for visualisation and alerts. The visualisation concerns outputs of individual sensors and higher-level quantities based on multiple SLMs. These elements are described in the following sub-sections.

5.4.2 Sound Level Meter

A Sound Level Meter (SLM) is generally composed of a high-quality microphone, connected to a board for signal conditioning, processing, local storage, and a display. For the MONICA project, the SLM is put inside a water-proof box, that includes as well a wireless router acting as a mobile Wi-Fi hotspot and a power bank for power supply to the SLM and the router. From the box, the microphone sticks out, protected by a windscreen against wind and rain. An image of the system is shown in Figure 5.18.

The SLM sends data through the router inside the box, and therefore relies on the telecommunication network (3G or 4G). The SLM can also connect to any existing Wi-Fi network in the vicinity.

The types of data that are sent depend on the requirements for the pilot test. For all cases, two different types of levels, the so called L_{Aeq} and L_{Ceq}, and spectrum data, are measured every second. L_{Aeq} and L_{Ceq} provide averaged levels, commonly used for environmental acoustic measurements. The spectrum decomposes the noise level into 1/3 octave frequency filters, from 12.5 Hz to 20 kHz. This represents a vector of 33 values. Sound levels and spectrum data occupy 116 bytes and are sent every second.

For advanced acoustic processing, audio recordings could be necessary. In this case, the amount of transferred data is much higher, since the sampling rate of the recordings could be relatively high. To reduce the required bandwidth, the data are compressed (but without loss) before sent out on the network. For the MONICA project, the required bandwidth is 1 Mbit/sec (since the compression of the audio recordings is dynamic, it varies, but it is generally less than 1 Mbit/sec).

The sound data are first sent to a dedicated cloud, called SLM-GW, for data management, storage and data reduction. The data are then accessible to the MONICA cloud through a RESTful interface. The data reduction tasks in the SLM-GW consists of processing the acoustic levels or the audio recordings, to provide averaged values and advanced data to the MONICA cloud. Two of the advanced calculations performed in the SLM-GW are the Annoyance Likelihood Index and the Contribution analysis. The Annoyance Likelihood Index is a metric between 0 and 10 indicating the level of noise annoyance [6] for the neighbourhood, 10 meaning a maximum annoyance. It is based on the comparison between initial measurements of background noise (before the event) using the levels L_{Ceq}, and the same type of levels during the event. This metric is available for every minute to the MONICA cloud. The purpose of the contribution analysis is to estimate the noise contribution of the venue in the neighbourhood. The noise level measured by a SLM located in the neighbourhood is often a mix of contributions: from the venue, but also from acoustic sources nearby such as cars. Therefore, to recover the contribution from only the event, specific processing needs to be performed. Two approaches are considered: one based on spectral data (implemented in the DSS module, see Section 5.4.3), the other is based on audio recordings (implemented in the SLM-GW). As mentioned earlier, the required transfer data rate between the SLMs and the cloud for audio recordings is relatively high.

The Sound Monitoring System has been tested during different pilot tests in Europe. To illustrate its implementation, we discuss experiences of two different pilot tests in 2019: KappaFutur Festival and Rhein in Flammen.

KappaFutur Festival. This event is located in Torino. It takes place every year in Parco Dora and is dedicated to electronic and techno music. It gathers about 50,000 participants during two days around 4 stages. Focusing on noise impact, low frequencies are pointed out by the neighbourhood as the main reason of annoyance, together with excessive levels during some music performances [6]. The noise monitoring system setup for 2019 edition in July consisted of 8 SLMs, of which four inside the venue, in front of each stage, and four outside, on balconies of dwellings.

The requested acoustic data are the same as for the previous case, so the data sent by each SLM was 116 bytes every second. For the SLMs in the dwellings, very small data losses have been noted. For the SLMs inside the venue, variations were seen in the quality of the communication network, causing some significant losses (not much the first day, but mostly the second day of the event). One explanation could be that the high number of people gathered in a relative limited area could generate situations where the cell phone network is overloaded.

Two additional SLMs have been set up, sending audio recordings, that require 1 Mbit/s data transfer rate. These SLMs were located outside of the venue, and the purpose was to try out the contribution analysis module based on audio recordings on one hand, and to support another system in the MONICA project called the Adaptive Sound Field Control (ASFC), on the other hand. For these two SLMs, the data loss rate was high because of insufficient network bandwidth outside the venue to support the required transfer data rate.

Rhein in Flammen. This event is located in Bonn, Germany. Here also audio recordings for six SLMs were used, relying on the telecommunication network. Based on local experiences concerning reliability and bandwidth, the local network operator was selected accordingly.

The bandwidth requirement for sending audio recordings is 1 Mbit/s/SLM which is equivalent to 429 MB/hour/SLM but the observed averaged data transfer was 316 MB/hour/SLM, thanks to the audio lossless compression performed in the SLMs. Very small amount of data loss has been noted. It demonstrates the feasibility of sending high sampling rate audio recordings, at the condition of a good telecommunication network support. However, one should be aware of the additional cost due to the high volume of transferred data.

5.4.3 DSS for Sound Monitoring

The DSS is the intelligence layer of the MONICA platform. It is used for a variety of applications, including sound monitoring and control. As an input, it uses data provided by the HLDF-AD and GOST components, such as sound heat maps, A- and C-weighted sound levels, sound spectra as well as location information from IoT devices such as wristbands. The output of the DSS is subsequently forwarded to the COP dashboard in order to be visualised in a clear and consistent way.

In a real-world scenario, information can be affected by uncertainty. In these cases, it is often not easy or desirable to group all pieces of information into crisp sets (sets with a binary membership where objects are either members of the set or not) based on precise parameters. Instead, fuzzy sets can be used for information classification based on imprecise criteria. For this reason, the DSS is developed as a Fuzzy Logic Device, producing deterministic output from deterministic input starting from a set of rules relating linguistic variables to one another using fuzzy logic. For the mapping to be performed, a subset of deterministic values are converted into fuzzy values, and vice-versa [7].

The fuzzy set theory was first introduced in [8] as an extension of classical fuzzy models. A fuzzy set is a collection of objects that do not have explicitly defined criteria of membership. Instead of a binary membership (i.e. 0 or 1) to a set, each object has a grade of membership [0, 1] instead, that indicates its degree of truth in a subjective way. In other words, its degree of membership indicates how that object "fits" into the set.

In complex sound management scenarios, there are two distinct types of problem knowledges that can usually be inferred for a situation, *objective knowledge* and *subjective knowledge* [9]. Objective knowledge refers to quantitative variables that can be used to accurately represent information. Subjective knowledge on the other hand corresponds to usually not-quantifiable knowledge in the form of verbal statements. Obtaining objective knowledge for a specific event, area or situation though is not easy and can often be inaccurate or not applicable; many times gaining subjective knowledge through the interaction with experts is better suited to capture the imprecise modes of reasoning that is essential for the ability of people to make decisions in an uncertain environment [10].

The DSS uses both types of information to create fuzzy antecedent-consequent "IF-THEN" rules that can be used to propose different courses of actions. One such example that describes a straightforward course of action is – *If the sound volume is too loud, lower the volume of the loudspeaker*. In this example, the notion of *too loud* is a linguistic variable that forms the antecedent part of the rule and can have a different meaning in different settings. The notion of *lower the volume of the loudspeaker* represents a crisp course of action and is the consequent part of the rule.

The DSS consists of four main modules: the fuzzification, knowledge base, decision making and defuzzification modules.

Input from HLDF-AD and GOST is received as crisp numerical values. A fuzzification module is then used, responsible for the mapping of such values to fuzzy sets. The resulting sets can be used in turn to activate all relevant rules by calculating their membership functions. The knowledge base is the process model of the system. It consists of a database that contains the rule structure for each different venue and event that the MONICA project is employed, constructed based on expert input. The fuzzy rules together with the input from the fuzzification module are combined to generate output by the decision-making module. The defuzzification module aggregates the rule consequents and selects the highest-rated one to produce a crisp output as the outcome of the Fuzzy Logic Device (FLD). This outcome is forwarded to the COP for visualisation. This allows stakeholders to have a real time assessment of the event with regards to local regulations.

Within the MONICA project, work has been carried out, as a collaboration between CERTH and Acoucité, to implement a DSS for sound monitoring for two pilots in Lyon. The French regulation regarding sound in festivals[4] was then taken as reference. The main requirements of the decree are:

- The sound pressure level (15 minutes average) should not exceed 102 dBA and 118 dBC at any place accessible to public in the audience area.
- For thresholds in the neighbouring area, the values are dependent on the background noise levels existing at the measurement point. A measurement campaign to determine the background noise levels was thus required. The requirement is that sound pressure level (15 minutes average) in the neighbouring area should not exceed the background noise level at the measurement points by more than the "emergence" values. Emergence values are defined in terms of Overall A-weighted Sound Pressure Level (L_{Aeq}) and in terms of octave bands from 125 Hz to 4 kHz. Emergence values depend on the period (day/night) and the duration of the event.

IoT SLMs of Noise Monitoring System (see Section 5.4.2) is set up to provide every second with the following acoustic data: A-weighted, C-weighted and 1/3 octave frequency bands sound pressure levels (1 second average).

---

[4]Decree no. 2017–1244 of 7th August 2017, in force since October 31st 2018.

Figure 5.19   Woodstower 2019, Left: IoT SLMs in audience area (highlighted in red), Right: Municipalities where SLMs outside the venue have been deployed.

This data cannot be directly used for assessing the event with regards to the regulatory limit thresholds. Thus, further computations are required at DSS. These computations are of two types:

- Obtaining 15 minutes average Sound Pressure Level from 1 second average Sound Pressure Level;
- Obtaining octave band spectrum from 1/3 octave band spectrum.

Once those computations are done, relevant data can be shown on the Noise Monitoring System Display and alerts can be set for monitoring threshold exceedances.

An example of the application of DSS for Noise Monitoring System at the Woodstower festival in Lyon, France during August of 2019 is presented here after.

5.4.3.1 DSS for sound monitoring at Woodstower festival 2019

A total of eight IoT SLMs have been deployed for the Woodstower festival, of which four in the audience area and four in the residential area located in municipalities around the venue of the festival (see Figure 5.19). The devices in the audience area have been located near the sound engineer's console at each of the four stages in the venue.

The IoT SLMs provided overall sound pressure levels and 1/3 octave band spectra and the connection was established using a Wi-Fi network specifically deployed for the MONICA project.

Positioning of the SLMs was planned to cover municipalities located to the north and south of the venue in order to include critical areas independently of meteorological conditions (i.e. wind direction varies from one year to another).

The devices sent overall sound pressure levels and 1/3 octave band spectra. Data were sent from IoT SLMs to the MONICA cloud using 4G connection through the MiFi device inside the IoT SLM box.

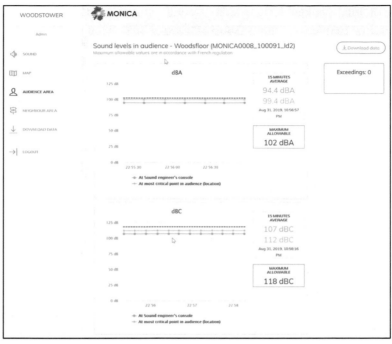

Figure 5.20 Sound levels in audience – Woodsfloor stage of the Woodstower festival.

Sound Monitoring Display for levels in the audience, as shown in the COP dashboard, is presented in Figure 5.20. Levels at the most critical point in the audience are estimated based on results from preliminary acoustic measurements done by Acoucité.

5.4.4 Sound Heat Map Module

The Sound Heat Map provides an estimate of the sound pressure level (SPL) at other positions than the one being measured by the SLMs. This is done using a forward sound propagation model, based on the existing sound propagation model "Nord2000". The model is built in 2D and requires the position of sound sources, positions of reflecting surfaces (walls) and a computation area (surface and grid size).

The computation model uses SPLs measured by the IoT SLM located in front of the stage to dynamically adapt the power of each source included in the model. Then, the map is calculated and becomes available in the MONICA cloud via a REST API.

It is nowadays common practice to use directive loudspeaker systems in large-scale concerts. Thanks to progresses made in the past few decades in terms of loudspeaker design, today systems allow getting directive stages even in low frequency sounds (sub-woofers). Thus, when computing a sound map of a venue, the possibility of controlling the directivity of sound sources is essential for getting accurate results in terms of spatial distribution of sound pressure levels within the audience.

The directivity can be expressed as a cardoid function. The basic equation of a cardioid pattern is:

$$D(\theta) = \frac{1}{2}(1 + \cos(\theta))$$

Cardioid polar pattern

where D is the directivity factor (from 0 to 1) between the source and the receiver, and θ is the angle with respect to the source's radiation axis. In a projected system (conic projection, ex: lambert 93), the angles are calculated from the Cartesian coordinates of the source and the receiver (Xs, Ys, Xr, Yr) as shown in the figure below:

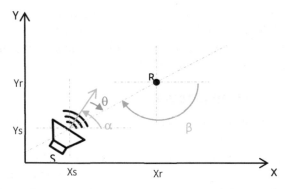

Cartesian coordinates of the source and the receiver.

$$D(\theta) = \frac{1}{2}(1 + \cos(\theta)) \quad \text{with}$$
$$\theta = (\beta - \pi) - \alpha \quad \text{and}$$
$$\beta = a\tan 2 \left(\frac{Ys - Yr}{Xs - Xr} \right)$$

The two inputs are the azimuth of the source direction (loudspeaker) and the azimuth of the path between the receiver and the source. Both are defined relatively to the X axis of the Cartesian extent. The angle β is calculated from the X and Y Cartesian coordinates. The computed directivity factors for cardioid source are used in the computation model for each source-receiver pair.

5.4.4.1 Sound Heat Map in Woodstower festival

Woodstower is a music festival taking place in a massive park in the northeastern part of the urban area of Lyon (France) at the end of August. During four days, musicians performed on four stages, namely: Mainstage, Woodfloor, Scène Saint Denis and Chapiteau. The event goers can walk around the venue and attend various performances.

The main challenge for the event manager is to respect the new French regulation (reported in Section 5.4.3). As explained in Section 5.4.2, IoT SLMs are an excellent way to control the levels at a long-term monitoring point (usually near the sound engineer's console). However, they do not allow getting information about the sound level distribution within the whole venue area. The Sound Heat Map gives that opportunity.

A computation model for the Sound Heat Map was built based on information provided by the Woodstower organisers such as venue layout and sound systems configuration, shown in see Figure 5.21.

The event manager chose only loudspeakers and subwoofers with a cardioid directivity pattern. The code for the Sound Heat Map computation should then be able to:

- Use cardioid sound sources.
- Calculate a Sound Heat Map for each of the four stages and to display the resulting global Sound Heat Map (i.e. the summation of the four stages) on the COP dashboard.
- Display the global level in dBA and dBC based on an energy summation of all the band frequencies.

Each stage has been modeled using a simplified model of the sound system (three cardioid sound sources per stage). The obtained directivity is shown in Figure 5.22.

Implementing the directivity pattern in the initial implementation of the heat map did not affect the calculation performance and allows a better identification of sound levels in different areas inside the festival site.

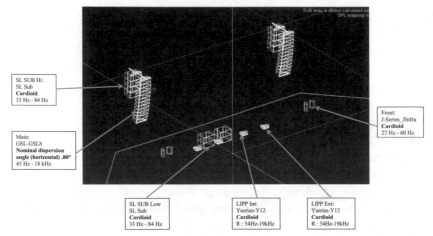

Figure 5.21 Sound system configuration for the main stage of Woodstower festival 2019 (d&B audiotechnik).

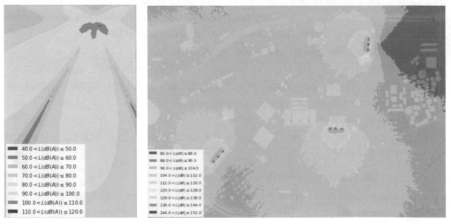

Figure 5.22 (Left) directivity of main stage at a frequency of 40 Hz. (Right) computed Sound Heat Map with three stages active (Scène Saint Denis, Chapiteau and Woodsfloor).

5.5 MONICA APP Layer

5.5.1 MONICA APIs

All applications and apps that need to access MONICA functionalities need to use the available MONICA APIs. There are three main MONICA APIs targeting different needs: Professional API, Public API, MessageHub.

5.5.1.1 Professional API

The Professional API is intended for components delivering data and services to the professional staff at the event. The main users of the API are the Common Operational Picture UI and the different apps that are part of MONICA.

The Professional API is based on the following technologies:

- MQTT interface for receiving messages for updating the COP status.
- Authorisation model based on Keycloak REST API.
- Odata-based API for retrieving resources from the IoT DB.
- Integrates SQL database with an OGC Sensorthings database for storage and management of time series of observations.

The Professional API provides the following main functionalities:

- Incident classification and management.
- Division of a geographical area into zones and subzones.
- Mapping of incidents, sensors, facilities and people/groups/crowds to zones.
- Fast retrieval of current status of the situational objects of interest.

5.5.1.2 Public API

The Public API is intended to be used by public apps and is basically a subset of the Professional API. The content that is accessible depends on what the event organiser makes public. Typically, the following items are available in the API:

- Public points of interest, for instance, position of medical services, toilets etc.
- MONICA collected data, which is made public. For instance, people count in different areas.
- Feedback collection, i.e. allowing the public apps to collect simple feedback of the event.

5.5.1.3 MessageHub

In addition to the rest-based API there is also a message hub implemented for pushing information to the apps and user interfaces. The main reason for supporting a push-based approach is to improve scalability, the clients will be notified of new available data rather than polling themselves. It will also increase the responsiveness of the application. The message hub is implemented using SignalR[5].

5.5.2 COP Dashboard

During an event monitored with MONICA, the CC will interact with event staff and build situational awareness using the main COP dashboard. The main interface is based on a map of the event venue with symbols marking different objects of interest. The main interface is an HTML-based app that can be used with any type of device that supports HTML 5. It is possible to run it on computers and tablets, but it is primarily designed to be used on devices that have a larger screen. It is possible to run the COP UI on a mobile phone, but it will have limitations due to the screen size.

Currently there are three main user interfaces that can be selected:

- Crowd/Security Monitoring that displays all the information related to Crowd/Security monitoring.
- Sound monitoring that displays all the information concerning sound levels etc.
- Staff view.

Depending on what is monitored at different pilot sites the COP dashboard is tailored to only show relevant information for the event. In the COP it is also possible to filter out information to unclutter the dashboard. Figure 5.20 shows an example of the COP view for sound monitoring while Figure 5.23 shows an example of the COP view for Crowd/Security Monitoring.

5.5.3 Professional Sound APPs

One example of an app built with the Professional API is the Professional Sound Application that is intended to be used by sound managers, sound engineers and event managers. With the application the users can follow the sound measured by the SLMs in real time. The application visualises 1/3 Octave spectra, LAeq and LCeq. It is also possible to combine two SLMs and compare them easily, see Figure 5.24. Furthermore, there is functionality to report sound feedback in different locations which will be shown as "sound incidents" on the COP dashboard.

---

[5]https://en.wikipedia.org/wiki/SignalR

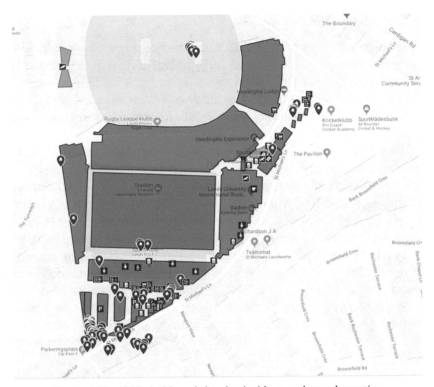

Figure 5.23 COP dashboard showing incidents and crowd counting.

Figure 5.24 Views of the MONICA professional sound app.

5.6 Conclusion

This chapter presented a set of IoT solutions, developed by the MONICA project, able to improve safety and security as well as to reduce the noise level for neighbours in open-air cultural events inside cities. These solutions are based on a large-scale deployment of innovative wearables and devices interconnected with closed-loop back-end services integrated into an interoperable cloud-based platform.

To support crowd management and monitoring use cases, MONICA has developed three types of wearables: a low-cost crowd wristband providing coarse localisation suitable for crowd monitoring, a more expensive wristband for security staff providing a more accurate localisation based on UWB and a third type of wearable with tracking capabilities based on GNSS and LoRa communication able to cover larger areas with a reduced installation cost. Finally, a smart glasses application has been developed to support the security staff to quickly share information about critical situations during an event. In addition to wearables, MONICA employed a set of crowd monitoring solutions relying on existing camera infrastructure. In particular, crowd analysis algorithms have been developed including crowd counting, crowd density estimation, crowd flow analysis, gate counting, and crowd anomaly detection.

Finally, concerning the sound monitoring use cases, MONICA has developed a sound monitoring system comprising IoT SLMs, the Sound Heat Map module and the DSS. For each pilot, these sound components have been used to cover two main areas of interest: the event venue and the neighbourhood. The sound data were processed by the DSS, which adopts a "fuzzy logic" approach to provide more elaborated results (e.g. 15 minutes average and sound contribution analysis) and noise-related alerts as output to be visualised in the COP dashboard. Furthermore, the implementation of the Sound Heat Map module provided a wider and graphic estimation of the sound levels by using interpolation algorithms and geometric modelling of the event area.

Acknowledgements

This work has received funding from the European Union's Horizon 2020 research and innovation programme as part of the "Management Of Networked IoT Wearables – Very Large-Scale Demonstration of Cultural Societal Applications" (MONICA) project under grant agreement No. 732350.

This work reflects only the authors' views and the commission is not responsible for any use that may be made of the information it contains.

References

[1] High Level Architecture (HLA), Release 4.0, AIOTI WG03 – IoT Standardisation, online at: https://aioti.eu/wp-content/uploads/2018/06/AIOTI-HLA-R4.0.7.1-Final.pdf

[2] OGC SensorThings Standard, online at: https://www.opengeospatial.org/standards/sensorthings

[3] Grafana, online at: https://grafana.com/grafana (retrieved: 2019-09-05).

[4] Github Repository: Grafana OGC Sensorthings Plugin, online at: https://github.com/linksmart/grafana-sensorthings-datasource

[5] M. C. Bor, J. Vidler and U. Roedig, "LoRa for the Internet of Things". In: EWSN, Vol. 16 (EWSN, 2016) pp. 361–366.

[6] K. Haddad, P. Munoz, E. Gallo, B. Vincent, M-H Song, "Application of Internet of Things technology for sound monitoring during large-scale outdoor events", Inter-Noise, Madrid, 2019.

[7] R. E. Haber, et al., "Application of knowledge-based systems for supervision and control of machining processes". In: Handbook of Software Engineering and Knowledge Engineering: Volume II: Emerging Technologies. 2002. pp. 673–709.

[8] L. A. Zadeh, "Fuzzy sets", Information and control, 1965, 8.3, pp. 338–353.

[9] J. M. Mendel, "Fuzzy logic systems for engineering: a tutorial", Proceedings of the IEEE, 1995, 83.3, pp. 345–377.

[10] M. J. Skibniewski, et al., "Quantitative constructability analysis with a neuro-fuzzy knowledge-based multi-criterion decision support system", Automation in Construction, 1999, 8.5, pp. 553–565.

[11] IoT European Large-Scale Pilots Programme (IoT-LSPs), online at: https://european-iot-pilots.eu/

6

IoT Technologies for Connected and Automated Driving Applications

Ovidiu Vermesan[1], Roy Bahr[1], Mariano Falcitelli[2], Daniele Brevi[3], Ilaria Bosi[3], Anton Dekusar[4], Alexander Velizhev[5], Mahdi Ben Alaya[6], Carlotta Firmani[7], Jean-Francois Simeon[8], Louis Touko Tcheumadjeu[9], Gürkan Solmaz[10], Francesco Bisconti[2], Luca Di Mauro[2], Sandro Noto[2], Paolo Pagano[2], Enrico Ferrera[3], Guido Alejandro Gavilanes Castillo[3], Edoardo Bonetto[3], Vincenzo Di Massa[7], Xurxo Legaspi[11], Marcos Cabeza[11], Diego Bernardez[11], Francisco Sanchez[11], Robert Kaul[9], Bram Van den Ende[12], Antoine Schmeitz[12], Johan Scholliers[13], Georgios Karagiannis[14], Jos den Ouden[15], Sven Jansen[12], Hervé Marcasuzaa[16] and Floriane Schreiner[17]

[1]SINTEF AS, Norway
[2]CNIT – PNTLab, Italy
[3]LINKS Foundation, Italy
[4]IBM, Ireland
[5]IBM, Switzerland
[6]Sensinov, France
[7]Thales Italia, Italy
[8]Continental, France
[9]DLR German Aerospace Center, Institute of Transportation Systems, Germany
[10]NEC Laboratories Europe, Germany
[11]CTAG, Spain
[12]TNO, The Netherlands
[13]VTT Technical Research Centre of Finland Ltd, Finland
[14]HUAWEI, Germany
[15]TU Eindhoven, The Netherlands
[16]Valeo, France
[17]VEDECOM, France

Abstract

The applications of the Internet of Things (IoT) technologies connect multiple devices directly and through the Internet. Autonomous vehicles utilise connectivity when updating their algorithms based on user data, interact with the infrastructure to get environmental information, communicate with other vehicles. They exchange information with pedestrians using mobile devices and wearables and provide information about the traffic attributes and data collected by the vehicle sensors. The connected and automated vehicles (CAV) require a significant quantity of collecting and processing data and through IoT applications and services the autonomous vehicles share information about the road, the present path, traffic, and how to navigate around different obstacles. This information can be shared between IoT connected vehicles and uploaded wirelessly to the cloud or/and edge system to be analysed and operated improving the levels of automation and the autonomous driving (AD) functions of each vehicle. This chapter gives an overview of the integration of IoT devices contributing to automated/autonomous driving, and the IoT infrastructure deployed and seamlessly integrated into the AUTOPILOT project use cases and pilot demonstrators, including the IoT platforms integration.

6.1 Introduction

The continuing advancement of intelligent connectivity can provide the responsiveness needed to make automated/autonomous vehicles a reality.

Automated/autonomous vehicles make the roads much safer as human errors can be reduced significantly. Technology makes automated/ autonomous vehicles possible to be deployed, and robust networks and powerful IoT solutions are essential parts to achieve this.

Intelligent connectivity enables new transformational capabilities in the mobility and transport sectors. The networks used for connecting IoT devices and vehicles must be ultra-reliable, as many critical tasks are executed remotely, and must rely on cost-effective edge infrastructure to enable low latency and scaling. Connectivity is, therefore, necessary for such services to work optimally. Intelligence enables the enhancement of user experiences through multi-access edge computing using, augmented reality (AR) and virtual reality (VR) technologies [15].

The automated/autonomous vehicles, IoT, and artificial intelligence (AI) connected systems are increasingly relying on information that is exchanged to perform and conduct their safety-critical operations. Keeping such systems (and the data within) trustworthy, secure, safe, and private for the required cases is a critical element for the acceptance and adoption of such autonomous systems.

IoT devices and technologies can support automated/autonomous driving functions in different ways and enhance these functions for different use cases. The combined autonomous vehicles and IoT ecosystems implemented in the AUTOPILOT use cases integrate the services provided by interoperable IoT platforms and IoT devices that provide additional information to the vehicles about the environment, surroundings and the dynamic events around the vehicles to enhance the automated/autonomous functions.

The AUTOPILOT project addresses automated driving progressed by IoT and is one of the IoT European Large-Scale Pilots Programme (LSPs) [1].

6.2 Automated Vehicles Connectivity Domains

A vehicle with automated features must have established reliable interactions with different domains that are interlinked through devices and systems. The whole ecosystem relies on the interaction among the onboard units (OBUs), roadside units (RSUs), and vulnerable road users (VRUs). Intelligent sensors and actuators in the vehicles, roads and traffic control units in the infrastructures collect various information to serve enhanced automated driving (AD).

6.2.1 Internet of Vehicles

The Internet of Vehicles (IoV) concept and the Vehicle-to-Environment (V2E) or Vehicle-to-Everything (V2X) connectivity applied for enhanced automated/autonomous transportation and mobility applications, requires ecosystems based on safety, security, privacy, reliability and trust to ensure mobility and convenience to consumer-centric transactions and services. The enhanced automated/autonomous vehicles and IoT applications cover several domains of interaction, connectivity, exchange of information and knowledge as illustrated in Figure 6.1 together with communications protocols may be used [1, 13]. Based on the ITS-G5 it can be implemented a process to secure infrastructure, by providing policies and services of strong authentication of both vehicle and infrastructures and by performing more and more Risk Assessments, mapping onto requirements of the ISA/IEC 62443 set of standards. The figure shows "all" the domains of interactions between the vehicle and the environment through communication and sensing capabilities.

Figure 6.1   Automated vehicle connectivity domains of interaction [1, 13].

- Vehicle-to-Infrastructure (V2I) communication is defined as the wireless exchange of information between vehicles and the roadside units of the infrastructure, such as traffic, road and weather condition alerts, traffic control, upcoming traffic lights information, or parking lot information.
- Vehicle-to-Network (V2N) communication is the wireless exchange of information between vehicles and cellular networks, used for value-added services such as traffic jam information and real-time routing or available charging stations for electric vehicles (EVs).
- Vehicle-to-Cloud/Edge (V2C) communication is defined as the wireless exchange of information between vehicles and the cloud or edge computing centres, for instance, used for tracking and usage-based insurance.
- Vehicle-to-Grid (V2G) communication is wired and/or wireless exchange of information between electric vehicles and the charging station/power grid for such as battery status and correct charging and energy storage and power grid load/peak balancing.
- Vehicle-to-Vehicle (V2V) communication is defined as the wireless exchange of information between vehicles about, for instance, speed and position of surrounding vehicles.
- Vehicle-to-Pedestrian (V2P) communication is the wireless exchange of information between vehicles and vulnerable road users (VRUs) for safety-related services.
- Vehicle-to-Home (V2H) communication is the wireless exchange of information between vehicles and a fixed or temporarily home, for instance, used for real-time routing.

- Vehicle-to-Device (V2D) communication is wired and/or wireless exchange of information between the vehicle and IoT devices either inside or outside the vehicle.
- Vehicle-to-Maintenance (V2M) communication is the wireless exchange of information between the vehicle and the vehicle condition responsible (automotive manufacturer or repair shop), including vehicle condition monitoring, predictive maintenance notification or alerts.
- Vehicle-to-Users (V2U) communication is the wired/wireless exchange of information between the vehicle and its current user, including situational details.
- Vehicle-to-Owner (V2O) communication is the wireless exchange of information between vehicles and its owner. Use cases may be vehicle rental, fleet management, freight tracking, etc.

The convergence of enhanced automated/autonomous vehicles, IoT and AI applications are accelerating the implementation of V2X concept and the move to mobility as a service (MaaS) and tier-one automotive companies, large technology companies and technology start-ups active involved in V2X, addressing first safety, security and privacy use cases to accelerate user acceptance and innovation. The overall interactions are covered under the name of V2X and consists of the following [1, 13]:

- The communication and sensing interactions between the autonomous vehicle and the dynamically changing environment.
- The communication and sensing interactions between the vehicle and its static environment.
- The communication and sensing interactions with different service providers.
- The communications with the owners, users, mobility service providers.

There are two key technologies considered for intelligent transportation systems (ITS), namely ITS-G5 and C-V2X, which are based on different design principles and radio interfaces [1, 13, 14]. However, the higher layers (above the PHY/MAC radio layers) can mainly share the same protocol stack. The two technologies are primarily intended for driver assistance warnings rather than autonomous driving but contribute to extending the line-of-sight limited operation of sensors such as cameras, RADARs and LiDARs.

ITS-G5 is defined by ETSI, and its radio air interface is based on IEEE 802.11p (DSRC in the US), which is an approved amendment of the Wi-Fi standard to add wireless access in vehicular environments (WAVE). ITS-G5 works independently of cellular networks, it supports V2V and V2I low latency short-range communication in the 5.9 GHz frequency band and uses orthogonal frequency-division multiplexing (OFDM) and a carrier sense multiple access (CSMA) based protocol in the MAC layer. ITS-G5 facilitates high reliability under high vehicle speed mobility conditions. Enhancements towards more advanced services such as autonomous driving are addressed by the IEEE 802.11 Next Generation V2X Study Group.

C-V2X is specified by 3GPP and is realised as LTE-V2X (3GPP rel. 14/15) for short- and long-range communication. The short-range mode can work independently of cellular networks, supports V2V, V2I and V2P communication, uses direct side-link communication over PC5 interface, uses orthogonal frequency-division multiplexing (OFDM) in the 5.9GHz frequency band, and its MAC layer is based on semi-persistent scheduling allowing deterministic sharing of the medium among multiple stations in a distributed manner. The long-range mode is cellular mobile network-dependent and supports V2N communication, i.e. up-/down-link communication between vehicles and base stations in a cellular LTE network over Uu interface. The next release 5G NR-V2X (5G New Radio V2X, rel. 16) addresses improvements such as lower latency, increased reliable communication, and higher data rates to support autonomous driving. 5G NR-V2X will complement LTE-V2X, i.e. not replace but co-exist with LTE-V2X.

LTE-V2X short-range mode and ITS-G5 are substitutes, but LTE-V2X has been shown in recent tests to have a superior performance in range/link-budget (reliability) [14]. However, ITS-G5 is not an equivalent replacement for LTE-V2X for providing C-ITS priority services. ITS-G5 provides different performance compared to LTE-V2X in direct side-link short-range communications and does not support long-range communications. The current, ITS-G5 cannot achieve the level of implicit compatibility between LTE-V2X and 5G-V2X, due to the different technological and design principles in the specifications of IEEE 802.11p (ITS-G5) and 3GPP C-V2X (LTE-V2X/5G-V2X). LTE-V2X is the natural precursor to 5G NR-V2X from the perspectives of the design and industrial ecosystem. The combination of these two V2X technologies can allow for the most cost-effective deployment of C-ITS services in EU [14].

A vehicle with automated features must establish interactions with different domains that are interlinked with one or more operational design domains like the use cases established in the AUTOPILOT project [1], namely the Automated valet parking, Highway pilot, Platooning, Urban driving, and vehicle sharing use cases. In this context for enhanced automated/autonomous vehicle applications, four connectivity domains are defined as essential connectivity building blocks of the IoT ecosystem in the AUTOPILOT project, namely V2V, V2P, V2D, and V2I. The IoT ecosystem relies on the interaction among the vehicles, the VRUs like pedestrians, a variety of devices and the infrastructure; to improve traffic management by increased efficiency, security and safety. Intelligent sensors and actuators in the vehicles, roads and traffic control infrastructures collect a variety of information to serve enhanced automated driving. These require robust sensors, actuators, and communication solutions, which can communicate with the control systems while considering the timing, safety and security constraints. Redundancy and parallel systems are required in all safety and security-critical applications. It is also worth mentioning that power saving mode, for example, for sensors and actuators, can be a barrier to real-time information. For battery-powered equipment, it will be a trade-off between power consumption and communication latency.

To mitigate the risks for the autonomous driving systems and infrastructures, security should be implemented at all layers and security policies should be defined following international standards and best practices. This can be achieved using existing frameworks like the ISA IEC 62443 [16].

6.2.2 Vehicle-to-Vehicle Domain

Vehicle-to-Vehicle (V2V) communication is defined as the wireless exchange of information (data) between vehicles. V2V communication facilitates the transfer of information and early warnings/control to ensure traffic safety, avoid traffic congestion, improve traffic flow and environment, etc. Interconnectivity between vehicles plays an important role in autonomous driving. High-speed environments and reliable real-time information are important issues for the V2V ad-hoc communication network, also referred to as VANETs (vehicular ad-hoc networks) or IVC (inter-vehicle communication) [11]. Examples of relevant standards are ETSI ITS-G5, IEEE 802.11p, IEEE 1609, and SAE J2735.

In the AUTOPILOT project at the French pilot site [1, 12], V2V communication is used in a platooning use case. This communication is made over ETSI ITS-G5 (IEEE 802.11 OCB) to allow the exchange of information between the vehicles forming the platoon. This information will be used to place the vehicles according to the others. The goal is to ensure that the platoon is not broken and that all vehicles are capable at any time to cross an intersection, for example. All the relevant information about the platoon is displayed to the driver on the embedded screen.

In the AUTOPILOT's Dutch pilot site, V2V communication is established via ITS-G5 connections and additionally a ultra-wide band (UWB) connection. An alternative approach for V2V information exchange is via IoT technology, using a MEC based local cloud service that forwards messages coming from one vehicle to another. In the AUTOPILOT's Spanish pilot site, V2V communication is performed by using the IoT in-vehicle platform system. To verify the impact of the fully IoT communication system, this V2V communication is achieved through the infrastructure and using standard oneM2M messages. These messages wrap the defined data models that allow IoT communication with the vehicle. Therefore, if a vehicle needs to send its information to other vehicles, that one will upload the corresponding message to the cloud, which later will be available for any other connected vehicle.

In AUTOPILOT's Italian pilot site, V2V communication is managed by an in-vehicle platform that implements an almost full ETSI stack. The main exchanged V2V messages are cooperative awareness messages (CAM). This information is used to provide a detailed look at the surroundings of the automated vehicle. CAM messages notify the information sensed by the in-vehicle sensors to the vehicles in the transmission range. This information is used by the automated driving data fusion algorithm to make decisions for both highway and urban driving use case scenarios. In the AUTOPILOT's Finnish pilot site, the vehicle is equipped with 4G/LTE communications. The vehicle ETSI ITS-G5 station transmits CAM messages regularly, and the vehicle position is also sent to the IoT platform, for use by other services.

6.2.3 Vehicle-to-Pedestrian Domain

Every year, many vulnerable road users (VRUs) are seriously injured or killed in accidents. Improved connectivity solutions can contribute to reducing the number of these fatalities thanks to an effective warning system for the involved actors. Vehicle-to-Pedestrian (V2P) connectivity is a field of research that studies the communications between vehicles and pedestrians. In the broadest sense, it typically considers bicyclists and motorcyclists, children in strollers, mobility-impaired people with wheelchairs, etc. However, these may also be classified as vehicles in V2V connectivity, if for example the communication unit is bicycle equipped and not used as wearables, mobile phones, etc. The goal of the V2P connectivity is to detect a pedestrian or more generally a VRU and notify information useful to avoid accidents.

The more intuitive device to warn pedestrians is through a smartphone or even a smartwatch, due to its increasing computational power, the availability of wireless connection and its widespread availability. Today's incumbent standard for vehicular communication is ETSI ITS-G5 that is based on the IEEE 802.11p amendment of Wi-Fi standard. Unfortunately, while Wi-Fi is supported by most smartphones, IEEE 802.11p is not implemented in commercial products yet [1, 12]. In 2014 Qualcomm and Honda, published a paper that describes a real implementation of this idea [2]. The prototype is made with a smartphone equipped with a Qualcomm Wi-Fi solution. As reported in the original paper: "The design goal was to provide an always-on, highly accurate and low latency pedestrian collision warning system, without introducing significant hardware or processing overhead to the smartphone". The paper states that good performances can be achieved, although some problems still need to be solved. Among the others, the accuracy of the position is given by the internal GNSS receiver of the smartphone, the congestion of the wireless medium and the certification of communications and application performances. Indeed, the certification procedure changes a lot depending on whether the application is considered as a supplemental alert or as a complete safety-critical warning system. A more recent prototype, created by Bosch, addresses motorcyclists is shown a video on YouTube [3]. An alternative approach is to exploit the cellular communication channel, owing to its complete availability on mobile devices. Waiting for the complete definition of LTE-V2X and 5G, several papers explored this idea showing good performances. In "Cellular-based vehicle to pedestrian (V2P) adaptive communication for collision avoidance" presented in 2014 [4], this approach is theoretically described, also taking into account the road-safety system in terms of energy consumption on the smartphone. In other papers, LTE communications are used, together with Wi-Fi, to exploit the advantages of both channels, i.e. the more extended communication range of LTE and the low delays of Wi-Fi direct communication. A further example is a study on the use of a pure Wi-Fi solution in "Vehicle to pedestrian communications for protection of vulnerable road users" presented in 2014 [5], which demonstrates the possibility of effectively using such a channel for safety purposes.

A different vision is to use different radio systems, with a dedicated transmitting device carried out by the VRUs. The system can be used to compute the distance and the position of the users without the need of a GNSS device and the related issues (e.g. accuracy in urban areas). One of the most important works in this sense is done in the Ko-TAG project (Security for vulnerable road users through Ko-Tag) [6], which uses an RFID-like approach.

In the AUTOPILOT project's Dutch pilot site [1, 12], a smartphone application is developed which connects to the Huawei OceanConnect IoT platform and the oneM2M platform. This smartphone application uses Global Positioning System (GPS) localisation to localise the VRUs in the area. This information is used to inform the vehicle of a possible VRU on the road where the vehicle is also driving. The vehicle must adapt its speed accordingly. Also, the other way around, the vehicle is sending its location to the smartphone using oneM2M, in order to warn the VRUs of an automated driving vehicle approaching. ITS-G5 beacons are used, in addition to the smartphone, to correlate the data transmitted from the smartphone with the location transmitted from the ITS-G5 beacons.

Finally, a Wi-Fi sniffer is used to detect surrounding Wi-Fi enabled devices, which can be used to detect crowdedness by detection of pedestrians and cyclists in the area, using their smartphone or other devices as trackers. While Wi-Fi detection applies to smartphones in the vicinity based on Wi-Fi sensing range (i.e. about 30 meters in outdoor scenarios), filtering mechanisms based on received signal strength indication (RSSI) levels may be used to detect only pedestrians closer to the vehicle. Due to the relatively low position accuracy utilising this technology, the output (number of devices detected and location of detection, logged by GPS) will only be used to map a crowdedness mapping of the area and not to individually position VRUs with smartphones. This information will then be used to inform other automated driving vehicles on how many VRUs are on a certain road, so they can adequately decide to take the less crowded routes.

6.2.4 Vehicle-to-Device Domain

In the AUTOPILOT project's French pilot site, the car/vehicle-sharing use case relies on a mobile application that allows the user to unlock and start one's vehicle. The system consists of an onboard unit that communicates with the mobile app via Bluetooth Low Energy (BLE). An embedded interface is also developed to display information about the vehicle to the driver, via an Android tablet that communicates with the vehicle through a serial and an Ethernet link. The serial link is used to communicate with low- level network in the vehicle and display failures, etc. The Ethernet one is used to communicate with the automated driving units and, for example, to guide the driver during the switch between manual and autonomous driving and autonomous and manual driving. During the car/vehicle-sharing use case, this interface displays its position on a map to the user. This position comes from the vehicle GPS through the Ethernet link. It also displays an alert when the vehicle enters a zone where the autonomous driving is allowed and when the vehicle is approaching the end of this zone. In the vehicle rebalancing use case, this interface is used to display the state of each vehicle in convoy.

In the AUTOPILOT's Dutch pilot site, V2D connectivity is used in platooning and automated valet parking use cases. In the platooning use case, the drivers are notified in their vehicle by a platoon manager service about the platoon status and related information. It uses an existing screen on the dashboard, which was modified for this purpose. Additionally, the lead driver of the platoon is informed about speed and lane advice via an Android on a dedicated smartphone. In the automated valet parking (AVP) use case, the vehicle communicates with an Android smartphone via an IoT platform using a 5G/LTE communication network. The AVP application running on the smartphone receives information such as vehicle state (e.g. current vehicle position, current AVP action and phase) and can send commands like "park" or "collect" to the vehicle. The data to be exchanged over the IoT platform has been defined in detail by the DMAG (data modelling activity group).

In AUTOPILOT's Italian pilot site, the vehicular IoT bridge should enable bidirectional semantic-full communications between vehicles and application entities both in-vehicle and in roadside infrastructure nodes. The bridge contains a processing unit able to manage all communication interfaces; an IEEE 802.15.4 wireless interface, an OBD/CAN interface, the IEEE 802.11p (ETSI ITS-G5) transceiver and the 4G modem for cellular communication. At the network layer, the bridge should be able to address 6LoWPAN destinations (to talk with the onboard WSN) and to address other C-ITS station using the GeoNetworking protocol. From outside, the bridge should be addressable as an IP node. The bridge should, at least, handle at the transport layer UDP (User Datagram Protocol) and BTP (Basic Transport Protocol) [2] and, at the application layer, CoAP communications. It also requires a bridge to abstract all the onboard generated data. This involves the abstraction of all the data that are shared on the OBD/CAN network. Therefore, it should be able to read the main messages in accordance with OBDII standards and to aggregate them with the information coming from a wireless sensor network. In the Italian pilot site use cases, the IoT bridge is an OBU developed so that it can manage several different devices (V2D): it will manage the connection to a tablet that will be used as HMI and as a sensor for vibration data. The OBU will also interface with an inertial measurement unit (IMU) via CAN and with a 6LoWPAN dedicated vibration sensor.

Moreover, to segregate the OBD/CAN from the V2X and sensors network traffic a secure gateway is implemented onboard preventing that undesired communication can happen between non-safety critical onboard zones and safety-critical onboard zones.

6.2.5 Vehicle-to-Infrastructure Domain

Within the platooning use case in the city centre of Versailles at the AUTOPILOT's French pilot site, the platoon must pass through two complicated crossroads. To do so, it is necessary to have V2I communication. When the platoon is approaching, the complicated intersection, the RSU detects the lead vehicle and passes the message on to the traffic light controller for it to change its phase in order to give priority to the platoon. The traffic lights interrupt their usual phase and switch specific traffic lights to the correct green/red combination so that the platoon can cross safely. Once the RSU has communicated with the cloud through the oneM2M server, the OBU is informed on whether the platoon can continue following its route. Once the platoon has gone past the junction, it goes back into its classic functioning mode. The traffic assist architecture for platooning is illustrated in Figure 6.2.

In the AUTOPILOT's Dutch pilot site, V2I connectivity is used in the highway pilot, platooning and automated valet parking use cases. Concerning

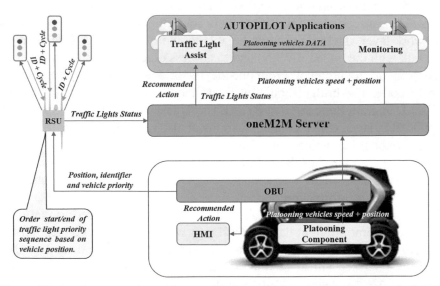

Figure 6.2 Traffic light assist architecture for platooning in complicated crossroads, (example from AUTOPILOT French pilot site) [1, 12].

the highway pilot use case, all exchanges to and from the vehicles go through the infrastructure. There are four major components of the system: detection (of anomalies by leading ego vehicles), reporting (of anomalies to the cloud), validation (or learning of hazards presence) and information (for the control of following vehicles). However, only the reporting and information components rely on V2I communication:

- For reporting, the vehicle communicates with the cloud with MQTT and HTTP over a 4G connection.
- For information, the vehicle communicates with the map provider with Web Socket and HTTP over a 4G connection.

The platooning use case uses V2I communication in the following different ways:

- Broadcasting CAM messages via ITS-G5 that are intercepted by the instrumented facility along the highway to support vehicle detection.
- Exchanging platoon status information with the cloud-based platoon manager service that involves IoT (oneM2M) and cellular (commercial 4G) technology.
- Publishing data to and retrieving data from an IoT-enabled (oneM2M) local dynamic map service deployed at the roadside. This concerns data that can be used to increase the environmental perception of IoT connected vehicles (platoon vehicles and other vehicles). The communication channel is realised through the Hi-5 pre-5G network, which provides coverage over a part of the road through a base station.
- Status information of four traffic light controllers controlled by RSUs; one on each successive junction on the road are received by the MQTT clients. The data in binary format is converted to JSON format with the ASN.1 decoder and published to the respective containers on the oneM2M platform. The binary data is also published to the oneM2M MQTT broker. All services and vehicles subscribed to this service can pick up this data.

The automated valet parking (AVP) use case:

- Parking spot occupancy and obstacle detection: The AVP use case features a stationary roadside camera and the micro aerial vehicle (MAV) as infrastructure devices. The MAV and the camera detect free parking spots and obstacles and send this information to the vehicle via the AVP parking management service (PMS) application. The vehicle communicates with the infrastructure devices using the IoT platform over the cellular network connectivity (e.g. 5G/LTE).

- The MAV detects the free parking spots and obstacle, processes the data and publishes the parking spot and obstacle status information to the IBM Watson IoT platform. The parking management service application, as an IoT application, registered by the Watson IoT platform, receives this data over MQTT and publishes it to the AVP vehicle.
- Roadside stationary camera: The traffic manager application is providing parking spot status and obstacle status update information, and deep learning algorithms send out parking spot status and obstacle status (along the access road to the parking lot). These algorithms are running in the servers and use an advanced message queuing protocol to communicate with the parking spot entity, which then publishes them to the containers' resources in the oneM2M platform. The data are updated to a Watson-specific format and published. The vehicle subscribed to this information gets these updates from the oneM2M platform. The data in the Watson-specific format is subscribed by an interworking proxy, which then forwards it to the IBM Watson platform. The vehicle or the parking management service application receives these data from Watson IoT platform over MQTT. For evaluation purposes, the parking spot entity also forwards the data directly to the IBM Watson platform. The data flow diagram is shown in Figure 6.3. The AVP data models follow the SENSORIS and DATEX data models, which are currently being standardised (for AUTOPILOT community) in the data modelling activity group (DMAG).

In the AUTOPILOT's Italian pilot site, V2I connectivity is managed using two different channels. The first one is the classic DSRC, used to exchange decentralised environmental notification messages (DENMs) with the RSUs and SPaT/MAP messages with the traffic lights. The second channel is LTE, mainly used to send information to the oneM2M platform. V2I connectivity is used in the highway, urban and highway driving use cases, and the DENMs are used to notify alert sensed by the IoT devices. More in details:

- Highway driving use case – For puddle detection, some dedicated 6LoWPAN sensors will detect a puddle. The sensors are connected to an RSU that sends the information directly to the surrounding vehicles using a DENM message. The same information is sent by the RSU to the oneM2M platform (via the cellular network) and then, through the cloud, it is validated from the Traffic Control Centre and sent back to the relevant RSUs and to the approaching vehicles, using both the LTE and the ETSI ITS-G5 channels. The information is sent in the

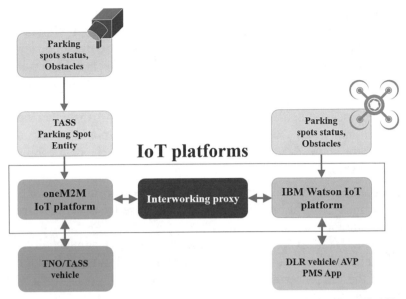

Figure 6.3   Interaction between the AVP devices and the IoT platforms [1, 12].

form of a different speed limit for the portion of highway affected by the puddles. In this way, both short and long-range communication are covered. As a further option, also NB-IoT water sensors are used: in this case, the information about the water level is transmitted straight to the oneM2M platform via a cellular network and then consumed from the applications that generate the alert messages. As for road works' warning, a notification is sent from the Traffic Control Centre to the oneM2M platform and then consumed by both the RSU, to notify the vehicles via ETSI ITS-G5, and the e-Horizon cloud application that dynamically updates the maps onboard the vehicles. The road works can be fixed or mobile.

- Urban driving use case – For pedestrian red-light violation, the pedestrians are detected thanks to a smart camera. The information is combined with the status of the pedestrian lights, and in case of violation, a message is sent using a DENM notification. The message is also sent to the oneM2M platform. For fallen bicycle detection, a bicycle is equipped with an IoT in-vehicle platform. This device will be equipped with sensors that permit to detect when the bicycle has fallen. This information is automatically sent via DENM by the bicycle. If the message is received by an RSU, this will be sent to the oneM2M platform.

- Highway and urban driving use cases – For pothole detection, the in-vehicle platform will act as a "virtual sensor" for vibration. The information can be taken by a 6LoWPAN sensor, a smartphone/tablet or an inertial measurement unit (IMU). The "virtual sensor" can work with only one source of information or combines different sources. When a pothole is detected, a message is sent to the oneM2M platform where it becomes available for subsequent usage by all the other vehicles. Additionally, the OBU reports to the oneM2M platform also an idea about the status of the road surface depending on the data coming from the sensors.

V2I communications are used to report relevant information coming from IoT to the oneM2M platform. These data are then used to give useful feedback to the autonomous driving function.

In the AUTOPILOT's Spanish pilot site, the V2I connectivity is supported with both cellular network connectivity and Wi-Fi. Through these channels, the bidirectional IoT communication will be performed, sending and receiving messages following the oneM2M standard in the urban driving and automated valet parking use cases.

Urban driving use case:

- Traffic Lights: In the pilot site, the different involved traffic lights will be connected to RSUs. These RSUs are monitoring the status of the traffic lights and publishing it to the IoT cloud platform (IBM Watson). These statuses are obtained by the in-vehicle IoT platform through an urban server, which will be providing and filtering this information to any connected vehicle.
- Vulnerable Road Users (VSUs): In order to detect VRUs, a smart camera is used, located in the surroundings of the road. This camera detects any pedestrian located in a relevant area and sends a VRU event message to the IoT cloud platform (IBM Watson). Afterwards, this information is collected by the mentioned urban server, which will provide and filter it to any connected vehicle.
- Hazards: In order to obtain the different hazard events that might occur, the control management system of the public authorities is used. By using a module that obtains the different hazard events and translates them to IoT messages, publishing them to the Watson IoT platform, these hazards are available to any vehicle connected to the same urban server.

Automated valet parking (AVP) use case:

- Drop-off and pick-up: A parking management system is developed. This parking management system can forward the user's command of pick-up and drop-off to the vehicle. Also, this system can detect VRUs that afterwards would be published to the IoT platform in the same way that it is done for the urban driving use case. The in-vehicle platform can then receive these commands and VRU events adapting its behaviour.

V2I communications are used to report relevant information coming from the IoT platform. This data is then used to give useful feedback to the autonomous driving functions.

In the AUTOPILOT's Finnish pilot site, V2I connectivity is supported with cellular 4G/LTE communications. In urban driving and automated valet parking use cases, communication is as follows:

Urban driving use case:

- Traffic Lights: Real-time information on signal state and the next phase is available both through cellular communications, through connection to the traffic light operator's server.
- Vulnerable road users (VRUs): In order to detect VRUs, a smart camera is used, which is installed at a mobile RSU. This camera will detect pedestrians and cyclists located in a relevant area and send a VRU event message to the IoT cloud platform. From there, the information will be made available to the vehicle.

Automated valet parking (AVP) use case:

- Traffic cameras, installed at the mobile RSU, monitor the parking area and detect objects and pedestrians either at the parking spaces or on the potential vehicle paths, and send the information to the IoT platform.

6.3 Automated Driving Use-Cases and Applications

In the AUTOPILOT project, five different use cases are developed and implemented in two or more of the pilot sites established in the project. An overview of the pilot sites and their respective use cases (denoted "+") are given in Table 6.1 [1, 12].

This includes connectivity with IoT devices, connectivity between vehicles, infrastructure and other sensors to enhance automated driving capabilities and technology that allows vehicles to monitor the state and availability of different services. As for the in-vehicle functions, three main groups of sensor systems such as camera-, RADAR- and LiDAR-based systems, together

Table 6.1 Pilot sites and use cases in the AUTOPILOT project [1, 12]

Pilot Sites vs. Use Cases	France (Versailles)	Italy (Livorno-Florence)	The Netherlands (Brainport)	Spain (Vigo)	Finland (Tampere)
Automated Valet Parking (AVP)	−	−	+	+	+
Highway Pilot	−	+	+	−	−
Platooning	+	−	+	−	−
Urban Driving	+	+	+	+	+
Car/vehicle Sharing	+	−	+	−	−

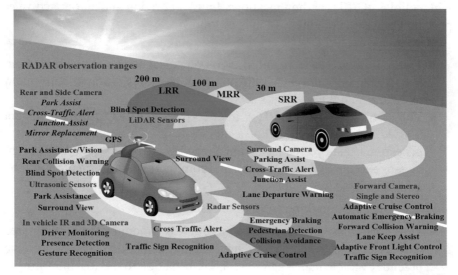

Figure 6.4 RADAR observation ranges, sensors, actuators and functions [7].

with ultrasonic sensors, are used for autonomous driving, as illustrated in Figure 6.4.

6.3.1 Automated Valet Parking

The aim of the automated valet parking (AVP) use case is to demonstrate how this functionality can benefit from different information sources, other than the onboard sensors, accessed via the principle of the IoT like parking cameras. Through the use of IoT, the IoT platform can monitor and/or

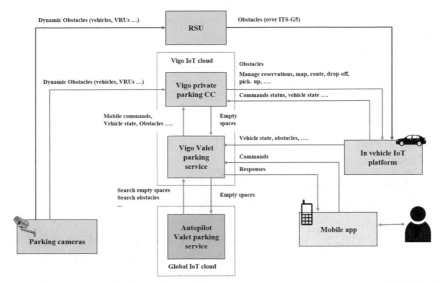

Figure 6.5 Automated valet parking (AVP) execution view, (example from the AUTOPILOT Spanish pilot site) [1, 12].

coordinate traffic on the parking lot and do efficient route planning based on real-time available traffic information. Hence, the IoT platform will exchange information about the dynamic and static obstacles in the parking lot and/or the route to be followed by the vehicle using the information provided by the parking cameras. The AVP use case has two main scenarios:

- Autonomous parking of the vehicle (drop-off scenario), after the driver, has left the vehicle at the drop-off point, that can be located near the entrance of a parking area.
- Autonomous collection of the vehicle (pick-up scenario), when the driver wants to leave the site, he/she will request the vehicle to return itself to the collection point, using, for example, a smartphone application.

In Figure 6.5, which is an execution view from the Spanish pilot site, the IoT devices and the functions that will be supported using the IoT platform are described. The following list is a detailed proposal of devices and functions to support the valet parking use case:

Private parking control centre:

- Informs when a parking spot is free or not.
- Manages reservations.

- Validates vehicle access.
- Manages maps and vehicle routes.

User's mobile device:

- Requests parking slots.
- Manages pick-up and drop-off events.

Smart cameras:

- Publish detected events (e.g. pedestrians or other objects on the parking place).

Connected automated driving (CAD) vehicle:

- Validates that the access to the vehicle is provided to the authorised driver.
- Informs when the vehicle is ready to move unmanned to the destination (parking place or collect point), e.g. when the driver has moved out of the proximity of the vehicle or has locked the doors.
- Manages pick-up and drop-off events and Navigation to the destination, following a route either determined by the IoT platform, while avoiding obstacles detected by either the vehicle sensors or the IoT platform.
- Informs when the vehicle goes into a low power consumption mode.
- Informs when an obstacle is detected.
- Informs about vehicle sensors values and position.

Interior parking areas are very controlled scenarios where the main challenges are the corners without visibility. IoT parking cameras can provide information on these blind spots, allowing the AD function to increase the SAE automation level from 3 to 4 [1, 12].

At the Dutch pilot site, the AVP use case story starts with the vehicle being manually driven to the drop-off point. After arriving there, the user activates the AVP function (e.g. by in-vehicle interface or smartphone app) and exits the vehicle. Services on the IoT platform determine an obstacle-free route to an available parking position based on information from IoT devices. The vehicle autonomously drives to the dedicated parking position. IoT devices involved in the use case are:

- Permanently installed cameras on the parking area that can detect free parking spots and obstacles.
- A micro aerial vehicle (MAV) that can provide information about free parking spots and obstacles, for areas the cameras do not cover.
- IoT-enabled vehicles with their own sensors.

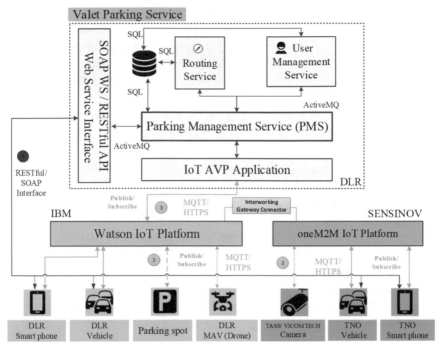

Figure 6.6 Automated valet parking (AVP) use case architecture, (example from the AUTOPILOT Dutch pilot site) [1, 12].

The primary goal of the IoT usage is, therefore, to gain an improved environment model that can possibly increase the efficiency and safety of the use case. Figure 6.6 depicts an overview of the IoT architecture of the AVP use case as deployed in the Dutch pilot site. Two IoT platforms from Watson IBM and oneM2M are used by the AVP, and the interoperability between the two platforms is realised through the bidirectional interworking connector that has been implemented for this purpose.

6.3.2 Highway Pilot Use Case

For the detection component of the Dutch pilot site system, three sensors in the vehicle are relied upon: a LiDAR, a front camera and an inertial measurement unit (IMU). An extra camera supports the use case for lane detection but is not directly involved in hazard detection. The LiDAR data is processed by a specifically developed algorithm, focusing on speed bump detection. The front camera data is processed by a specifically developed algorithm, focusing on potholes. The IMU data is processed by a specifically developed algorithm, capable of detecting anomalies without specific classification.

For the information component of the system, one actuator relies upon the active cruise control (ACC) unit. Moreover, turning lights are controlled to support lane changes scenario. The way all these are interconnected is illustrated in Figure 6.7.

It is worth noting that the raw data from sensors are indeed passed directly to the runtime environment where the real-time detection algorithms runs. However, everything else is coordinated through an in-vehicle IoT platform (here an MQTT Broker) that ensures the coordination between the results from all other software modules. In addition to the IoT devices within vehicles, the use case also takes advantage of a roadside camera that monitors the road for anomalies too (e.g. static objects like fallen cargo). The detection from this camera is passed through onto the oneM2M IoT platform.

The use case carried out in the Italian pilot site involves vehicles with IoT enhanced automated driving (AD) functions, driving on a "smart" highway. The test vehicles are equipped with an onboard IoT open vehicular platform enabling IoT triggered AD functions, like speed adaptation, lane change, and lane-keeping. Some vehicles also have special sensors, such as an IoT-based pothole detector.

The "smart" highway is a highway where a pervasive IoT ICT system is deployed based on a network of roadside sensors or other sources capable of collecting information and making it available to cloud-based applications. In the use cases, connected vehicles and the traffic control centre (TCC) also have an important role. For safety reasons, the connected vehicles precede and follow the AD vehicle driving in convoy.

The goal is to show how the combined use of IoT and C-ITS can mitigate the risk of accident for an AD vehicle when at a certain point, the road becomes dangerous because of the two kinds of hazardous events: Wet road (puddle) and road works. In the following, the functions of the different IoT devices are described.

6.3.2.1 Hazard on the roadway (puddle)

Puddle IoT sensors:

- In the Italian pilot site, two kinds of such sensors are deployed, using different communication technologies: 6LowPAN and NB-IoT. They continuously monitor the highway in critical locations sending two kinds of signals; a low-frequency heartbeat and a high-frequency alert triggered by the rising of the water level.

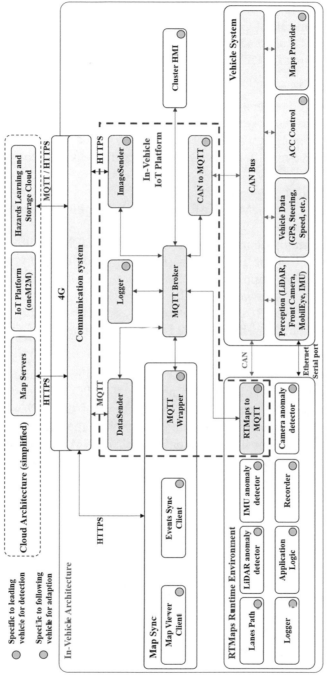

Figure 6.7 SW and HW IoT architecture, (example from the AUTOPILOT Dutch pilot site) [1, 12].

- The 6loWPAN puddle sensors send the messages to the roadside ITS-Station by means of CoAP.
- The NB-IoT puddle sensor sends the message straight to the oneM2M platform using the LTE cellular network and CoAP protocol, as well.

Roadside ITS station:

- Roadside ITS station is a programmable gateway with multi-access technologies (notably 6LowPAN, ETSI ITS G5, LTE, and Ethernet). It is an RSU, compliant with ISO/TC204 WG16 standards, able to exchange information over different networks, using different protocols, including the IoT ones.
- The RSU always listens to the 6loWPAN sensors and sends the measurement to the oneM2M IoT platform of the pilot site with a certain frequency.
- When a hazard occurs, the RSU broadcasts a DENM with the lowest quality level of the information (i.e. not yet validated by the TCC), toward both the approaching vehicles via the ITS-G5 network and the oneM2M platform via LTE cellular network.
- Furthermore, the RSU publishes on the oneM2M platform the CAMs collected from the vehicles in the ITS-G5 communication range.

The traffic control centre (TCC):

- The TCC implements a DATEX II node that is allowed to supply information from the whole highway network. The TCC is also responsible for managing ITS on the oneM2M platform of the Italian pilot site. Two kinds of services are provided leveraging the subscription to the oneM2M platform: hazard validation and DENM forwarding. It also publishes to the oneM2M platform the relevant traffic information from the DATEX II node, to be consumed by the highway infotainment service (FI-PI-LI App).
- When a hazard like flooding on the road occurs, the TCC is notified by the subscription to the oneM2M platform. After assessing the severity of the danger, it validates the hazard and broadcasts a DENM with the highest quality level of the information (i.e. validated by the TCC) to the RSUs along the highway, using the cabled LAN.
- The TCC subscribes to the CAMs of the vehicles published by the RSUs on the oneM2M platform. The information is combined with the Bluetooth and Wi-Fi transit data loggers to perform the travel time analysis and live overview on the TCC video wall.

- The TCC subscribes to the AD vehicle's sensor data on the oneM2M platform in order to provide ITS services to the users of the highway.

The automated driving (AD) vehicle:

- The AD vehicle broadcasts CAMs over the IEEE 802.11 OCB (ETSI ITS-G5) network; at the same time, the AD vehicle publishes data from its sensors to the oneM2M platform.
- The AD vehicle is approaching the hazard on the road; the in-vehicle application (Connected e-Horizon (CeH)) subscribes to the alert from the oneM2M platform.
- The in-vehicle IoT platform combines the information obtained by the CeH with that obtained by DENM via the IEEE 802.11 OCB (ETSI ITS-G5) network and then feeds the appropriate autonomous functions that perform either the necessary adaptation of the driving style in a smooth way, if sufficiently in advance.
- In a case when a vehicle is close to the hazard and for some reason (i.e. the warning from the IoT services was not received, or the warning was received just by the safety channels of ITS-G5 (DSRC), etc.), an emergency braking is needed, and this event is registered by the in-vehicle application and sent to the oneM2M IoT platform of the pilot site.
- At the same time, the cloud monitors the performance of the vehicle, checks that the in-vehicle application feeds the appropriate autonomous functions, and sends a notification to the in-vehicle HMI.

Connected vehicles:

- The connected vehicles lead and follow the AD vehicle; they continuously broadcast CAMs over the ETSI ITS-G5 (IEEE 802.11 OCB) network; at the same time, they publish its sensor's data the oneM2M platform.
- The connected vehicles are approaching the hazard on the road; the in-vehicle IoT platform receives the information from both the RSU along the track and the oneM2M IoT platform.
- The in-vehicle application pre-alerts the driver about the hazard using the information obtained by the oneM2M IoT platform of the pilot site and by the DENM.

An overview of the demonstration storyboard is shown in Figure 6.8: IoT sensors placed along with to the highway monitor continuously the presence of puddles and if a warning condition has been detected, send an alert to the

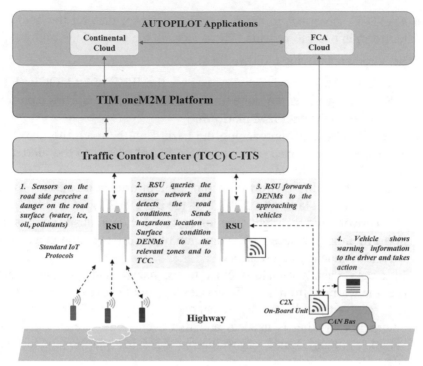

Figure 6.8 Hazard on the roadway (puddle) execution view, (example from the AUTOPILOT Italian pilot site) [1, 12].

RSU that broadcasts this information to vehicles (DENM) and to the TCC. It validates the alert, forwards the DENM message to farther away RSUs and feeds the IoT oneM2M cloud platform with alert related data.

The information on the presence of puddles generates a temporary update of the speed limit in the interested area, which is transmitted from the cloud to the CeH installed inside the prototypes. The in-vehicle application feeds the appropriate autonomous functions that perform a smooth speed adaptation (IoT-enabled speed adaptation for AD vehicle) in combination with information obtained from DENM. In consequence, IoT technology assists the rising of the SAE automation level from 3 to 4 [1, 12].

6.3.2.2 Roadworks warning by traffic control centre (TCC)

A roadworks event is planned by traffic/road operators, and a temporary speed limit is associated with the event. Two IoT-assisted AD manoeuvres are expected:

- The AD vehicle has to reduce its speed approaching the roadworks area, travel at the temporary speed limitation and increase the speed again at the end of the roadwork area.
- The AD vehicle has to stay on the current lane without any human steering action. Moreover, in the presence of a lane closed due to roadworks, it has to perform a lane change and avoid the obstacle.

An overview of the demonstration storyboard is shown in Figure 6.9:

- A sensor node is attached to the road works trailer and announces the presence of roadway works to an RSU.
- Then the RSU triggers DENM messages, broadcasting information about available lanes, speed limits, geometry, alternative routes, etc.
- The TCC broadcasts the DENM messages to farther away RSUs. At the same time, the TCC feeds the oneM2M platform with roadworks related data.

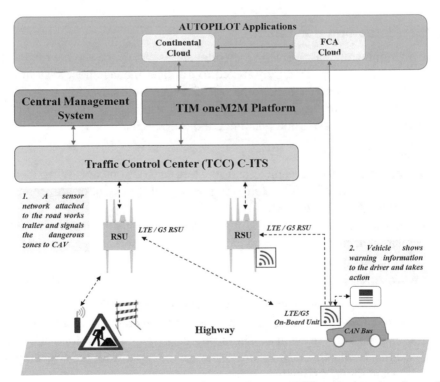

Figure 6.9 Roadworks warning by traffic control centre (TCC) execution view, (example from the AUTOPILOT Italian pilot site) [1, 12].

- Then the information is consumed by the CeH application and transmitted to the cloud as a modified dynamic speed limit that considers the generated dynamic event.
- The cloud immediately notifies to the prototype vehicles the updated information for the onboard CeH device. Thus the in-vehicle application feeds the appropriate autonomous functions that perform the necessary adaptation of the driving style in combination with information obtained from DENM. A notification/warning can be generated through the in-vehicle HMI.

Expected benefits; IoT can provide to the AD vehicle information in advance on the presence of obstacles, roadworks or other vehicles in the rear blind spot. With that information, the in-vehicle application can instantiate both smooth IoT-enabled speed adaptation and lane-change manoeuvres. In such a way, IoT technology assists the rising of the automation level from 3 to 4 [1, 12].

6.3.3 Platooning Use Case

The platooning use case of the French pilot site is part of the vehicle rebalancing business case and is closely linked to the fleet management system that indicates which vehicles have to be transferred from one station to another. The added value of the IoT in the platooning use case is illustrated in the following aspects of mission planning, and traffic light assist:

Mission planning:

- Choose the leading vehicle and its start/end stations according to data collected via IoT objects (e.g. the position of the operator, the charging level of the vehicle, etc.).
- Choose the follower vehicles, the start/end station and their order in the platoon according to data collected via IoT sensors in each vehicle and in the parking spots.

Traffic light assist:

- Suggest a reference speed to the operator in order to minimise the waiting time (red light) at each intersection that counts with a traffic light along the entire itinerary. (The traffic assist architecture for platooning is already illustrated in Figure 6.2 as an example from the French pilot site and the V2I domain).

The main scope in the Dutch pilot site is to show how increased flexibility in platoon navigation and manoeuvring capabilities can be realised, and how it can benefit from the use of IoT technology. For instance, platoon forming is done under the control of a platoon manager service that calculates the estimated time of arrival and rendezvous point of platoon vehicles based on the actual positions and speeds of those vehicles.. An additional function of the service is to guide the platoon after successful formation. Guidance involves speed and lane advice to the lead vehicle, based on the traffic situation on the road ahead. For example, the platoon service receives regulatory information from the road operator (max speed and lane access/or closure) and takes data from the IoT platform (oneM2M) concerning vehicle traffic conditions and traffic light status data. In order to minimise the probability of platoon break-up, the platoon service provides specific speed advice. After (an unlikely) break-up of the platoon, the service will support reformation of the platoon. The platooning use case utilises various communication channels (V2V and V2I). V2V concerns operational driving of the Cooperative Adaptive Cruise Control (CACC) while the bidirectional V2I channels are mainly used for exchange of data related to tactical driving (such as lane or speed choice). Relevant IoT data are the road operator originated info, the actual traffic state data (from road-side surveillance cameras), platoon state data and traffic light data. Logging takes place on the vehicle (vehicle state and control) and on the IoT platform.

The execution view of the systems and processes involved during the platoon formation stage gives some insight into the system architecture implemented for platooning. The intended procedures in Figure 6.10 are:

- Traveller steps into the vehicle and starts the vehicle-sharing application.
- Traveller defines whether he/she wants to be leading or following in platoon.
- Traveller defines the destination.
- Vehicle sharing application already knows about existing platoons and can match.
- Vehicle sharing app gives route to the Watson IoT platform, which sends it to the oneM2M IoT platform.
- Traveller presses the vehicle GUI to put the vehicle in platoon formation mode.
- Platoon service receives a message from the vehicle that it wants to platoon.
- Platoon service application receives a message from the vehicle-sharing app that matches has been made.

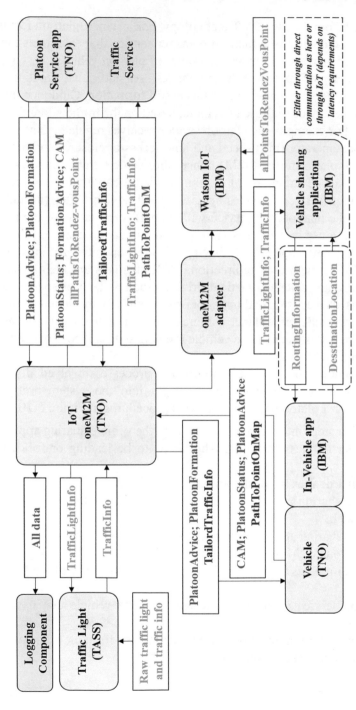

Figure 6.10 Platooning use case execution view of platoon formation (example from the Dutch pilot site) [1, 12].

- Platoon service application gives route(s) to the planner; fill platoon formation message with info from the planner and send to vehicles.
- Vehicle receives a platoon formation message containing platoon ID and planner information.

The platoon service listens to the cloud-based Traffic Manager application, which delivers regulatory road information. The traffic operator can update the traffic management info such as speed limits, lane status, etc., using the GUI and publishes this information to a respective container in oneM2M. The operator can also publish road map information (usually static) wherever there is any change to the otherwise static map. The platooning vehicles subscribed to these containers in oneM2M get these updates and adapt their driving accordingly.

6.3.4 Urban Driving Use Case

Urban driving assisted by IoT has the main objective to support connected, and automated driving (CAD) functions through the extension of the CeH of an automated vehicle. The vehicle can process data from external sources that enrich those provided by its own sensors (Camera, LiDAR, RADAR, etc.). In Figure 6.11, which is an execution view from the Spanish pilot site, the IoT devices and the functions that will be supported using the IoT platform are

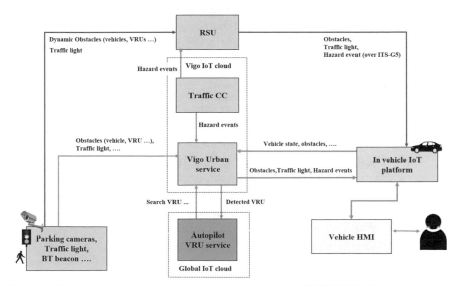

Figure 6.11 Urban driving execution view, (example from AUTOPILOT Spanish pilot site) [1, 12].

described. The following list is a detailed proposal of devices and functions to support the urban driving use case:

- Traffic control centre: Informs when there is a hazard on the road (e.g. accident, traffic jam, road work warning).
- Traffic light: Informs about the traffic light status and time to change.
- Smart cameras: Publish detected events (e.g. pedestrians or other objects on the intersection).
- Connected AD vehicle: Informs when an obstacle is detected, and about vehicle sensors values and position.

Considering all the information provided by IoT devices, the CAD systems will adapt their behaviour accordingly. The complexity of urban scenarios makes it essential to have as much redundancy information as possible. IoT platform provides data about the traffic lights and road events through 3G/4G. Furthermore, the frequency with which the IoT platform sends the data is higher than other advanced V2X communication. For the case of VRUs, the information received by IoT complements the data from the AD sensors, so it provides more reliable and accurate results. There are other objects that could not be detected if there were no IoT services (IoT camera or sensor information from other vehicles). As a result, IoT technology allows for increasing the SAE automation level from 3 to 4 [1, 12].

The urban driving use cases concern IoT-assisted speed adaptation in the common urban scenario, considering traffic light, presence of bicycles, pedestrians and other vehicles. An Italian pilot site overview of the execution is shown in Figure 6.12, including pedestrian detection with a camera, connected bicycles, and potholes detection:

Pedestrian detection with a camera:

- An AD vehicle is approaching an intersection regulated by a "smart" traffic light.
- A smart camera detects a pedestrian or an obstacle on the lane. The information is processed locally and notified to the RSU via IoT protocols. The stereo-camera used for the implementation can even provide information about the position of the VRU or obstacle with a good confidence degree. Moreover, a connected traffic light sends to the RSU via SPaT/MAP messages information about the time to green/red.
- The RSU receives the information transmitted by both devices (smart camera and traffic light), fuses the data and sends it by DENM messages to all the interested actors on the roads.

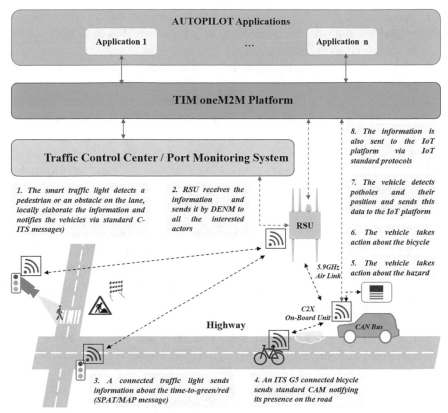

Figure 6.12 Urban driving execution view, (examples from AUTOPILOT Italian pilot site) [1, 12].

- The OBU of the AD vehicle receives the information and smoothly adapts the speed to the situation. The detection of VRUs and the traffic light status is also displayed on the HMI of the vehicle.
- The information is also sent to the oneM2M platform and can be retrieved by other vehicles in the same area via cloud applications.
- At the same time, the area monitoring centre consumes the information from the oneM2M platform and displays a new advisory speed limit for the interested area to avoid possible problems.

Connected bicycle:

- An AD vehicle is moving in an urban scenario with other road users, including a connected bicycle.

- The connected bicycle is equipped with battery-powered communication modules and dropout sensors: currently, it sends CAM messages to other vehicles and to the infrastructure.
- At a certain point, the bicyclist falls while the AD vehicle is approaching and a DENM is triggered.
- The AD vehicle, informed by IoT of the dangerous situation, smoothly decreases its speed and stops before reaching the accident area.
- The information is also sent to the oneM2M platform and can be retrieved by other vehicles in the same area via cloud applications.
- At the same time, the area monitoring centre consumes the information from the oneM2M platform and displays a new advisory speed limit for the interested area to avoid possible problems.

Potholes detection:

- A wireless vibrations sensor installed on the vehicle collect the data of the raw signal accelerations on the three axes and notifies to the OBU, via 6LowPAN or MQTT protocol, the occurrence of a vibrational shock above a certain level (threshold), due to a pothole presence on the road.
- The OBU combines this information with other data coming from the CAN bus (speed, odometer, etc.) and GPS and sends this data to the oneM2M IoT platform, by using MQTT and/or HTTP as application protocols.
- An upcoming AD vehicle consumes the information and can arrange its speed accordingly.

In such complex scenario, the IoT inputs to AD functions are many: IoT information about the traffic light phase and remaining time can be used from AD vehicles to adapt their speed in order to cross the intersection with green traffic light; and if not possible, to safety stop at the traffic light or queue behind other vehicles. Moreover, a smart camera on the test site can provide information on pedestrian traffic light violation. AD vehicles can use this information to stop at the traffic light even if the traffic light on its side is green. IoT enabled speed adaptation for AD vehicle is also related to the bicycle presence and if a fallen bicycle is detected, and to the road conditions. What is more, in this scenario, the IoT technology enhances the rising of the automation level from 3 to 4 [1, 12].

6.3.5 Car/Vehicle Sharing Use Case

A car/vehicle sharing service is intended as a service to enable different customers to make use of a fleet of vehicles (either self-driving or not) which

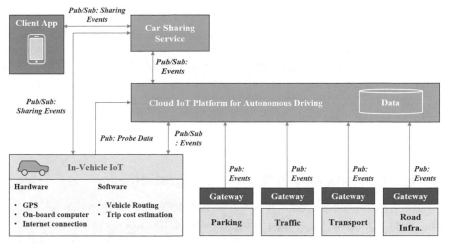

Figure 6.13 Vehicle sharing use case architecture (example from the Dutch pilot site) [1, 12].

is shared amongst them. Vehicle sharing can be interpreted as a service that finds the closest available vehicle and assigns it to a single customer or drives the closest available vehicle to the requesting customer. Vehicle sharing can also be intended as ridesharing when multiple customers that possibly have different origins and destinations share a part of the ride on a common vehicle (either self-driving or by driving it themselves). Finally, vehicle-sharing services can also be thought of as services that allow customers to specify pick-up and drop-off time-windows to increase flexibility and planning.

Figure 6.13 shows the target architecture for the vehicle-sharing use case at the Dutch pilot site. The focus here is on the interaction between the various vehicle-sharing actors and components and the open IoT platform common services, represented as one box.

The users should book vehicles and manage (modify, cancel, etc.) their bookings using the central vehicle-sharing service through a mobile or desktop application, referred to as the client app. The proposed architecture requires that shared vehicles should be equipped with the necessary hardware and software to:

- Communicate their probe data (GPS location, speed, etc.) to the open IoT platform common services and the vehicle-sharing service.
- Compute optimum routes and their costs (distance, energy consumption, etc.) given an assigned destination. These may be fully implemented inside the vehicle itself or may be delegated to external web services.

IoT enabled devices and vehicles of the IoT ecosystem should publish relevant events (traffic, accidents, weather, parking spot availability, etc.) on the open IoT platform. In order for the vehicle-sharing service and shared vehicles to be notified about events that may affect their planned trips, they should subscribe to the open IoT platform for relevant events. The open IoT platform should be responsible for collecting data from the various IoT devices, storing them and communicating the relevant pieces of data (events) to subscribers.

6.4 IoT Devices and Platforms Integration

IoT devices for autonomous driving applications are deployed through IoT platforms that offer integrated services where the IoT devices interact and exchange information. The integration of IoT devices into IoT end-to-end platforms provides the hardware, software, connectivity, security and device management tools to handle the different IoT devices used in the different use cases across the AUTOPILOT project's pilot sites. Different sections provide information on how some of the integration is implemented presenting the managed integrations, device management, cloud connection, cellular modem, etc., to manage and monitor the IoT devices in different use cases. Table 6.2 gives an overview of the communication infrastructure in the AUTOPILOT project [1, 13]. Fields denoted "+" means that the communication technology is implemented in the respective use case in one or more of the pilot sites.

6.4.1 French Pilot Site in Versailles

The oneM2M standard defines two mechanisms to integrate oneM2M and non-oneM2M IoT devices into the IoT platform:

- Integration of oneM2M devices: The IoT devices are called application dedicated nodes (ADN) and can interact with the oneM2M platform directly via the Mca, one of the oneM2M standard interfaces. The IoT devices send requests and receive notification using the oneM2M RESTful API.
- Integration of non-oneM2M devices: The oneM2M standard is highly extensible and allows the integration of non-oneM2M devices and applications, regardless of their vendor or provider. A dedicated software component called Interworking Proxy Entity (IPE) shall be developed and deployed for this purpose. The IPE provides interworking between the oneM2M platform and specific IoT device technologies or protocols.

Table 6.2 Communication technologies in the AUTOPILOT project [1, 13]

Use Cases vs. Technologies	Urban Driving	Automated Valet Parking	Highway Pilot	Platooning	Car/Vehicle Sharing
Long Range Wireless Communication Networks:					
3GPP 4G (LTE)	+	+	+	+	+
3GPP 4.5G (LTE advanced)	+	−	−	+	+
IoT Wireless Communication Technologies:					
IEEE 802.15.4	+	−	+	−	−
IEEE 802.11	+	+	−	+	+
IETF 6LoWPAN/LP-WAN	+	−	+	+	+
LoRaWAN	+	−	−	+	+
Bluetooth/BLE	+	+	−	+	+
RFID	+	−	−	+	+
3GPP NB-IoT	−	−	+	−	−
Intelligent Transport Systems wireless technologies:					
ETSI ITS G5	+	+	+	+	+
IEEE 802.11-OCB	+	+	+	+	+
LTE Cellular-V2X-Release14	+	−	+	−	−
IP Communication:					
IP-V4 TCP/UDP	+	+	+	+	+
IP-V6 TCP/UDP	+	−	−	+	−
IoT Protocols:					
DDS	+	+	−	−	−
MQTT	+	+	+	+	+
oneM2M standard	+	+	+	+	+
Facilities, Transport and Application Protocols:					
ETSI CAM	+	+	+	+	+
ETSI DENM	+	+	+	+	+
ETSI SPaT	+	+	−	−	−
ETSI MAP	+	−	−	−	−
CEN/TS 16157 DATEX II	−	−	+	−	−
DIASER NF P 99-071-1 G3	−	−	−	+	−

Figure 6.14 illustrates the main components and interactions of the AUTOPILOT's French pilot site. The connectivity within the vehicle is handled by the vehicle connectivity module (VCM) developed.

Vehicle remote control data is not exposed to the IoT platform. It is pushed to the OEM platform (via the VEDECOM Broker) using a separate interface available on the vehicle. In addition, a separate communication

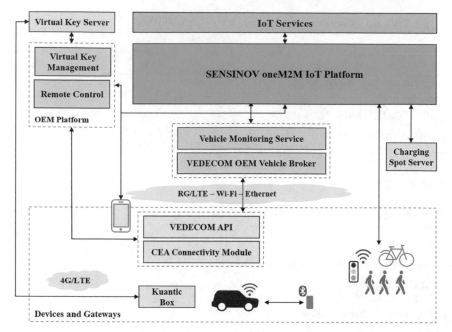

Figure 6.14 AUTOPILOT's French pilot site IoT platform integration [1].

channel for virtual key management is established between the Kuantic server, deployed on the cloud, and the Kuantic box, deployed on the vehicle.

Vehicle monitoring data is exposed to the OEM vehicle Broker using the API service available on the vehicle. Data is pushed to the IoT platform via an IPE to make it available for high-level IoT services using a generic data model and Mca the oneM2M Mca interface.

Other IoT devices, including traffic lights, bicycles, charging spots, passengers' devices are considered as oneM2M-enabled devices and will interact with the IoT Platform using Mca interface.

6.4.2 Dutch Pilot Site in Brainport

There are several use cases implemented and rolled out on the AUTOPILOT's Dutch pilot site. The case implementations are being developed by various project partners using different IoT platforms and technologies:

- oneM2M interoperability integration platform provided by Sensinov.
- FIWARE IoT Broker [17].
- Watson IoT platform.
- Huawei IoT platform.

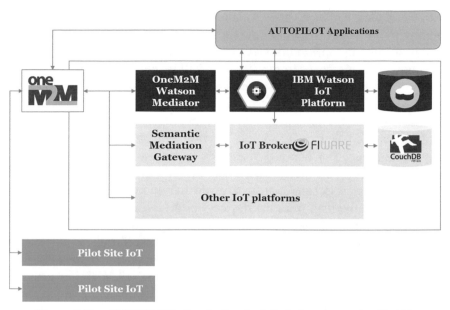

Figure 6.15 AUTOPILOT's Dutch pilot site IoT platform integration [1, 12].

Since the platforms generally perform similar tasks and provide comparable interfaces (e.g. device management, discovery, message brokers, etc.), it has been a challenging task to make all components work together. Moreover, the pilot site devices can connect to one of the platforms, i.e. the platforms are able to discover the devices and communicate with them. The goal was to make the platforms and devices interoperable, and Figure 6.15 illustrates the integration between the platforms and devices in the AUTOPILOT's Dutch pilot site:

- AUTOPILOT applications that implement the use cases.
- An oneM2M platform that all devices connect to by default to "hide" the complexity of the communications between the platforms and applications.
- A set of IoT platforms that should either be able to communicate with the oneM2M platform or implement support to the oneM2M communication protocols by itself.
- The IoT devices connected to the oneM2M platform of the pilot site.

In this scenario, there are two platforms involved in the communications from the bottom up to the top; from the device to the oneM2M interoperability platform, then to the target IoT platform and finally, an application that deals with the device, and vice versa. Another scenario is that the devices connect directly to the target IoT platform (e.g. Watson IoT platform) to reduce the burden of the interoperability between the platforms. This may be useful if one knows that messages from the device will be consumed only by one IoT platform. In this case, there is no need to build a hierarchy of the platforms and pass the messages emitted by the device through the full stack. The drawback of this approach is that the interoperability platform does not know all the connected devices. To address this problem, an announcement process may be introduced. When a device is connecting to an IoT platform, this platform makes an announcement to the interoperability platform to convey that a new device is connected to a given IoT platform, and if somebody wants to consume data from the device via the interoperability platform, it must lookup for the device and message at this IoT platform.

The various platforms and applications are interfaced, as depicted in Figure 6.16. FIWARE focuses on a common data model and powerful interfaces for searching and finding information in IoT. FIWARE is using the OMA Next Generation Service Interface (NGSI) data model as the common information model of IoT-based systems and the protocol for communication. NGSI-9 and NGSI-10 are HTTP-based protocols that support JSON and XML formats for data. Let us shortly describe these two interfaces.

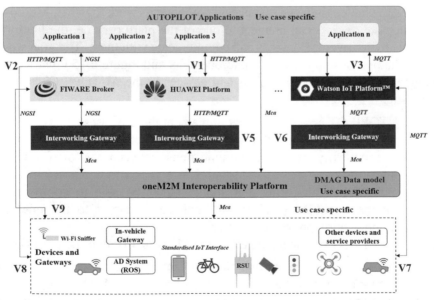

Figure 6.16   The platform interfaces in AUTOPILOT's Dutch pilot site [1, 12].

- NGSI-9 is used to manage the availability of context entity. A system component can register the availability of context information, and later on, the other system component can issue either discover or subscribe messages to find out the registered new context information. Detailed specifications can be found in the FIWARE NGSI-9 Open RESTful API Specification [9].
- NGSI-10: is used to enable the context data transfer between data producers and data consumers. NGSI10 has a query, update, subscribe and notify context operations for providing context values. A context broker is necessary for establishing data flow between different resources as well as consumers or providers. Detailed specifications can be found in the FIWARE NGSI-10 Open RESTful API Specification [10].

The micro aerial vehicle (MAV) and its ground station computer act as one single IoT device in the AVP use case. The communication between the MAV and the ground station is based on a local IEEE 802.11n Wi-Fi connection that guarantees high-data bandwidth and continuous local availability. Small Open Mesh OM2P routers are used on both sides. The ground station computer connects to the Watson IoT platform via 4G/5G. The Huawei OceanConnect IoT platform is an open ecosystem built on IoT, cloud computing and Big Data technologies. It provides over 170 open APIs and serial agents that enable application integration, simplifies and accelerate device access, guarantees network connection and realises the seamless connection between upstream and downstream products for Huawei partners. The used communication protocols are MQTT and HTTP. Applications requiring access to the Huawei OceanConnect IoT Platform need to be authenticated first. Once an application is successfully authenticated, it may perform the following actions:

- Collect device data from the IoT connection management platform using either an active query or data subscription.
- Issue commands to a specified sensor through the IoT connection management platform.
- Issue rules to the IoT connection management platform, allowing response events and commands to be triggered based on the rules.
- Subscribe to device information (events) from the platform.

6.4.3 Italian Pilot Site in Livorno-Florence

At the AUTOPILOT's Italian pilot site, the IoT devices are integrated into the IoT oneM2M platform according to the oneM2M standard, as illustrated in Figure 6.17:

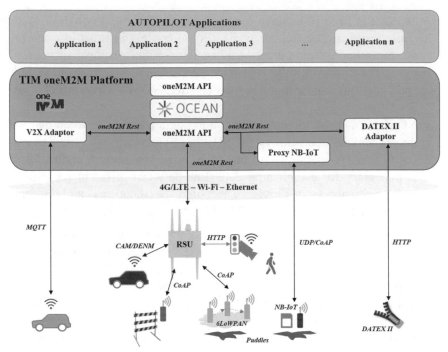

Figure 6.17 AUTOPILOT's Italian pilot site IoT device integration [1, 12].

- IoT oneM2M platform: A federated model where several heterogeneous IoT platforms are interconnected. A central IoT platform includes various modules: big data management and storage, real-time and batch analytics, security and privacy, semantics, etc. Interoperability between the central IoT platform and the pilot site IoT platforms is addressed in this platform.
- In-vehicle IoT platform: An in-vehicle component that provides communication with the cloud IoT platform and the interfaces to other in-vehicle components.

The in-vehicle software architecture for the IoT platform integration is illustrated in Figure 6.18. In this scheme, it is possible to represent the IoT in-vehicle platform and also the interconnections between this container with other onboard sensors in the host vehicle and the cloud system and/or other vehicles and RSUs.

The IoT in-vehicle platform is composed by a quad-core ARM processor. The platform runs an optimised version of Linux and provides

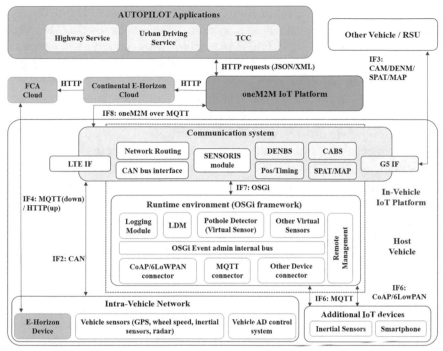

Figure 6.18 AUTOPILOT's Italian pilot site software architecture for OBU IoT platform integration [1, 12].

several interfaces like IEEE ETSI ITS-G5, Wi-Fi, Bluetooth, Ethernet, CAN, 6LoWPAN, and LTE. The board implements CAM, DENM and SPaT/MAP standards with the possibility to send messages both over ETSI G5 and LTE channels. The board will manage the lane level computation of the surrounding vehicles' position. Finally, it mounts a GNSS (Galileo + GPS) receiver that is used for positioning and synchronisation. The unit can synchronise other hosts (within 10 ms) using the NTP standard or other protocols. The IoT in-vehicle platform is modular software including application container and communication system, which are deployed on the OBU. The runtime environment part of the OBU is composed of several software modules.

The functionality of the remote management is implemented by software, which allows configuring the platform by adding/removing bundles, introducing the idea of remote monitoring and control of external application based on an OSGi platform. Through the event admin internal bus, the connectors have the same communication interface to the bundles, which they interfaced in the application container. The application container also encases the functionality of data management, with the modules of local dynamic maps (LDM) and the pothole detector. LDM is a database that achieves integrated management of map and vehicle information (functional requirement of context awareness). It contains information on real-world and conceptual objects that have an influence on the traffic flow. The bundle of the pothole detector represents the implementation of the pothole detection algorithm. It is based on data fusion techniques in order to implement the concept of "virtual sensors". This module collects data from multiple sensors on the vehicle (IoT in-vehicle components or OEM in-vehicle components), processes the various data and sends the results of this elaboration to the cloud oneM2M platform, RSUs or other vehicles via the communication system.

Regarding the IoT device adaptation, the target is to support different IoT communication protocols with the devices. The IoT connectors showed in Figure 6.18 are used to integrate with 6LoWPAN data coming from additional IoT devices (i.e. inertial sensors), which are used by edge applications on the OBU (CoAP/6LoWPAN connector). They are also used to integrate with MQTT protocol data coming from additional IoT devices (e.g. smartphone), which are used by edge applications on the OBU (MQTT connector).

The communication system part of the OBU manages different high-level capabilities. The module CAN bus interface reads data coming from the CAN bus and decodes important data coming from the in-vehicle sensors that are sent directly to the oneM2M platform or used by edge applications on the OBU. The module Pos-Timing reads the positioning data and timing information through the GPS hardware module to set the position on CAM and DENM messages. CABS and DENBS modules take data from the CAN bus, position and time from Pos-Timing and create a CAM/DEN message as described in the proper ETSI standard [8]. They also receive CAM/DEN messages coming from other vehicles and save them on the LDM. The SPaT/MAP messages in the communication system are generated from a traffic light, and SPaT/MAP module decodes them, saving the relevant information in the LDM for further use. SPaT/MAP offers a potential channel for detailed information exchange between traffic systems and road users.

The capability of message routing is assigned to network routing, which manages the connectivity of all the in-vehicle modules that need network connectivity. Moreover, it manages the channels where CAM and DENM messages are sent. In the OBU, they can be transmitted on the ETSI G5 radio channel and/or on the cellular way towards the oneM2M platform or for debugging.

As far as the interoperability part is concerned, it should be considered that the in-vehicle IoT platform should work with heterogeneous devices, technologies, applications, without additional effort from the application or service developer. OEM-specific components relate to components such as actuators for power steering and brakes, inputs to gearbox or vehicle sensors needed for the normal vehicle functions (MAP, MAF, ABS, etc.). Software modules implementing drivers to virtualise such OEM-specific components into vehicle IoT platform are needed, so as to satisfy the OEM systems communication functionality. The OBU can also exchange data with additional IoT devices such as inertial sensors or the motion sensors of, for example, the smartphone. These data are interfaced with the IoT in-vehicle platform using CoAP/6LoWPAN or MQTT protocols as already described, and better implement the concept of "virtual sensors" added to pothole detection.

In order to have a complete vision of Italian pilot site architecture, external components should also be mentioned. The AUTOPILOT applications interface the IoT platform and implement the AUTOPILOT functions in the cloud. Each application communicates with the vehicle via the IoT platform. An application can also comprise a component that runs in the vehicle platform. These components can be either an IoT application or an in-vehicle application, depending on the level of integration with the IoT platform. The IoT platform implements the IoT functions at the cloud or edge level. It also comprises other vehicles and roadside elements.

For example, in the use case of urban driving, a smart traffic light detects a pedestrian or an obstacle on the lane. The information is processed locally and notified to the RSU using the IoT protocols to the vehicles via standard C-ITS messages. Moreover, a connected traffic light sends information about the time to green/red (SPaT/MAP messages). The RSU receives the information, fuses the data and sends it by DENM to all the interested actors on the roads. The information from RSUs and OBUs is also sent to the IoT data platform via IoT standard protocols, and it can then be processed by the area monitoring centre for real-time risk assessment and safety services.

Figure 6.19 illustrates the RSU software architecture for IoT platform integration. The peculiarity given by the modularity and configurability of the designed software, it is possible to customise it depending on the context in which it is inserted (i.e. OBU or RSU IoT platform). In this case, the runtime environment contains a bundle related to pedestrian detection. The module Jaywalking Detector represents the implementation of the algorithm that notifies this event when a pedestrian crosses the strip while the traffic light is red. In these conditions, the IoT platform of RSU is interfaced with a

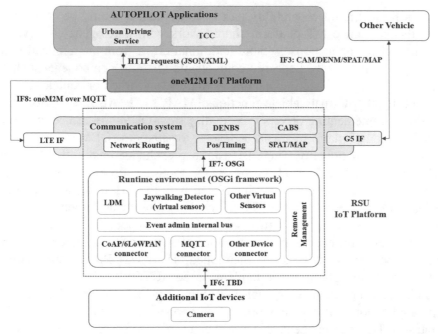

Figure 6.19 AUTOPILOT's Italian pilot site software architecture for RSU IoT platform integration [1, 12].

camera that may register the wrong crossing of pedestrians and send data to the IoT platform. In this bundle, the data are elaborated and the notification of "detected jaywalking pedestrian" is sent to oneM2M IoT platform exploiting HTTP request (JSON, XML) via oneM2M protocol, or to other vehicles using the CAM/DENM/SPaT/MAP interfaces.

6.4.4 Spanish Pilot Site in Vigo

AUTOPILOT's Spanish pilot site is composed of three main IoT platforms, as illustrated in Figure 6.20:

- IoT platform: A federated model where several heterogeneous IoT platforms are interconnected. A central IoT platform includes various modules like big data management and storage real-time and batch analytics, security and privacy, semantics, etc. Interoperability between the central IoT platform and the pilot site IoT platforms is addressed in this platform.
- In-vehicle IoT platform: An in-vehicle component that provides communication with the cloud IoT platform and the interfaces to other in-vehicle components.

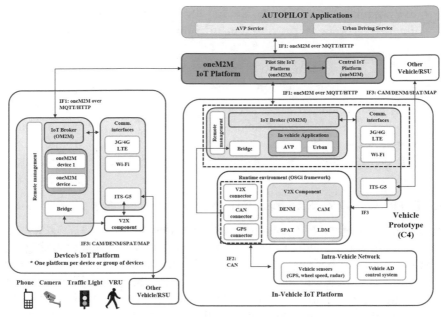

Figure 6.20 AUTOPILOT's Spanish pilot site IoT platform integration [1, 12].

- Devices IoT platform: The devices can be new devices or existing devices adapted to become IoT devices able to be integrated into the IoT ecosystem.

The in-vehicle platform and device platform software components are described in Figure 6.20. The more important software components are:

Communication interfaces: The component responsible for providing connectivity to the device.

- The supported interfaces are cellular (3G/4G LTE), Wi-Fi and ITS-G5 wireless interface.

IoT Module: This module translates the information that comes from the different devices into oneM2M messages and translates oneM2M message into understandable information for the vehicle.

- IoT Broker: OM2M based ASN-CSE (oneM2M) that acts as an IoT gateway. It provides the HTTP and MQTT connectivity to the cloud IoT platform.
- Bridge: Responsible for translating all the information from the vehicle into oneM2M and for publishing and providing any needed methods to obtain this data.

- IoT Applications: Responsible for the interaction with the physical devices in order to provide the full functionality expected in the use cases.

Runtime environment: OSGi framework that contains the stack that enables the V2X communication.

- V2X Component: Contains several modules that are able to process data coming from V2X communication through ITS-G5. The component provides the encoding/decoding for the SPaT/MAP, CAM and DENM messages. Includes the connectors that give access to the IoT module.

6.4.5 Finnish Pilot Site in Tampere

Figure 6.21 illustrates the communication architecture of the AUTOPILOT's Finnish pilot site, including the two use cases; automated valet parking and urban driving. For both use cases, the same infrastructure is used; the prototype vehicles and a traffic camera installed on the mobile RSU. The data of the traffic camera are processed locally, and information on objects is transmitted to the IoT platform. Both the vehicle and the mobile RSU internal network are

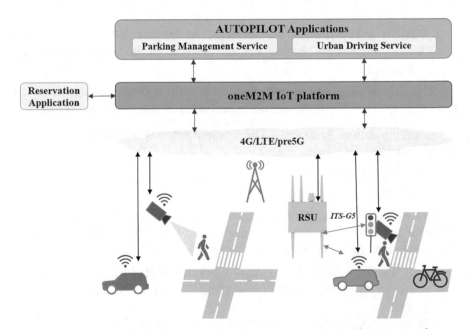

Figure 6.21   AUTOPILOT's Finnish pilot site communication architecture overview.

based on the data distribution service (DDS) network. Information exchanged between the vehicles, RSU and the services, like the parking management service, is based on MQTT and is being sent to an open oneM2M IoT platform. Communication between vehicle, mobile RSU and the IoT platform uses available mobile commercial network (4G/LTE) or a Pre-5G innovation platform, which is installed in the city of Tampere. Vehicles receive signal phase information from traffic lights also from the traffic light operator's server over MQTT. Services for parking management and for management of the urban driving use case are developed and integrated with the IoT platform.

Figure 6.22 illustrates the IoT platform architecture integration of the Finnish pilot site and is composed as follows:

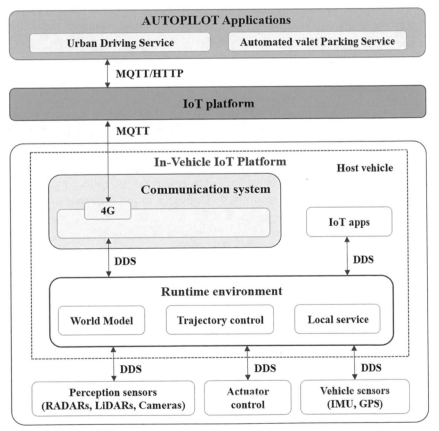

Figure 6.22 AUTOPILOT's Finnish pilot site IoT platform integration [1, 12].

- An open IoT platform for connecting the different devices based on oneM2M. The main purpose of the IoT platform is to act as a broker.
- The in-vehicle IoT platform provides communication with the IoT platform, and with the different devices and applications in the vehicle. Data is exchanged between the different applications using data distribution service (DDS).
- The mobile roadside unit has a similar architecture as the vehicle unit. The mobile roadside unit processes the information from the traffic camera and makes this information available through the IoT platform to the vehicle and the parking management system. The system also has a storage process for assuring that all data needed for evaluation are made available.
- In addition, there is a connection to the traffic light server. Information on the traffic signal phases is available in real-time over the cellular network as MQTT messages.

6.5 Conclusion

The IoT technologies used in different AUTOPILOT use cases have demonstrated that it is possible to support automated/autonomous driving functions as defined by SAE levels and IoT technologies and platforms embedded in vehicles and infrastructure, enhancing the automated/autonomous driving functions. There are five different IoT platforms used for collecting, exchanging and processing the data from the IoT devices in the different use cases in the different pilot sites presented:

- FIWARE IoT Broker.
- IBM Watson IoT platform.
- HUAWEI OceanConnect IoT platform.
- Telecom Italia (TIM) oneM2M IoT platform.
- Sensinov oneM2M platform.

The integration of IoT technologies and platforms is adapted to the infrastructure of the pilot sites. The use cases map the AUTOPILOT architecture, and the IoT technologies are integrated into different architecture components and interfaced/connected to the infrastructure of each pilot site. The IoT technologies used were adapted to the autonomous driving function requirements in terms of speed of access (latency), availability and range (covered area).

The vehicles used in the different AUTOPILOT use cases starts at level 2 with internal systems that take care of the different aspects of driving, such as steering, acceleration and braking. The driver is able to intervene if any part of the vehicle system fails. Examples of level 2 include use cases helping vehicles to stay in lanes and self-parking features, with more than one advanced driver assistance system (ADAS) aspect. Tesla's Autopilot and Nissan's ProPilot are examples of level 2, as the vehicles can automatically keep you in the right lane on the road and keep you at a safe distance from the vehicle in front when in a traffic jam.

The results from the AUTOPILOT projects show that the IoT devices and technologies can support the autonomous driving functions in use cases such as urban driving/vehicles rebalancing, highway driving and automated valet parking.

Acknowledgements

The work presented in this chapter has been supported by the European Commission within the European Union's Horizon 2020 research and innovation programme funding, project AUTOPILOT [1] under Grant Agreement No. 731993. Special thanks to all involved project partners, whose names do not appear on the author list, but who also contribute significantly to the development and testing of the automated driving applications presented in this chapter at various pilot sites.

References

[1] The AUTOPILOT project – Automated driving progressed by Internet of Things, online at: https://autopilot-project.eu/
[2] X. Wu et al., "Cars Talk to Phones: A DSRC Based Vehicle-Pedestrian Safety System," 2014 IEEE 80th Vehicular Technology Conference (VTC2014-Fall), Vancouver, BC, 2014, pp. 1–7.
[3] Bosch Motorcycle-to-vehicle communication, online at: https://www.youtube.com/watch?v=BXXlodI9gO0
[4] M. Bagheri, M. Siekkinen and J. K. Nurminen, "Cellular-based vehicle to pedestrian (V2P) adaptive communication for collision avoidance," 2014 International Conference on Connected Vehicles and Expo (ICCVE), Vienna, 2014, pp. 450–456.
[5] J. J. Anaya, P. Merdrignac, O. Shagdar, F. Nashashibi and J. E. Naranjo, "Vehicle to pedestrian communications for protection of vulnerable

road users," 2014 IEEE Intelligent Vehicles Symposium Proceedings, Dearborn, MI, 2014, pp. 1037–1042.

[6] The Ko-TAG project, online at: https://www.iis.fraunhofer.de/en/ff/lv/l ok/proj/kotag.html

[7] M.S. Greco. Automotive Radar. IEEE Radar Conference, Atlanta, May 2012.

[8] O. Vermesan and J. Bacquet (Eds.). Cognitive Hyperconnected Digital Transformation Internet of Things Intelligence Evolution, ISBN: 978-87-93609-10-5, River Publishers, Gistrup, 2017.

[9] "FI-WARE NGSI-9 Open RESTful API Specification", FIWARE Forge, 2017, online at: https://forge.fiware.org/plugins/mediawiki/wiki/fiware/index.php/FI-WARE_NGSI-9_Open_RESTful_API_Specification

[10] "FI-WARE NGSI-10 Open RESTful API Specification", FIWARE Forge, 2017, online at: https://forge.fiware.org/plugins/mediawiki/wiki/fiware/index.php/FI-WARE_NGSI-10_Open_RESTful_API_Speci fication

[11] Handbook to the IoT Large-Scale pilots Programme, online at: https://wiki.european-iot-pilots.eu/index.php?title=HANDBOOK_TO_THE_I OT_LARGE-SCALE_PILOTS_PROGRAMME

[12] The AUTOPILOT project – Automated driving progressed by Internet of Things. Report on development and integration of IoT devices into IoT ecosystem. D2.4, July 2018.

[13] The AUTOPILOT project – Automated driving progressed by Internet of Things. Final specification of communication system for IoT enhanced AD. D1.8, June 2019, online at: https://autopilot-project.eu/deliverable s/

[14] Alliance for Internet of Things Innovation (AIOTI), IoT Relation and Impact on 5G, (Rel. 2.0). AIOTI WG03 – IoT Standardisation. February 2019.

[15] Intelligent Connectivity, GSMA Report, 2018, online at: https://www.gs ma.com/IC/wp-content/uploads/2018/09/21494-MWC-Americas-repo rt.pdf

[16] The AUTOPILOT project – Automated driving progressed by Internet of Things. Final specification of Security and Privacy for IoT-enhanced AD. D1.10, June 2019.

[17] F. Cirillo, G. Solmaz, E. L. Berz, M. Bauer, B. Cheng and E. Kovacs, "A Standard-Based Open Source IoT Platform: FIWARE," in IEEE Internet of Things Magazine, vol. 2, no. 3, pp. 12–18, September 2019.

7

IoT in Brazil: An Overview From the Edge Computing Perspective

**Marcelo Knorich Zuffo[1], Laisa Caroline Costa De Biase[1],
Pablo César Calcina-Ccori[2], Catherine Pancotto Portella[3],
Adilson Yuuji Hira[3], Gabriel Antonio Marão[4],
Irene Karaguilla Ficheman[3], Geovane Fedrecheski[1]
and Roseli de Deus Lopes[1]**

[1]Escola Politécnica da USP, Brazil
[2]Instituto de Matemática e Estatística da USP, Brazil
[3]LSI-TEC, Brazil
[4]Fórum Brasileiro de IoT, Brazil

Abstract

In 2019, the Brazilian Government published the National Plan for the Internet of Things, which works towards fostering the development of this area in four priority environments: agribusiness, health, smart cities and industry. The plan also states that to establish a basis for solutions in these domains, investments in the following strategic fronts are needed: human capital, innovation, technology, and regulation.

We here summarize important scientific and technological advances in IoT, conducted by Brazilian institutions in the aforementioned strategic fronts: Code IoT education platform (human capital); Caninos Loucos SBC family and SwarmOS (technology); telecommunication regulation and the General Law for Personal Data Protection (LGPD) (regulation, security and privacy); and five applications (innovation) – Smart traffic lights, Smart surveillance, health monitoring of childhood cancer patients, Sleep apnea diagnosis, and Internet of Turtles. We also discuss regulatory aspects towards flexibilizing IoT services, and a recent law that protects the privacy of citizens in Brazil. These efforts clearly show a growing development of IoT in Brazil, particularly in areas that solve urgent problems, such as health and the environment.

Additionally, the existence of a national IoT platform leverages the massive creation of high-impact applications in the near future.

7.1 Introduction

The Internet of Things will dramatically change our lives, spreading connected computers with sensors and actuators, generating all sorts of smart things and producing enormous quantity of data, generating a whole set of new services. The potential socioeconomic impact of the IoT on economic productivity and improvements in public services in Brazil was estimated by McKinsey Consulting to be up to $200 billion – equivalent to approximately 10% of the 2016 Brazilian GDP. For example, in freight transport, real-time monitoring of goods could reduce costs up to 25% while intelligent choice of routes could reduce costs up to 20% [19].

Taking into account this context, the Internet of Things is considered an opportunity to the Brazilian industry to be positioned as a relevant solution provider in this segment. Since 2007, Brazilian stakeholders have monitored the segment transformations that started the IoT movement and have prepared to support the IoT ramp up.

Small technology-based companies are making moves to take advantage of that market. In 2016, there was a significant increase in the number of projects involving innovations in the Internet of Things submitted by startups to FAPESP[1] Innovative Research in Small Business (PIPE) Program. Just as an example, there are currently 21 projects led by startups from the state of São Paulo engaged in developing IoT solutions applied to things, such as health services, vehicle tracking, livestock management, building automation and energy management. Among them, we could mention Exati, a company in Curitiba, which has developed an IoT platform for a street lighting management system used in 200 Brazilian cities [19].

In 2019, the Brazilian Government published Decree 9854/19 establishing the Brazilian National Plan for the Internet of Things, in which an action plan was designed for the sector. This plan defined four fields of action: human capital, innovation, regulation and technology. This chapter presents some selected works in each of these fields from the edge computing perspective. The next section describes the Brazilian National Plan for the Internet of Things; one section is then presented to each of the fields of action. Section 7.3 presents the Human Capital field, presenting the Code IoT platform. Section 7.4 describes 5 IoT applications. In Section 7.5 we present the Caninos Loucos (hardware) and SwarmOS (software) platforms for IoT.

---

[1]The State São Paulo Research Foundation.

Section 7.6 shows some advances in IoT regulation. In Section 7.7 we discuss and present some concluding remarks for the chapter.

7.2 Brazilian National Plan for the Internet of Things

The Brazilian National Plan for the Internet of Things is a public policy that has been created by the Ministry of Science, Technology, Innovations and Communications (MCTIC) to sponsor the expansion of the IoT in Brazil. It refers to a set of strategies and public policies that seek to involve companies, the government and research institutions to disseminate the use of Internet-linked devices in Brazilian industry and services [1].

7.2.1 Priority Environments

The national plan selects technological niches and economic segments in which Brazil could have more ability to compete. As shown in Figure 7.1, four environments were given priority status for investments: Agribusiness, Health, Smart Cities and Industry. These environments were selected according to the existence of well-established companies in Brazil and in which there are good opportunities for developing innovations, with huge national demands.

The goal of each priority environment is described as follows:

- **Agribusiness:** to increase the Brazilian productivity and relevance in the global trading of agricultural products, with high quality and

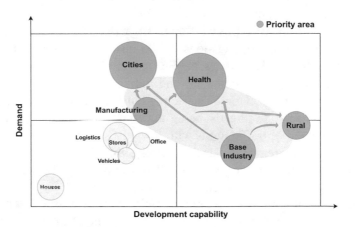

Demand x Development capacity x Offer (size of the circle)

Figure 7.1   Demand × Developing capacity of IoT in Brazil.

Source: National Plan of IoT

sustainability, positioning the country as the largest tropical exporter of IoT technology.

- **Health:** to expand access to high quality health in Brazil, through health monitoring decentralization and increase in efficiency of health centers.
- **Smart Cities:** to enhance the quality of life of residents through technologies that allow integrated management of resources, and to enhance mobility, public safety, and resource usage.
- **Industry:** to foster the production of more complex items and to improve the national productivity through innovative business models and greater cooperation among various productive chains.

7.2.2 Main Structural Fronts

To achieve the goal of each environment, the plan is further structured in four main fronts: Human Capital, Innovation and International Insertion, Technical Infrastructure and Interoperability, and Regulation, Security, and Privacy. In the first front, **Human Capital**, the goal is to raise the potential for building IoT solutions, while directly benefiting the population through courses, grants, and other public policies. Some challenges to be addressed in this front include the enhancement of basic education and better integration of industry and academy, and how to quickly train and to attract high-quality professionals to serve the demand that IoT will incur. A particularly successful initiative in this respect is the CodeIoT program, which offers six free Massive Open Online Courses (MOOC) and has taught over 50,000 students, most of which were not previously introduced to programming or electronics, on how to build IoT solutions.

The **Innovation and International Insertion** front seeks to develop new IoT platforms and applications, while also making them stand out in the global landscape. It includes public financing to develop pilot projects in the most relevant environments, and enhancing competence centers to develop IoT enabling technologies, such as hardware and software platforms. Some pilot projects are already being developed, including a platform for smart management of traffic lights with low cost communication, and health monitoring of children with cancer. Finally, this front will also create an IoT Observatory, to facilitate tracking IoT development in Brazil and sharing news and other initiatives carried by the national plan.

Another front comprises **Technical Infrastructure and Interoperability**, and its main goal is to foster the development of open IoT platforms to support the creation of advanced and interoperable applications, as well as to facilitate the development of connectivity solutions. Regarding the development of open platforms, the Caninos Loucos program stands out as the official platform for Single Board Computers within the national plan, while the SwarmOS is a new software platform for decentralized IoT applications. Finally, the **Regulation, Security, and Privacy** front aims to adjust the regulation to facilitate IoT adoption, while keeping risks of new technologies as low as possible and protecting the personal data of its users. A significant challenge resides in minimizing the impact of impositions set by existing telecommunication regulations over new business models and services offered by IoT applications. Advances in this regard are being made by the National Agency of Telecommunications (Anatel), which seeks to flexibilize regulations and encourage IoT adoption. Data privacy and security also faces challenges, as the IoT significantly increases the points for data collection. While the General Law of Personal Data Protection (LGPD), issued in 2018 by the government, may be a starting point for protecting privacy, more work has to be done, especially regarding security risks and certifications.

To enforce the execution of the national plan, a Chamber for Management and Monitoring of Machine-to-Machine and Internet of Things Communication Systems Development (Câmara IoT) was created [1]. The new chamber is composed by members of the MCTIC, the Ministry of Economy, the Ministry of Agriculture, Livestock, and Supply, the Ministry of Health, and the Ministry of Regional Development. Members from other public and private associations may also be invited to contribute to the Chamber.

The Brazilian National Plan for the Internet of Things is serving as a catalyst to foster value generation throughout the country, which ultimately benefits companies, users, and society as a whole. The following chapters describe selected scientific and technological advances in the fronts and environments aligned with the plan.

7.3 Human Capital: The Code IoT Project

Considering the global importance of IoT and the way it is already changing our lives, it is important to encourage new generations of engineers and computer scientists to learn and to study different aspects of Internet of Things. This stimulates the solution of problems and development of solutions that improve agribusinesses, health, cities and industries. This topic is widely addressed in undergraduate and in graduate courses, but not necessarily in K-12 (basic education) environments, especially with high school students who can understand the concepts, start developing simple solutions, eventually choose STEM careers and, in the future, contribute to creating intelligent IoT applications.

Seeking to bring basic education students closer to the concept of Internet of Things and programming tools, physical computing and application development, we created the Code IoT platform with free online MOOCs in different aspects of IoT.

7.3.1 Methodology

Initially we conducted a face-to-face activity with 32 high school students and their teachers when we offered a 16-hour IoT workshop. The activities were structured to engage students in creative construction and learning processes based on project-based learning strategies and aimed to modify their understanding of the technologies they already use, as well as their ability to create conceptual IoT solutions and to implement simple initial prototypes.

Activities related to programming, physical computing, and robotics have been offered to K-12 students due to the development and dissemination of new tools suitable for use by children and teenagers. Working with these themes enables engagement in interactive, dynamic, and multidisciplinary learning activities that can contribute to increased motivation and assimilation of scientific, technological, mathematical, artistic, and engineering concepts in solving real-life problems.

During the workshop, we filmed, photographed and observed the students and their teachers. We also interviewed and collected feedback from the participants.

Based on the IoT workshop experience and the feedback analyses, we created the Code IoT platform available at www.codeiot.org.br with six online courses that present and discuss several aspects of IoT:

1. **Introduction to IoT.** This is a theoretical course that presents basic concepts of Internet of Things, explains what it is and how it works, shows some of its applications that are already part of our daily lives and explains tendencies in this area.
2. **Learning to code.** In this course, users take first steps in the universe of programming. Using the programming language Scratch, they have the opportunity to create projects involving stories, animations and games, to interact with the online Scratch community and to learn important programming concepts in a practical way.
3. **Electronics: concepts and basic components.** This course explains how an electric circuit works and how to create circuits with electronic components that are easily found. Users get familiar with electronics concepts that will help them understand how things work. They also

learn how to assemble a basic electronic kit to take first steps in projects construction.

4. **Physical computing.** In this course, users learn how to create projects using microcontrollers capable of interpreting information from the environment and able to execute actions in the physical world. They understand how some of the electronic devices that we see around us work and learn how to create intelligent objects, integrating programming and electronic circuits.

5. **Apps for mobile devices.** This course approaches apps creation. Users practice and explore concepts that are behind the operation and creation of Smartphone apps. They learn to develop programs and interfaces, using AppInventor. They get familiar with design and usability aspects that are important in mobile application development, and they create their own apps and see them running on their Smartphone or tablet.

6. **Intelligent Connected Objects.** This course integrates electronics, programming and Internet of Things knowledge to create solutions for real world problems. Users have the opportunity to use what they have learned in previous courses to create solutions for real world problems connecting various technologies. They create and develop intelligent objects able to communicate with Smartphones and interact with the environment.

7.3.2 Implementation

The Code IoT platform was launched in September 2017. The six courses are free online MOOCs and have been offered several times since then, typically 2 to 3 times each year. The courses are six-week long and demand a 4 to 6-weekly-hour effort. All the courses are based on Problem-Based Learning paradigms and always end with a project the users have to create and to submit. Participants that conclude all the tasks proposed receive a certificate for each course. Courses are intended for basic education students and teachers; however, anyone interested in the subject can enroll.

In 2017, the courses were all in Portuguese, and in 2018, the courses were translated into Spanish and English and have been offered simultaneously in the three languages since then. In three years (September 2017–September 2020), the platform has had 105,579 registered users, counts on 212,233 enrolments in the courses, issued 9,414 certificates totalizing 169,796 training hours.

After analysing the users profiles, their age and occupation, we observed that most users were university students (undergraduate and graduate) and professionals interested in the IoT subject. We then started to work with high school teachers in face-to-face workshops, for them to be acquainted with the Code IoT platform, to conduct interactive activities with their students and to encourage them to engage in the online courses. In two years (September 2017–December 2019), we conducted 21 workshops in two Brazilian cities, in which 1,228 teachers participated and have impacted 9,233 high school students.

7.3.3 Future Work

We intend to continue our effort in promoting human capital development by intensifying the promotion of the Code IoT platform with High School teachers and students and conducting face-to-face workshops.

We are also planning to attend undergraduate and graduate students creating new MOOC courses on IoT and the Caninos Loucos hardware platform as well as the IoT middleware software called SwarmOS.

7.4 Innovation: IoT Applications in the Brazilian Industry

In this section, we describe some ongoing efforts towards the solution of problems of high interest in Brazil, such as health, traffic, urban surveillance for security, and environment. The first four applications are the result of a consensus amongst several research, industrial, and government institutions under the program of IoT National Pilots developed by the Brazilian Development Bank (BNDES). Those projects will take place in the beginning of 2020 and will have a great impact on the solution of the above listed problems. The *Smart traffic lights* project proposes installing edge devices in the traffic lights of São Paulo city, for better traffic light control. The *Smart surveillance* project will put mobile cameras with computing abilities for capturing hazardous scenes through computer vision algorithms. The *Health monitoring of cancer patients* and *Sleep apnea diagnosis* projects aim to use IoT devices to monitor the health of patients in order to perform early diagnosis and save lives. The rest of this section presents further details about these projects.

7.4.1 Smart Traffic Lights

The project aims to implement and to evaluate a network of smart traffic lights with remote programming through a Fixed-Time Traffic Light Control

Center, aiming to offer tools to improve the effectiveness and efficiency of urban traffic management. IoT devices built into the fixed-time traffic light controllers will be used to establish a wireless communication network between the control center and the controller, allowing their remote reprogramming and monitoring.

Smart transport is a priority for São Paulo city, especially in improving the modal fluidity. Brazil currently loses approximately $156.2 Billion with traffic jams in the city of Sao Paulo. Traffic light controllers have a fundamental contribution to traffic. In the city of São Paulo, there are about 6,000 traffic lights. Among them, 4,500 operate in fixed time; in other words, configured to operate following a temporal schedule. The remaining 1,500 traffic lights are real-time, remotely controlled by a fibre optic network, determining the state of the traffic lights at each instant.

Between 1993 and 1997, five real-time Area Traffic Centers (CTAs) were implemented in the City of São Paulo. Five CTAs were required instead of one because of a restriction that each manufacturer's semaphore controller model could only be installed in their respective CTAs. This restriction not only increase the required investments to operate the system but prevents integration. Thus, CET began a search for solutions that would allow semaphore controllers from any manufacturer to be connected to CTAs with standardized and open communication protocols.

Currently, the São Paulo Traffic Engineering Company (CET) does not yet have a Fixed Time Traffic Light Control Center; configuration changes or problem identification and remediation require CET teams to be relocated for reprogramming or recovery, resulting in a high operating cost and low agility in problem-solving. There are great opportunities for system improvement, since the identification of non-working traffic lights is mostly received by citizens' complaints, and the average time between receiving a notification and troubleshooting is 9 hours. Thus, remote access to fixed-time semaphore controllers is currently a demand, as it would lead to improving the system availability and the fluidity of transport modes, especially in adverse or peak usage situations.

Taking advantage of IoT technologies for remote programming and diagnosis of traffic lights, the project aims to connect Fixed-Time Traffic Light *Connected Controllers* to a Fixed-Time Traffic Light *Control Center* through a standardized and open communication protocol. The IoT solution is divided into three main layers: devices, network, and application; a fourth layer permeates all the others: security.

- Device layer: comprises the Fixed-Time Traffic Light Connected Controllers, which are able to receive remote updates to its local database of traffic light schedules. These devices can also notify the Control Center about any emergency or problem detected in the traffic light function;
- Network layer: pursuing the flexibility of semaphore controllers manufacturers, the use of a standardized and open communication protocol is mandatory. The solution will use an event-oriented IoT messaging protocol and a long-range, low-cost, IoT-suitable communication, such as LoRaWAN, which uses sub-gigahertz frequencies for extremely energy-efficient data transmission over distances of up to 10 km. One of the technical and scientific challenges faced in this project is the potential size of the configuration packages, which must be optimized to fit within the communication mechanism bandwidth requirements;
- Application layer: comprises the Fixed-Time Traffic Light Control Center, which will be developed using open source software, commissioned by the municipality but not yet validated in the field.
- Security layer: appropriate and well-established protocols will be implemented for each device, network and application. As the traffic management, more specifically traffic lights control, is a critical mission, security is of major importance.

7.4.2 Smart Surveillance

According to the Numbeo[2] ranking, Brazil is the 7th country in the world with the highest rate of criminal occurrences. In Latin America, Brazil is only behind Venezuela, which leads the global ranking. In addition to direct damage to the impacted population, the effect of violence on the country's economy can cost 3.14% of the GDP, according to estimates by the Inter-American Development Bank in 2014. In order to be more effective in combating crime, law enforcement agencies focused on the use of technologies, among which, the use of fixed cameras installed in several cities in Brazil to reduce costs and to increase effectiveness in combating risk situations to citizens. However, despite the benefits provided by fixed cameras, their static nature limits their coverage of specific areas. In addition, criminals can simply change the place of operation. If safety for all is a goal, solutions that are more effective have to be introduced.

---

[2]www.numbeo.com

A major benefit of the Internet of Things is that it can transform any object in the physical world into an information retriever and transmitter. In this context, this project proposes creating a mobile sensing network by installing IoT devices in vehicles, since they provide natural mobility, increasing the coverage area and reducing the chance of being predicted by criminals. Another benefit of vehicular sensing lies in the possibility of not simply detecting crimes, but detecting life-threatening situations in general, such as accidents.

If the effectiveness of police in fighting crime depends on a widespread surveillance, success in emergency response is particularly impacted by its speed, because the faster the help is provided, the greater are the victims' survival or criminal apprehension chances. Currently, the State of Sao Paulo emergency notification system is based on telephone calls, which depend on people contacting, explaining the occurrence and giving the location of the incident. Only after this process is completed is the emergency service able to allocate a resource for the call. Since individuals involved in emergency situations may be disoriented or unconscious, an automatic, geolocalized, and reliable notification may save lives.

The pilot will use a device capable of reading and notifying vehicle license plates (Sentinel), which will be installed in a car fleet to perform the system evaluation in the field. The pilot fleet used in this pilot will consist of 50 cars from the Police force and 10 volunteers' cars. A module to receive the notifications from IoT to be integrated in the PM Operations Center (Central), will also be implemented; it will include the notification of accidents involving vehicles and the license plate readings. For the accident reporting assessment, the Sentinel will simulate the incidents. The IoT solution is divided into three main layers: devices, network, and application; a fourth layer will permeate all the others: security.

- Device layer: comprises the Sentinel, which will build on a pre-existing computing platform, the Labrador Single Board Computer (SBC), and incorporate the necessary communication sensors and modules;
- Network layer: the project will use LoRa, a low-cost and long-distance protocol for IoT communications;
- Application layer: comprises data management, storage, and analysis tools, and includes APIs for receiving incident notifications;
- Security layer: considering the sensitivity of the transmitted information, such as license plates and data about criminals, the confidentiality and integrity of the exchanged messages is critical.

7.4.3 Health Monitoring of Childhood Cancer Patients

The spread of Mobile Devices has sparked a new era of possibilities for IoT-based healthcare solutions. Future generations of IoT solutions promise to transform the healthcare industry by enabling breakthrough computing and communication capabilities, where individuals are monitored online by connected wearable sensors, enabling interoperability of personalized health and wellness-related information, patient's vital parameters, as well as data on physical activity, behaviors, and other critical parameters that affect daily quality of life.

Infections represent the main immediate cause of death among children undergoing cancer treatment. As the first symptoms of infection appear, immediate referral to their Health Service Center is absolutely vital, fever being the main symptom.

The context of the project is to monitor the patient's body temperature remotely, transmitting its readings through Bluetooth Low Energy to a smartphone, which stores and sends this data to a cloud-hosted web service via 4G wireless network. This allows the treatment center and the treating physician to receive alerts and monitor the temperature of their patients in real time from a computer, tablet or smartphone. The main objective is to develop and to analyze the use of wearable sensor-based IoT technologies for monitoring vital signs of people, specifically temperature in Child Cancer patients.

The focus of this study is to evaluate the mentioned platform as a precise tool for detecting the infectious condition, allowing the notification of patients and caregivers, which will allow immediate referral to the clinical treatment service in the emergence of fever symptoms, and to analyze their evolutionary impact in this process.

This project uses an IoT platform called Caninos Loucos Pulga, being developed under the National Microelectronics Program of MCTIC, for constructing microsensors based on micro PCB (Printed Circuit Board) Shield. The clinical study will be conducted at the ITACI – Child Cancer Treatment Institute, HC-FMUSP – Pediatric Oncology Treatment Service at the University of São Paulo, accredited by the State of São Paulo Health Department (SES-SP). This project will be implemented in the 2020–2021 biennium.

7.4.4 Sleep Apnea Diagnosis

Currently, for diagnosing sleep disorders, there is the all-night polysomnographic study performed in laboratory, which is the gold standard method for diagnosing sleep disorders. The polysomnographic study allows recording several parameters: Respiratory effort through Inductive Plethysmography Chest Strap, Nasal Flow for Pressure Measurement, Oxygen Saturation (O_2), and Heart Rate.

Although polysomnography is considered the gold standard method for diagnosing sleep disorders, it is necessary to expand diagnostic methods, since not all patients have access to polysomnography, as it is an expensive exam. Moreover, there are Brazilian cities that have no doctors trained in sleep medicine and no laboratories or sleep clinics.

Apnea is currently the most prevalent sleep disorder in about 32% of the population. Thus, providing an affordable Sleep Apnea diagnostic test would bring enormous social and public health benefits as an alternative to polysomnography. In addition, an IoT-based Sleep Apnea diagnostic test would allow patients from remote locations to be monitored and diagnosed remotely.

The objective of this work is developing a sleep quality monitoring system to provide a solution for diagnostic test Sensors with IoT technology, directed to diagnosing Sleep Apnea. An important effort to enable proper monitoring in this work is Signal Characterization and Pattern Recognition for Apnea Diagnostic Calibration by IoT Sensors in relation to Polysomnography measures.

IoT Sensor exams represent an alternative to polysomnography for its ease of access and dissemination, provided there is a network connection, as well as its low cost, besides being an appropriate approach to meet a public service demand. While sleep disorders, such as Apnea, currently affect a large portion of the population, these patients do not currently have access to adequate services due to the lack of an affordable diagnostic test. This project also will use a platform called Caninos Loucos Pulga[3].

This clinical study is conducted at the Sleep Institute (Instituto do Sono), a reference center for sleep disorder care. It is associated to the Federal University of São Paulo (UNIFESP).

7.4.5 Internet of Turtles

In the last 50 years, the growing expansion of coastal cities in Brazil increased sea pollution and exploitative hunting, which threatens some turtle species.

---

[3]Pulga literally means *flea* in Portuguese.

Monitoring those individuals helps to protect them; however, several technical challenges arise. First, turtles spend most of their time under the ocean, which deteriorates the materials of specialized equipment. Second, seawater high conductivity is a problem for electronic equipment. Third, the considerable distances traveled by turtles require long-range antennas to cover such large areas.

Some initiatives, such as the Tamar Project[4] and the Guajiru NGO seek the preservation of turtles in critical regions on the Brazilian coast, such as Ubatuba (SP), Salvador (BA), and João Pessoa (PB). Actions by these institutions include protection of new-born turtles, preservation of their ecosystem, and sustainable development of local communities. Monitoring living specimens is paramount, and technology plays an important role in this task.

To provide a technical solution for the monitoring problem, the Internet of Turtles project is an ongoing effort developed by the University of São Paulo in collaboration with the Tamar Project and the Guajiru NGO. The objectives of the Tamar project include developing an electronic device to monitor turtles that satisfies the environmental constraints. These constraints include minimum size, low cost, lightweight, and long-range communication capabilities.

Two key technologies to implement the Internet of Turtles project are: the Caninos Loucos Pulga single board computer and the LoRaWAN network protocol. The Caninos Loucos board family is described in detail in Section 5.1; it includes the Pulga chip, whose physical size and energy consumption are low. Accordingly, the LoRaWAN network protocol provides long-range transmission of small data packages, besides using Sub-GHz communication, which is highly efficient for distances over 10 km. The Internet of Turtles project also comprises the research of software-defined radios to be integrated into the Caninos Loucos Pulga chip.

Turtle monitoring will be achieved by embedding the developed device into turtle shells and sea buoys. The support from the Tamar Project and from the Guajiru NGO will be of great importance for deploying the system.

7.5 Technical Infrastructure and Interoperability

This section describes two complementary efforts towards a unified IoT platform for the Brazilian market. It consists of a hardware platform called

---

[4]https://www.tamar.org.br

Caninos Loucos[5] and a software IoT middleware called SwarmOS. Together, both technologies provide a full stack platform that leverage local innovation in diverse IoT application areas.

The Caninos Loucos is a program to design and to deploy a family of single-board computers, whose development was the result of several factors, among which are the high taxes for importing hardware technology in Brazil. The Caninos Loucos has evolved since then, from the Labrador SBC to a whole family of boards ready for industrial applications.

The SwarmOS is a bio inspired IoT middleware that creates a decentralized network of heterogeneous devices, mediates the communication, and facilitates resource sharing across devices. The Swarm constitutes the natural complement for the Caninos Loucos hardware platform, as both work in tandem to provide a full platform for the future IoT.

7.5.1 The Caninos Loucos Hardware Platform

Single Board Computers, or SBCs, are complete computers integrated into a single printed circuit board. They usually have very small dimensions, close to the size of a credit card, and affordable cost, of the order of a few dozen dollars. Despite their reduced size and cost, SBCs are very powerful computers at low power consumption, incorporating multiple input and output interfaces of General Purpose I/O (GPIOs), wireless and wired Internet connection, USB ports, sensors and actuators. With the evolution of technology, SBCs have become an essential platform for developing Internet of Things and Industry 4.0. In recent years, there has been a proliferation of marketing models and volumes that exceed millions of units sold in many countries, such as SBC Raspberry PI designed and manufactured in the UK. In this context, we present the Caninos Loucos Program, which aims to: specify, design, develop, manufacture and market a family of Open SBCs fully developed in Brazil and in Latin America. It also aims to develop SBCs focused on the needs of the maker community and, at the same time as the local industry, to promote the development of IoT initiatives in Industry 4.0, from the availability of a development platform that can be easily modified and adapted to the needs Latin American regions. This project is part of the current National Plan for Internet of Things, promoted by the Brazilian Federal Government. Moreover, the Caninos Loucos Family has covered all levels of application in edge computing, starting from smart sensor nodes to

---

[5]Caninos loucos literally means *mad dogs* in Portuguese.

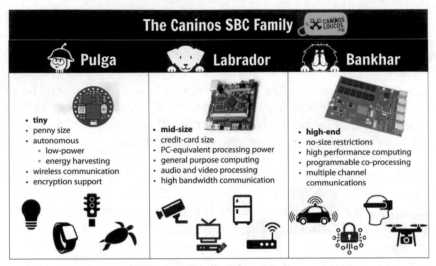

Figure 7.2   The Caninos Loucos single-board computer family.

high end performance computers. Figure 7.2 provides an overview of each member of the family with their main features and usages.

7.5.1.1 Hardware requirements at the edge

Edge computing consists in having most, or even all, of the processing work happening at the site, instead of in the cloud, to improve responsivity and reduce bandwidth use. To meet this requirement, the edge-based systems, or smart objects, as they are called, need to have embedded electronics with a considerable processing power, memory capability, reliability and scalability. In this sense, the hardware has to be robust in terms of both energy and security, to be able to communicate with other systems ensuring interoperability and, finally, being able to process the data in a reasonable time. Also, power consumption is very critical at the edge since many of the applications involve energy constraints, especially the ones related to agriculture and medical applications. In this sense, having a long battery life option is essential for the platforms that work at the Edge. Lastly, it demands both short- and long-range communications for most of its applications and a large variety of communications are used at the edge, such as Wi-Fi, LoRaWAN, BLE, Sigfox, Ethernet and others.

Therefore, the supply of smart objects will depend on the design of electronic systems. Electronic system designs may be developed with a variety of strategies that will impact: the cost of the project (non-recurring engineering cost) and the time to market. The system design can range from a totally dedicated project, even including the design of a new electronic component (ASIC), to the use of a ready-to-use computational module. This also facilitates the integration of computing devices to smart objects by companies not previously familiarized with electronics and the demand for personalized products, with specialized batches in lower quantity.

7.5.1.2 The need for an open SBC platform

To meet the demands of edge-based systems, the single board computers present the opportunity to consolidate products with smaller time-to-the-market, since SBCs are essentially compact computational modules, that are expandable, scalable and with a variety of configurations, processing performances and costs. Therefore, it can also significantly reduce the non-recurring engineering costs involved in an IoT project. These costs represent the investment on the development itself, including hardware and software, in cases of embedded electronic products; by using the SBC one can minimize the hardware project cost by using a known platform and focus on the software design. This approach will save both the time and money spent in development. In this sense, SBCs establish a paradigm of computer as a device, which allows it to be embedded in practically anything anywhere.

Hence, SBCs are ideal platforms for IoT and edge computing solutions, since edge computing consists in taking most of the processing work to the edge instead of the cloud. The latter requires a combination of powerful processing and low power consumption, which SBCs already have, since they are largely used in embedded systems; for that, they need to have extended working life as well as the capability of having communication according to the application of use. Moreover, the extended fields of usage of the SBCs demand a high number of peripheral possibilities and variable memory capability. Therefore, hardware requirements vary with the application but in general the SBCs have enough processing capability to handle most, or even all, of the computing load of the system. Besides, SBCs that implement DSP instructions and cryptography hardware acceleration ensure the security in edge computing systems.

Moreover, since each application has its own requisites, the versatility of SBCs is important. They have as many peripherals as possible, such as wireless communication, I2C, SPI and other interfaces to communicate with other components or with the external world. Finally, the strategy of using SBCs in edge computing allows a variety of peripherals and communication protocols needed in smart systems development with the advantage of low development costs, low time to market. Besides, they ensure the security of the system by not having backdoors or allowing industrial espionage, since they can be customized for each application, provide economy to the overall project due to the lower prices and taxes, and also integrate it on the IoT wave.

7.5.1.3 Caninos Loucos family as a platform for edge computing

The SBC Caninos Loucos presents several innovations, such as internal processing, low energy consumption, optimized communication protocols for the Internet of Things, adaptability to different processes, high concern for information security, ease of use, and an open and collaborative approach to projects. For greater versatility and appropriateness to this concept, the family uses a two-board strategy: an IoT core module consisting of the computational (CPU and Memory) unit and a base board with interfaces and peripheral support. This flexibility of the proposed platform is a differential; it will generate a standardization of pinning and a printed circuit board architecture that will allow meeting various demands by business, start-ups and inventors.

The two-board strategy enables the Caninos Loucos SBC to work as platform for edge computing since its main board contains a powerful processing module associated to a base board that can be customized, allowing versatility on power and communication, the two main restraints in edge computing applications. Moreover, the standardization of the pinning permits users to change the computing module according to the processing needs of the application while maintaining the peripherals on the BaseBoard or vice-versa.

The Caninos Loucos Family has different boards for different uses, while maintaining the two-board concept. In this sense, the development of the Caninos Loucos family aims to cover three categories of applications, according to computing capabilities: SBC-tiny, SBC-mid e SBC-high. Figure 7.2 summarizes the main characteristics of the Caninos Loucos board family.

The first SBC in the Caninos Loucos Family is the Labrador, which focuses in the *SBC-mid* category and includes credit card-sized boards (8.5 cm × 5.5 cm) or 46.75 cm^2. Applications for this board family include communication gateways, microservers, personal computers, embedded boards in white-good appliances, educational toys, among others, since it can run Linux and Android and access the internet via cable or Wi-Fi. It has enough processing and communication power to process high resolution audio and video. Thus, it is a miniaturized generalist computational platform, with computing power equivalent to a low-performance personal computer and size close to a credit card. These platforms bring versatility and agility to a wide range of applications, including thin clients, home appliances, security cameras, set-top boxes, gateways, etc.

The IoT module, called Labrador Core, contains the processing unit, the power management unit and the volatile and non-volatile memories. These components are highly sensitive to impedance variations and demand high speed signals, which implies a more complex and sophisticated project. The baseboard, called Labrador Base, has a simpler design, with lower frequency signals and simpler components. This strategy is appropriate to the open hardware approach, since it consolidates two projects with very different redesign difficulty level and with software compatibility, since a single module can be used with different baseboards customizable according to each application. Figure 7.3 shows a Labrador SBC with the IoT module and the BaseBoard, designed and produced in Brazil.

The second SBC in the family is the Pulga, the Tiny SBC that comes to meet a demand regarding the trend of distributed processing using the edge computing approach since many IoT designs are strongly centered in the cloud, with low processing power being demanded by the device itself. It constitutes a single board computer, with considerable processing capacity, yet at low cost, small size, which is versatile in communication mechanisms and sensors but with high autonomy. The device high autonomy is achieved using low-power components associated with an energy-harvesting circuit and optimized software for low-power consumption. Its design adopts

Figure 7.3   The Labrador single board computer.

Figure 7.4   The Flea single board computer.

new communication standards for low-range communications allowing the constitution of mesh networks of more than one thousand nodes, which is particularly interesting for sensing and monitoring applications. The *SBC-Tiny* family is targeted to last-mile edge computing, where connectivity, computing capabilities, and energy consumption are extremely limited. The size of these boards is less than 2 cm^2 Applications for these boards include sensors and actuators for diverse areas, such as agribusiness, home and industrial automation, health, fitness, entertainment, and wearables. Figure 7.4 shows a Flea SBC in its initial version, manufactured by LSITEC in Sao Paulo, Brazil.

Finally, the *SBC-high* family proposes high network computing performance as the main characteristic, cable of reaching a 1 Teraflop of processing power with much lower consumption than other high-end solutions. Potential applications of this family are autonomous vehicles and virtual reality engines, among others.

7.5.2 SwarmOS

The term swarm was first proposed [2] to refer to sensory found at the edge of the cloud, and identified the opportunity for materializing/serving areas such as cyber-physical systems, cyber-biological systems, immersive computing and augmented reality. Subsequent work led to a more concrete definition of the Swarm, particularly the proposal of an initial architecture

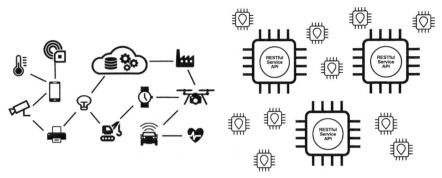

Figure 7.5 The SwarmOS architecture.

for the Swarm [3] in the context of a larger project called TerraSwarm. They also outlined a common framework for devices to communicate and to share resources, called SwarmOS. The architecture of the SwarmOS framework was further developed [4], complementing the already existing distributed storage system (*data plane*) with a module responsible for sharing and managing resources (*control plane*).

The Swarm is a self-adaptive network for autonomous smart objects. Devices do not rely on the cloud for storage and processing; instead, part of this work can be performed in the device itself. The Swarm is a heterogeneous network constituted of different kinds of devices, with variable computing power and energy capabilities. In Figure 7.5, we illustrate the general structure of the Swarm. Making a parallel with swarms of bees, with specialized bees contributing to a common goal, the Swarm is composed of specialized devices whose interaction solves a common problem. The Swarm network behaves as an organism and shows an organized behavior resulting in an emergent collective intelligence.

7.5.2.1 The Swarm architecture

Device functionalities are exposed to and shared with the network as *services*. Thus, the Swarm can be seen as a large network of interacting services. Every service is specialized in a specific functionality, and many services can perform similar or equal functionalities. The true potential of the Swarms resides in the composition of services, which dramatically extends the functionalities offered by the network. Given the intersections with the *microservice* architecture style, we adopted many of its concepts for the Swarm, such as the use of services as the main building block; loose coupling and high cohesion;

decentralized governance; decentralized data management; and evolutionary design [5].

Interaction among devices is performed opportunistically, with no prior agreement. The connection among devices is established in real-time, based on the availability of devices in the network. As a response to an external event, devices in the network form groups to perform an action or to give an answer. Although those groups formed are transient, the success of each interaction is recorded in the network and serves to build a measure of *reputation* for each device. The Swarm platform is based on a lightweight middleware installed in every IoT device called SwarmBroker, which acts as a communication facilitator. Some functions provided by the Swarm-Broker include registry and semantic discovery of services in the network, enforcement of policies for access control, a decentralized mechanism for service contracting and reputation, based on blockchain. The transaction model creates contracts between service consumers and providers which are chosen by a combination of price and reputation, thus creating an economic model for resource sharing in the IoT.

7.5.2.2 The SwarmBroker

The actual software framework that implements our Swarm vision is called *SwarmBroker*. It acts as a facilitator of communication among services. We define two categories of services in the Swarm: *platform service*, which constitutes the core functionalities of the Swarm network; and *application services*, all other services that participate in the Swarm. Platform services include *discovery* of other services; *registry*, a distributed catalog of services; *access control*, which determines the access to resources; *binding*, which translates commands among protocols; *policy management*, a repository of policies used by access control; *contracting*, which establishes service-level agreements for the use of services; *mediation*, which offers a semantic support for discovery service; and *optimization*, which analyses data generated by device interaction to tune network parameters and policies. The Broker can be seen as the collection of platform services. Figure 7.6 illustrates the landscape of platform services that constitute the Broker.

Every device participating in the Swarm has a Broker installed in it. Since the Swarm network is composed of heterogeneous devices, the Broker number of platform services varies in the devices, according to their capabilities. Less powerful devices will provide fewer platform services. Accordingly, devices with minimum computing capabilities that do not allow installing new software have an external software *proxy* to translate communication.

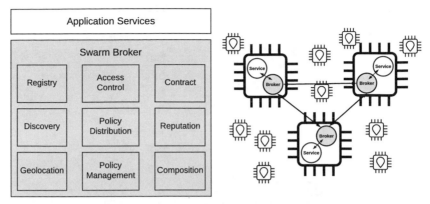

Figure 7.6 The Swarm OS broker organization.

Different implementations of the Broker are expected, to cover a wider range of devices. Currently, we have four implementations, using different programming languages: C, Lua, Java, and Elixir.

7.5.2.3 Semantic discovery in the Swarm network

Finding a suitable service to interact in the Swarm is a problem of major importance. The Swarm architecture, shown in Figure 7.7 proposes a functionality-based search of services. A requester service searches for a service that matches the expected functionality. Initially, an exact functionality-matching was implemented, based on string comparison, which poses severe limitations, such as not being capable of matching equivalent functionalities that use different names.

Several initiatives were devoted to exploring the use of semantics to the problem of service discovery. More recent work highlighted the opportunity

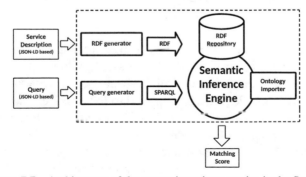

Figure 7.7 Architecture of the semantic registry service in the Swarm.

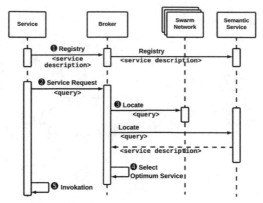

Figure 7.8   The semantic discovery process in the Swarm.

of applying those technologies to the Internet of Things. We here introduce a framework to enable semantic discovery of services in the Swarm, as shown in Figure 7.8. We describe the benefits of semantics for the aforementioned problems and present an architecture and implementation for our solution.

The authors in [6] propose a novel architecture for semantic discovery in a decentralized and heterogeneous environment, and a novel document format for service description and service request, focused on human friendliness and ease to use.

7.5.2.4 The Swarm economy

The distributed and decentralized nature of the Swarm network poses new challenges to security and transaction models. Traditional technologies, such as the public key infrastructure (PKI), are not suitable since they require a centralized certificate authority (CA). To overcome this challenge, the Swarm adopted the blockchain technology, used in the Bitcoin cryptocurrency, to create a decentralized mechanism for trust in the economic model of the Swarm network [7].

The economic model of the Swarm aims to regulate the transactions of services in the Swarm network. This model includes trust, a rewarding mechanism, billing, reputation, and a full virtual economy system. The model based on microeconomic principles applied to IoT services, has the following components: a *transaction* is a trade of *computing resources* between a *customer* and a *provider* of different *owners*.

The Swarm Broker is responsible for linking the parties and for facilitating transactions. The price of a resource is the number of credits necessary for the service provider to grant access to the service consumer. Credits are the owner's asset; they are used by the service to contract or to purchase any service on behalf of its owner.

The Swarm economic model is based on the price of a service and on the reputation of both service consumer and provider; hence, it is called *price-reputation* model. A service provider defines the number of credits necessary to allow a third party to use it. On a service request, candidate providers are ranked by the lowest price according to the formula:

$$P = \begin{cases} P_{\min} + \frac{P_{\max} - P_{\min}}{T_{th}}(T_{th} - T_{pc}), & T_{pc} < T_{th} \\ P_{\min}, & T_{pc} \geq T_{th}. \end{cases}$$

During the transaction process, reputation points evaluate the success of the operation. The price-reputation transaction is the simplest transaction defined for the Swarm framework: the consumer gets the service by paying a number of credits settled by the provider, depending on their behavior, they both get reputation points during the process.

As the Swarm is an organic network of heterogeneous participants, a fair set of rules is necessary to guarantee a fair trading of resources. We created an economic model, following principles from microeconomics, as previous efforts did. We identified the participants of the model, created a taxonomy and proposed a microeconomic model for resource trading in the Swarm. The economic model describes how transactions take place in a distributed environment. The implementation of the price-reputation model takes advantage of the blockchain technology to store information credits and reputation of the participant devices.

The advances in an economic model for the Internet of Things go in the same direction of a growing trend in world economy called sharing economy. As in the physical world, the Swarm favors the digital sharing of resources over the acquisition of dedicated devices. As a consequence, it produces a reduction of device consumption and a better global use of resources.

7.5.2.5 Security and access control in the Swarm

The resource-sharing vision of the Swarm can only be implemented if security is built into the system. This includes both the use of appropriate algorithms and protocols to protect exchanged messages, and a flexible access control mechanism to govern which interactions are allowed. For example, the owner of a street-facing security camera may make it available for sharing during daylight, and a smart building will need different policies to control access to different devices on different floors, which are rented to different stakeholders. Thus, managing the access among large quantities of devices becomes a significant challenge.

The Swarm approach to access control uses Attribute-Based Access Control (ABAC), a flexible and comprehensive system in which subject requests to perform operations on objects are "granted or denied based on assigned attributes of the subject, assigned attributes of the object, environment conditions, and a set of policies specified in terms of those attributes and conditions" [8]. As ABAC is still in its maturation phase, a new model called ABAC-them was introduced. It focuses on combining simplicity and expressiveness, and its main characteristics are:

- Enumerated policies: attribute enumeration allows creating policies that are easy to parse and to embed into small devices.
- Hierarchical attributes: allow creating high-level policies that are easier to write and to understand. During execution time, low-level attributes present in access requests benefit from attribute hierarchies, which allow them to match with the high-level policies.
- Typed attributes: provides a counterbalance to policies that can grow large when using enumeration, such as those involving numerical ranges. Multiple attributes: very specifically, this feature allows easily creating conjunctions when using enumerated policies.

As an example, Figure 7.9 shows a policy written according to the ABAC-them model. It states that "any security appliance can be accessed and modified by an adult family member" and is serialized using Javascript Object Notation (JSON).

The ABAC-them model was implemented within an architecture based on the NIST recommendation for ABAC systems [8]. It comprises four main

```
{
  "user_attrs": [
    ["string", "Role", "AdultFamilyMember"]
  ],
  "operations": ["read", "update"],
  "object_attrs": [
    ["string", "Type", "SecurityAppliance"]
  ],
  "context_attrs": [
    ["time_interval", "DateTime", "* * 8-18 * * *"]
  ]
}
```

Figure 7.9   An access control policy example.

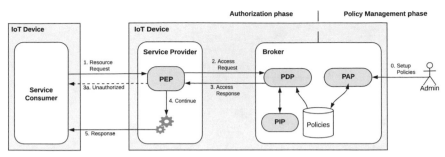

Figure 7.10 Security and access control architecture on the Swarm using ABAC policies.

points. The Policy Decision Point (PDP) evaluates policies managed through the Policy Administration Point (PAP), while the Policy Information Point (PIP) accounts for gathering context and other attributes, and the Policy Enforcement Point (PEP) intercepts requests and verifies their permission with the PDP. While the original NIST architecture proposes that the PDP, PIP, and PAP reside in an authorization server, the proposal within the Swarm puts all points inside the IoT device, thus enhancing its autonomy and security. One challenge emerging from this modification is that the policies are now distributed, and a policy-sharing mechanism must be developed. In a previous work, a policy distribution algorithm was implemented, which allowed devices to gather policies from surrounding devices, which would be edited by a human user, and then pushed the policies to the appropriate devices again [16]. Figure 7.10 shows the architecture of the access control module.

7.5.2.6 Resource-constrained devices: The Swarm minimum broker

The Swarm Broker is the software agent installed in each device to mediate the interaction with the emergent and complex network of devices. The Swarm Broker turns the device into a *swarm-insect*, i.e., a member of the Swarm. To overcome the challenge of heterogeneity in the IoT, particularly the integration of resource-constrained devices, a Minimum Broker (MB) has been proposed [9], which contains the core features necessary for a device to be part of the Swarm. Figure 7.11 depicts the interaction between Minimum Broker and common Swarm Broker, focusing on the discovery process.

The simplest possible behavior of a Swarm participant is to be able to be discovered and to provide information, as a simple sensor does. It is not hence necessary to support the creation of new locate requests, and only a simplified

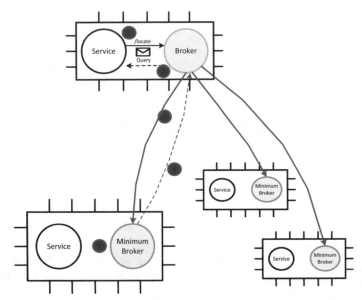

Figure 7.11   The discovery process in the Swarm minimum broker.

implementation for answering locate requests suffices. While the complete process for responding to location requests considers queries that arrive via either unicast (e.g. HTTP) or multicast (e.g. SSDP) and supports forwarding queries, the locate service in the MB only supports queries arriving via multicast. The reception of unicast locate messages was not considered an essential feature, as it is superseded by multicast in local networks, and would only work with the pre-requisite that a remote broker already knows the address of the Minimum Broker.

7.5.2.7 The future of platforms for the Internet of Things

In [10], G. Rzevski foresees the IoT as a complex network of devices, characterized by the seven properties of a complex system. *Connectivity*, with heterogeneous devices richly interconnected in a global network; *autonomy* of behavior, as edge devices become more intelligent; *emergent behavior* that results from device interaction; *nonequilibrium*, a common characteristic from markets is applicable to the future IoT when an economic model for resource trading is widely adopted; relations between participants are *non-linear*; thus, a small input can result in a large event (butterfly effect); *self-organization* is the ability to change the behavior or structure to adapt to unexpected events; *co-evolution* states that as the IoT network changes, its environment also changes in an inevitable and irreversible way.

The characteristics above describe a common scenario that platforms for the future IoT should consider. As devices become more capable, richer interactions lead to the need of an adaptable, self-organizing platform for device cooperation. The common cloud-centric architecture of most IoT applications will be superseded by decentralized and distributed platforms, where edge devices will have a greater protagonist. Interaction models based on multi-agent systems will be the basis for the future IoT. The SwarmOS constitutes a step in this direction.

7.6 Regulation, Security and Privacy

The regulatory field in Brazil has been mainly affected by the actions of the National Agency of Telecommunications (ANATEL), and the LGPD. While Anatel is working towards reducing barriers to IoT large-scale adoption, the LGPD paves the legislative ground to protect the privacy of customers in all economic sectors, including the IoT.

7.6.1 Telecommunication Regulation

Anatel, an independent organization that has regulated and supervised telecommunication services in Brazil since 1997, is working towards the flexibilization and reduction of barriers to the expansion of IoT and M2M applications. In August 2018, and then again in August 2019, Anatel released a public consultation to receive inputs from society for 45 days, which are to be considered during the creation of new regulatory policies for IoT in Brazil [11]. These policies are expected to make the exploration of telecommunication services more flexible, facilitate the setup of roaming, and provide consumers with easy access to details regarding the service level they have contracted. The new regulation is expected to be approved by the end of 2020, and it is aligned with the IoT National Plan, which seeks to implement and push forward the IoT in Brazil.

A topic discussed recently within the agency is whether a specific service type for IoT applications should be created by the agency, with the ultimate decision being that it should not [12]. The main reason is that Anatel considers that the existing regulations for radiofrequency and telecommunications already allow a vast range of applications, and it would be easier to perform small changes to existing rules than to create a completely new regulation.

7.6.2 General Law of Data Protection

The use of IoT solutions that collect data in large scale raises concerns about privacy. Aligned with global concerns over the topic, such as the European General Data Protection Regulation (GDPR), the Brazilian Congress passed the General Law of Data Protection (LGPD), whose main goal is to enhance the privacy of personal data and to allow greater control over it from a consumer perspective. It also creates clear rules for how data should be treated by organizations and strengthens the power of regulatory agencies to perform control [PR2018]. According to a comparative analysis, the differences between LGPD and GDPR are minor, and the LGPD can be referred to as a "GDPR à la Brasileira" [13].

Approved in August 2018, the actual law enforcement is predicted to begin only on February 2020, so as to give companies an 18-month interval to adjust to the new regulation. The LGPD concerns all economy sectors and applies to every company that collects data in Brazil, independently of its source country. Therefore, every IoT company with operations in Brazil will need to comply with it. The law also provides that companies can only collect personal data with the consent of the users, which can request access to their data and demand its complete erasure at any time. Violations of the law may entail warnings, fines, and even partial or full suspension of operations, depending on the severity of the case. Regarding fines, the values may vary from 2% of the past year revenue to R$ 50 million, with the addition of daily penalties [14, 17].

7.7 Conclusions

The Brazilian National Plan for the Internet of Things helped to formalize and to converge ongoing IoT initiatives and to promote new ones. In this work, we summarized some selected work towards the accomplishment of the premises settled by the National Plan. Those initiatives were carried out by the University of São Paulo, the National Telecommunication Agency (Anatel), and the Brazilian Government comprehending strategic areas. The Code IoT education platform constitutes an investment in human capital, which has a direct impact on future technical developments. The Caninos Loucos Single Board Computer family and the SwarmOS IoT platform together constitute a national software and hardware platform for IoT. The five applications: Smart traffic lights, Smart surveillance, health monitoring of childhood cancer patients, sleep apnea diagnosis, and Internet of Turtles, are representative examples of innovative application for smart cities, health and environment domains. Finally, the flexibilization of regulations by Anatel and the creation of the General Law of Data Protection (LGPD) constitute the advances in regulation, security and privacy.

The examples above constitute concrete efforts towards an extensive adoption of IoT technology in strategic areas. Although all these use cases have substantial results, their development continues as there is a clear demand for further advances.

Acknowledgements

The content of this chapter summarizes the work of different projects, coordinated by Brazilian institutions, such as the Centro Interdisciplinar de Tecnologias Interativas (CITI) from the University of São Paulo, and LSI-TEC.

Also, a number of institutions provided financial support for developing these projects, such as SMART Modular Technologies[6], the Ministry of Science, Technology, Innovation and Communications of Brazil (MCTIC)[7], the Brazilian Development Bank (BNDES), the Coordination for the Improvement of Higher Education Personnel (CAPES)[8], the National Council for Scientific and Technological Development (CNPq)[9], the University of São Paulo (USP)[10], Samsung[11], LG Electronics[12], and Santander Bank[13].

Further acknowledgements include:

- Caninos Loucos: Mr. August R. Machado, Mr. Tadeu M. Frutuoso, Eng. Sílvio Dutra, Dr. Casimiro de A. Barreto, Eng. Edgar Righi, Eng. Marcelo Ordonez, Eng. Guilherme Garcia, Eng. Sérgio de Paula, Mr. Mário Nagamura. Special thanks to Jon 'Maddog' Hall for his contribution and inspiration to the project, including the project name (*Caninos loucos* means *mad dogs* in Portuguese).

---

[6]https://www.smartm.com
[7]http://www.brazil.gov.br/government/ministers/science-technology-innovation-and-communications
[8]http://www.capes.gov.br/
[9]http://www.cnpq.br/
[10]http://www.usp.br
[11]https://www.samsung.com/us/
[12]https://www.lg.com/br
[13]https://www.santander.com.br/

- SwarmOS: Mr. Gabriel M. Duarte, Mr. Phillipe S. Rangel, Mr. Guilherme C. Marques, Mr. Gustavo Rubo, Mr. Carlos E. Laschi, Mr. Matheus B. Guinezi, Mr. Renan Oliveira, Mr. Douglas Navarro, Mr. Rafael C. Sales, Mr. John Esquiagola, Dr. Flávio S. C. da Silva.
- CodeIoT: Dr. Ana G. D. Correa, Dr. Marcelo A. José, Eng. Alexandre A. Martinazzo, Eng. Cassia Fernandez, Eng. Isabela Angelo, Eng. Leandro Coletto Biazon, Ms. Elena Saggio, Mr. Erich P. Lotto, Mr. Fábio G. Durand, Mr. Mário Nagamura, Mr. Rodrigo Suigh, Ms. Letícia Lopes, Mr. Renatto O. M. Domingues, Mr. Charles R. Silva, Ms. Julian Lepick, Sra. Lídia Chaib, Mr. Yohan Takai, Sra. Ohanna J. do Amaral, Mr. Migyael G. T. Vieira.

References

[1] Ministério da Ciência, Tecnologia, Inovações e Comunicações (MCTIC). "Decreto que institui o Plano Nacional de Internet das Coisas é publicado". online at: https://www.mctic.gov.br/mctic/opencms/salaI mprensa/noticias/arquivos/2019/06/Decreto_que_institui_o_Plano_Nac ional_de_Internet_das_Coisas_e_publicado.html

[2] J. M. Rabaey, "The swarm at the edge of the cloud-a new perspective on wireless."2011 Symposium on VLSI Circuits-Digest of Technical Papers. IEEE, 2011.

[3] E. A. Lee, et al. "The swarm at the edge of the cloud." IEEE Design & Test 31.3, 2014, pp. 8–20.

[4] L. Costa, et al. "Swarm os control plane: an architecture proposal for heterogeneous and organic networks." *IEEE Transactions on Consumer Electronics* 61.4, 2015, pp. 454–462.

[5] J. Lewis and M. Fowler. "Microservices." *martinfowler. Com*, 2014.

[6] P. C. Calcina-Ccori, et al. "Enabling Semantic Discovery in the Swarm." IEEE Transactions on Consumer Electronics 65.1, 2018, pp. 57–63.

[7] L. De Biase, et al. "Swarm economy: a model for transactions in a distributed and organic IoT platform." *IEEE Internet of Things Journal*, 2018.

[8] V. C. Hu, et al. "Guide to attribute-based access control (ABAC) definition and considerations (draft)." *NIST special publication* 800.162, 2013.

[9] L. De Biase, et al. "Swarm Minimum Broker: an approach to deal with the Internet of Things heterogeneity." 2018 Global Internet of Things Summit (GIoTS). IEEE, 2018.

[10] G. Rzevski and P. Skobelev. "Managing complexity". Wit Press, 2014.

[11] Agência Nacional de Telecomunicações (ANATEL). "ANATEL aprova consulta pública para diminuir barreiras à expansão de IoT e M2M no Brasil". online at: https://www.anatel.gov.br/institucional/noticias-destaque/2333-anatel-aprova-consulta-publica-para-diminuir-barreiras-a-expansao-de-iot-e-m2m-no-brasil

[12] Agência Nacional de Telecomunicações (ANATEL). "CONSULTA PÚBLICA N° 39", online at: https://sistemas.anatel.gov.br/SACP/Contribuicoes/TextoConsulta.asp?CodProcesso=C2268

[13] C. Perrone and S. Strassburger. "Privacy and Data Protection-From Europe to Brazil." *Panorama of Brazilian Law* 6.9-10, 2018, pp. 82–100.

[14] G. Camargo. "LGPD: 10 pontos para entender a nova lei de proteção de dados no Brasil" Computerworld, online at: https://computerworld.com.br/2018/09/19/lgpd-10-pontos-para-entender-a-nova-lei-de-protecao-de-dados-no-brasil/

[15] L. De Biase, et al. "Swarm Minimum Broker: an approach to deal with the Internet of Things heterogeneity." *2018 Global Internet of Things Summit (GIoTS)*, IEEE, 2018.

[16] G. Fedrecheski, et al. "Attribute-Based Access Control for the Swarm with Distributed Policy Management." *IEEE Transactions on Consumer Electronics* 65.1, 2019, pp. 90–98.

[17] Presidência da República. "Lei Geral de Proteção de Dados Pessoais (LGPD)", online at: http://www.planalto.gov.br/ccivil_03/_ato2015-2018/2018/lei/L13709.htm

[18] L. De Biase, et al. "Swarm economy: a model for transactions in a distributed and organic IoT platform." IEEE Internet of Things Journal, 2018.

[19] F. Marques. "Brazil's Internet of Things". Pesquisa FAPESP, 259, 2017.

8

IoT Technologies and Applications in Tourism and Travel Industries

M. Dolores Ordóñez[1], Andrea Gómez[2], Maurici Ruiz[3],
Juan Manuel Ortells[1], Hanna Niemi-Hugaerts[4], Carlos Juiz[3],
Antonio Jara[2] and Tayrne Alexandra Butler[1]

[1]AnySolution, Spain
[2]HOPU, Spain
[3]University of the Balearic Islands, Spain
[4]Forum Virium, Finland

Abstract

Tourism is one of the most dynamic industries in the world, but it has also an industry to be greatly and directly affected by the 4th revolution, as can be evidenced by the impact of the internet in the evolution of this industry. Disruptive technologies, including IoT play a crucial role in the way of understanding and managing this industry and especially in how the offer and demand are linked. The great diversity of IoT applications in the tourism industry defines the competitiveness not only of the private companies involved but also of the destinations which are being transformed into Smart Destinations as a natural evolution from Smart Cities, which are influenced by the tourism sector. Smart Cities are characterised by smartly managing different areas. Smart Destinations require this smart management as well as the integration of the stakeholders' value-chain throughout the entire process. In this process, IoT has a crucial role in enhancing the experiences of tourists, to more efficiently manage the destination, and to offer a channel of information exchange. As a result, a more efficient destination, with better and more personalised experiences will improve the competitiveness of the tourism destination and of the quality of life of its citizens.

8.1 Introduction

Europe is one of the world's leaders when it comes to tourism, welcoming 713 million tourists in 2018 [1]. Europe's top position in worldwide tourism can be maintained and enhanced by using new technologies such as IoT and artificial intelligence to automate different processes in the industry according to the needs of a new kind of visitor. The visitor's profile needs to include capabilities that manage tourist experiences by use of Smartphones, including transport, tour guides, bookings and information in real-time. Companies in the tourism sector also need to use this technology to better understand the needs of the visitor, as a channel to discover and recover data to build future smart strategies for the tourism industry's entire value chain. A consequence of the industry's globalisation is the emergence of new tourism destinations, and an extraordinarily high level of development in tourism activity, especially in urban environments, a clear example being European capitals. This rise in the number of people visiting different tourism destinations and cities, together with the increasing importance of sustainability in the management of tourism destinations and urban areas, has generated the pressing need for new technological developments that can help to make destination and city management more efficient and productive.

New disruptive technologies, such as IoT, artificial intelligence, and distributed ledger technologies (DLTs), are crucial to achieving more efficiency and productivity in Tourism and Travel Industries. The ability to collect data from different types of sources and transport it to platforms, where it is then analysed and used to improve decision-making processes thanks to Big Data, is the first step to highlighting the importance of IoT in Smart Cities. To a high extent, this has been automatically transferred to tourism destinations, therefore generating what we call Smart Destinations. Here it is essential to point out that Smart Destinations are no more than Smart Cities but with an extra layer of complexity when it comes to management, which is simply tourism activity. So, all Smart City applications using IoT and other technologies are directly applicable in Smart Destinations. Beyond this, the fact that tourism activities are carried out in Smart Cities increases the need for disruptive technologies and places people even more at the centre. And, if Smart Cities intend to improve the quality of life of its citizens, Smart Destinations aim to improve the quality of life of every single person living or spending time in a tourism destination.

This entire process would not be complete without mentioning the importance of transforming tourists themselves into digital tourists. These tourists have changed their travel habits, and now stay in places for shorter periods, but travel more frequently and use mobile devices, especially Smartphones during almost every stage of their trip (before/preparation – during/stay over – after/memories). These tourists who are and want to be permanently connected, generate and consume data during the entire life cycle of their trip, and they also require a whole range of different digital services to help make their experience better.

The personalisation of experiences is another issue where IoT has a key role, and this also has a direct consequence on the competitiveness level of the different destinations.

The changes generated by the tourism industry's new trends, together with the tourists' new needs and requirements and the demand for tourism destinations to continue being competitive, have been the basis for the concept of 'Smart Tourism Destinations' emerging in Spain [2].

This concept is supported by the Smart Tourism Destinations (STD) UNE 178501 standard that promotes standardisation by establishing governance, accessibility, sustainability, innovation, and technology as the fundamental pillars for a destination to be considered *Smart*. Therefore, a STD or Smart Destination is not a tourism destination that only sets up digital infrastructures capable of improving interaction with digital tourists, but a destination where the final goal is to make the territory itself Smart by promoting sustainable and efficient management together with economic development to therefore improve the quality of life of every person living/staying/visiting the tourism destination.

In the EU and without having regulations in this sense, the European Commission has launched the European Capital of Smart Tourism initiative [3] that recognises the achievements of cities as tourism destinations in 4 areas: sustainability, accessibility, digitalisation, and cultural heritage and creativity.

Some of the most relevant definitions of *Smart Destination* are:

- UNE 178501 Standard for STDs "An innovative area for tourism, accessible to everybody, consolidated on the basis of cutting-edge technological infrastructure that guarantees the sustainable development of a specific territory, facilitates the interaction and integration of visitors with their surroundings and increases the quality of their experience while in the tourism destination, as well as the quality of life of the people living in the area.
- Organisations such as SEGITTUR defined the concept of Smart Tourism Destination as [4]: "A Smart Tourism Destination is an innovative tourism destination, consolidated and based on cutting-edge technological infrastructure and the tourism territory itself. An area highly committed to environmental sustainability, culture and socio-economic

progress, provided with a smart system that collects information in a procedural way, analyses, and understands events in real-time, making it easier for visitors to interact with their surroundings and for managers to make decisions, increasing their efficiency and substantially improving the quality of the experience".

8.2 IoT Technologies and the Tourism Sector

The tourism industry consists of multiple stakeholders and impacts global economy. Customer experience and personalised travel in the tourism industry are at the top of the list when it comes to IoT improvements for the industry. This is only possible if local infrastructure allows for the deployment of devices that provide information and the necessary data to make decisions in real-time, as well as information that helps to carry out simulations for the prediction of future scenarios. It is therefore essential for tourism destinations to implement comprehensive information collection, analysis, and distribution systems among all of the actors included in the tourism destination's value chain, thereby facilitating the decision-making process for each of them, in real-time.

According to Antonio López de Ávila, former president of SEGITTUR: "Intending to become a Smart Destination means that the tourism destination will need to implement a strategy that will increase its value by using innovation and technology. This process will help to increase competitiveness, not only because tourism resources will be used better but also because other resources will be identified and created; improving efficiency in marketing and production processes; and using renewable energy sources. Everything

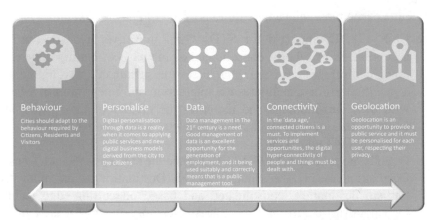

Behaviour
Cities should adapt to the behaviour required by Citizens, Residents and Visitors

Personalise
Digital personalisation through data is a reality when it comes to applying public services and new digital business models derived from the city to the citizens

Data
Data management in The 21ˢᵗ century is a need. Good management of data is an excellent opportunity for the generation of employment, and it being used suitably and correctly means that is a public management tool.

Connectivity
In the 'data age,' connected citizens is a must. To implement services and opportunities, the digital hyper-connectivity of people and things must be dealt with.

Geolocation
Geolocation is an opportunity to provide a public service and it must be personalised for each user, respecting their privacy.

Figure 8.1   Personalisation of services in Smart Tourism Destinations.

must be focused on boosting sustainable development in tourism destinations based on three aspects (environmental, economic, and socio-cultural elements), and subsequently, by improving the quality of the visitors' stay and the quality of life of its residents. In this way and in the short-term, the marketing and production processes will improve and be more efficient, employment and tax revenues will rise, and most importantly, overall satisfaction levels will also see a clear improvement" [4].

One of the most critical aspects of using IoT in tourism is the ability to personalise what's on offer based on the information collected and obtained from the tourists' connectivity itself.

The fact of knowing what tourists like and don't like will help to personalise what is offered to them, by sending them 'custom-made' information and therefore fulfiling their expectations, and subsequently making them loyal customers. If we know one of our clients likes vegan food, we can send them information about culinary workshops they can go to or restaurants specialised in this kind of food. In fact, if he/she were walking near to one of these restaurants and we knew they liked eating at a particular time, we could send them an alarm telling them what restaurants are in the area, help them to book a table and tell them how to get there; and all of this without realising it, but thanks to IoT.

The range of applications of IoT technologies in SDs is extensive, and every day it is enriched with new devices and functionalities. Table 8.1 shows the primary sensors that would be applicable.

Table 8.1 Types of sensors and applications in the frame of IoT and tourism

Type of Sensor	Objectives	Applications
Temperature	Measure the amount of thermic energy	Cooling system control. Applications: manufacturing processes, production of agricultural systems
Proximity	Identify the presence or absence of nearby objects. They can also identify the object's properties	Presence control. Applications: Movement of people, the presence of tourists, alarm system, and safety. Online information on offers for tourists. Available parking spaces
Pressure	Identify pressure and changes.	Maintenance of water and heating systems
Water quality	Identify water quality and monitor ions (chlorine, organic carbon, pH, etc.)	Controls water distribution systems Quality of water for swimming

(Continued)

Table 8.1   Continued

Type of Sensor	Objectives	Applications
Chemical/Gas	Identify changes in chemical composition in liquids and in the air	Environmental monitoring, contamination control (CO_2, CO, O_3, O_2, NO_2, NO_3, particles, etc. Safety and security systems
Smoke	Smoke detectors	Safety and security systems from fires
IR	Infrared sensor that identifies environmental characteristics	Health care (blood flow and arterial pressure, etc.)
Level sensors	Determine the level of the amount of fluid, liquid flowing in an open or closed system	Water deposit, fuel levels
Image sensors	Take pictures and videos	Safety. Car, medical . . . images
Movement sensors	Detect physical movement in a certain area	Safety. Intruders, tolls, manual driers, lights, Air conditioning, parking, fans
Accelerometer sensors	Measure the physical acceleration of objects. They detect vibrations, inclination, and acceleration	Anti-theft systems. Sports monitoring
Gyroscope sensors	Measure the angular speed Velocity around an axis	Car, drones . . . navigation
Humidity sensors	Very similar to temperature sensors	Air conditioning control. Hospitals and pharmaceutical industries
Optic sensors	Number of rays of light and electromagnetic energy	Electric energy control. Environmental and energy control
Positioning sensors	Provide geographical coordinates to position objects	

8.2.1 IoT Applications in the Tourism Sector

Some of the most important IoT examples in the travel industry are the following:

Personal Control

One of the most widespread uses of IoT technology within the travel industry so far, has been the possibility of enabling a higher degree of personalisation in hotels, and on flights, and this is primarily achieved by allowing customers to control more appliances or services through a centralised device, such as a tablet or even their own Smartphone.

By implementing internet-enabled heating, lighting, and television, customers can turn these appliances on or off from wherever they are. They may even be able to choose a specific temperature and light level and have the devices maintain those levels automatically. Similar technology can also be used on flights, regulating seat temperatures and air conditioning.

Seamless Travel

Another excellent use for IoT involves streamlining customers' experiences as much as possible, across all areas of the travel industry. In airports, this may mean using sensors and sending information to passengers' Smartphones, warning them when their luggage is nearby, and allowing them to locate it faster.

In hotels, the check-in process can be made seamless, with hotels sending electronic key cards to guests' phones, which, when used, automatically check them in without them even having to stop at the front desk. Sensors may also be used to warn restaurant staff when guests arrive, and automatically send them the right table number.

Smart Energy Saving

While IoT can enable personalisation, it can also offer businesses financial benefits through automated or smart energy saving. In a hotel, for instance, internet-enabled devices and sensors can allow for the room temperature to be adjusted continually, meaning heating is only used when it is really needed.

A similar principle can also apply to lighting. Some hotels already use IoT technology to control when lights are turned on and off. Sensors automatically detect the levels of natural light in the room, reducing the power of light bulbs in the process, meaning less energy is wasted, and high-powered lighting is only used when natural light is not enough.

Location Information

Companies working in the travel industry can also use the Internet of Things to send location-specific information to customers and to also gather other valuable data. By combining Smartphone capabilities with beacon technology or other sensors, messages can be sent to tourists when most relevant and based on where they are.

For instance, messages about local attractions could be sent with information on the times when they are least busy, or messages pointing out nearby public transport services could also be sent, as well as messages and alarms depending on when people are using specific hotel facilities at different times so that the amount of staff needed is adjusted.

Maintenance and Repairs

Finally, the Internet of Things can also be used to directly benefit IoT devices by providing valuable, real-time information about their current status and working order. This can be vital for many of those working in the travel and tourism industry, allowing for essential devices to be repaired or replaced before they stop functioning.

For example, hotel staff can be warned if a radiator or light bulb starts to deteriorate. As well as in hotels, the Internet of Things can also be deployed to allow airlines to fuel airplanes more efficiently or replace parts at the right time, striking the ideal balance between gaining maximum value and maintaining safety.

Ultimately, IoT involves adding internet connectivity to everyday devices and appliances, allowing them to communicate with one another, and this offers numerous benefits for those working in the travel industry, including the ability to deliver a greater customer experience and to optimise internal procedures.

8.2.2 User/Tourist Experience

As a context of Smart Tourism Destination, it is essential to understand the role of Smart Cities as a domain where cities seek to build sustainable and smart strategies for city growth, taking into consideration its people, institutions, and finally, the technology. Because of the relevance of the tourism economic activity, the sector also needs to use smart strategies and the technological infrastructures of Smart Cities, having given rise to the Smart Tourism Destination as a new domain of research, solutions, and tools [5].

In the domain of Smart Tourism Destinations, technology could be used in different sectors depending on the aspects it covers. Smart Tourism brings together all of the tools and deployments to interact with the physical environment as tourist guides, Virtual Reality (VR) and Augmented Reality (AR), among others. On the other hand, the e-tourism cluster joins technological enablers that can be used prior to, during, and after a visit/trip as a planning tool (booking of hotels, restaurants, etc.) [6].

The impact that Smartphones have on people's lives, being a device used for everything to do with our daily lives, has led to the emergence of what is called mobile tourism. This concept summarises the current situation of tourism, where visitors do everything with their Smartphones, for example making reservations and getting information on places related to the heritage sites visited [7].

Under this new approach in the tourism sector, which places ICTs as a critical element, the use of Smartphones has become the main point of analysis in companies, it is the point of contact between visitors and the Smart Tourism Destination. Hence, the needs and preferences of users who visit a certain area can be known, and therefore supply can be adapted to demand [8–10].

As another important aspect in the new visitor profile, in addition to the use of technology, it should be noted that with the active incorporation of the Millennial, Z and hashtag generation to the tourism sector as clients, destinations must start to offer personalised and exclusive experiences, forgetting about "tourist packages" as such that are currently marketed by travel agencies. This generation, that have the need to visit new places, want to go to places where they feel like they are at home (short time citizens) and where they will be able to live once-in-a-lifetime experiences [10].

8.2.3 Disruptive Technologies in Smart Tourism Destinations

ICTs have opened up a multitude of markets in all sectors. Cities and tourism destinations, as already mentioned, have also been affected, and currently, managers and institutions require technological solutions that generate the sustainable strategies that cities need.

Although ICTs are one of the great stars of this new market (Smart Cities & Smart Tourism Destinations), the authors in [11] stress that all deployments and technological infrastructures used must be based on the needs of people and institutions.

The technological pillar of Smart Cities and Smart Tourism Destinations could be summarised into three types of technologies: IoT that refers to the connection between objects and humans, Big Data, as structured and non-structured data, generated by IoT infrastructure and other processes and Smart Cities Platforms that manage the Big Data and IoT in cities [12, 13].

The IoT concept was born in 1999 in the Technological Institute of Massachusetts (MIT) by the hand of Kevin Ashton and referred to the situation that was beginning to be seen with respect to the number of objects and people that were already connected to the network [12].

The work of Pathak defines, in this concept, four points to understand it: a device that connects objects to the Internet, wireless networks, data collected in the cloud and capacity for analysis of this data, also seen as a key feature: The bi-directionality of the information, that is to say, the possibility of establishing communication with an object that is connected and

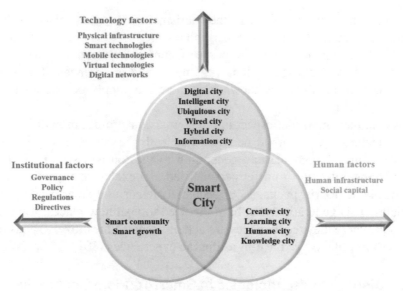

Figure 8.2   The main pillars of a Smart City.

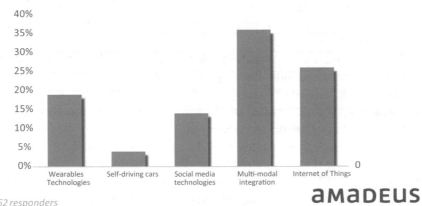

Figure 8.3   Disruptive technologies in tourism.

Top Five Emerging Technologies in 2017

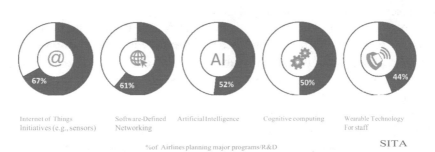

Figure 8.4   Top five emerging technologies in 2017.

to receive an answer. One could conclude that the definition of the concept such as the conversion of the things or traditional objects into intelligent ones: Smart [12, 14].

IoT [15] has the potential to have a significant impact on businesses, automating processes without any human-to-computer or human-to-human interaction. Based on the sheer volume of adoption and the ability to offer advanced interconnection and communication among devices, systems, and services, the Internet of Things is expected to disrupt the travel industry. According to SITA 2015 Airline IT Trends Survey, 86% of all airline companies believe that IoT will provide "clear benefits over the next three years." IoT is streamlining the end operations of hotels, airlines, and other travel companies by connecting smart devices, systems, and processes. By taking advantage of IoT technology, the travel industry can implement increased operational efficiency and more personalised guest experiences.

8.3 IoT Applications for the Tourism Value Chain

8.3.1 IoT and Hotels

As stated in an article by Dr. Ajay Aluri West Virginia University, "The IoT platform is the answer to scientific management in the digital life – a "shortcut" to get things done efficiently and effectively for both consumers and businesses" [16]. The boom in IoT Technology will boost the future of the hospitality industry; it will provide a competitive edge in the market and through the interconnection of devices, (sensors, actuators, identification tags, mobile, etc.) through the internet. IoT is no longer just a concept, it is very much a part of the industry and statistics are growing because IoT enables processes, data, and outcomes [17].

An example of a real use case is NADIA and its applications for the tourism sector. The introduction of IoT in the field of new technologies has not gone unnoticed in the tourism sector, which has probably been the sector to have most benefited from the age of digitisation. However, there are still very few specific IoT applications in this sector.

NADIA was launched in 2017 with the idea of exhibiting the great potential IoT has in the tourism sector. An example of this is '*Hotel Room' by NADIA.* by installing different types of sensors in hotel rooms, this platform helps to monitor the use of the sensors at all times.

It may seem that this is just another simple *Domotics* project, but NADIA's great potential is based on two aspects:

On the one hand, **hotel customers** who can use an App or a browser to set the temperature in their hotel rooms, turn on the television, chose the colour of the LED lights, etc. Allowing them to customise their room and enjoy a unique experience.

On the other hand, **hotel management**, this is a powerful tool to remotely control facilities by turning off lights or air conditioning systems . . . in empty rooms.

Figure 8.5   Hotel Room by NADIA.

Figure 8.6   Sensors and IoT devices used for environmental monitoring in hotels.

But, NADIA's real potential comes from its excellent analysis capacity. For the first time, hotel managers will be able to know how long customers are in their rooms for, their schedules, what they like, etc. From here on, the options available are endless and go from identifying things customers are unsatisfied with before they even happen, adapting the times for cleaning their rooms, or fixing something; identifying things that are not working properly, blown bulbs, cooling systems . . . ; planning times to enter the rooms and fixing things, knowing that the room is empty. If to all of these new features, we add the possibility of making the system smart, we are taking a giant step towards the optimisation of energy consumption without having to affect the comfort and well-being of guests.

8.3.2 IoT and Airports

Airports are the main points of entry for tourism destinations, and they are places where IoT has more direct applications, in the buildings themselves, to improve efficiency and generate savings as well as in relation to users, gaining information, improving services, and personalising supply. But it is also important to remember that airports are hubs that connect the whole world, and this has direct implications related to safety and security.

The authors in [17] state that "Airport council international ACI confirms the potential of IoT in airports and the aviation industry through the operational improvements and data exchange among the stakeholders.

Moreover, the data sharing among the collaborative stakeholders will enable them to make better decisions leading to better customer service at passenger screenings, checkpoint management, and identity management by real-time processing in the lanes and to the border and security agencies. Technology is creating new opportunities at a low cost for the air transport industry and it is ready to transform many novel techniques such as improved connectivity to airplanes and baggage tracking etc. available globally and also easy to deploy.

8.3.3 IoT and Tourist Attractions

In Smart Tourism Destinations, it is important to include a smart structure that allows interaction among visitors and sectors that make it up, as an innovation that opens the frontiers to a greater diversity of the public, as well as institutions and companies that market them [18].

Analysing Smart Tourism Destinations, the perception of visitors must be taken into account in the same way as in Smart Cities. People a primary key pillar. According to relevant authors in this field, we must approach the tourism destination as a brand, including the destination's companies and offers. Everything must be presented as a global, agile and unified experience for visitors, facilitating their interaction with it [12, 19].

The visit to a destination should be understood as a way to obtain new experiences. Visitors seek to consume a product that can be considered as emotional. Therefore, in the case of Smart Tourism Destinations, technology must contribute to an immersion where visitors get information and interact with the environment as part of the network.

Another critical factor is that visitors seek a fluid experience where all the information they need during their stay is easily accessible (such as obtaining information, reservations, purchases, etc.). When the user enjoys a territory the rupture that the lack of access to information could cause the rupture of that immersion, decreasing their sense of satisfaction [20].

For this reason, technologies such as IoT, that allow for the creation of contextualised experiences in Smart Tourism Destinations and which interact with the users' Smartphones, enable the generation of products and services that adapt better to this new visitor profile.

Among the new tools that use IoT in Smart Tourism Destinations, one to underline is the *Be Memories* project, an innovative tourist guide that disseminates the intangible heritage of a destination with the use of IoT and through the collaboration of the residents as content creators. This solution has been deployed in Ceutí (Spain) and Bristol (United Kingdom). *Be Memories* is based on two lines of innovation [10]:

- An Agile channel to interact (IoT): Due to devices called Smart Spots, which create an open Wi-Fi network in different cultural points and tourist attractions of the city, *Be Memories* generates smart areas known as Smart POIs (Smart points of interaction). Visitors can interact with *Be Memories* using their Smartphones. This Wi-Fi network automatically opens a Web-App stored in the device with the cultural content about the area.
- Intangible heritage content for visitors making them feel like short term citizens: *Be Memories* disseminates the widespread knowledge of residents (trailer of the Be Memories content: https://bememories. hopu.eu/#/main), integrating visitors into the city's network. Also, the content is presented in a short video with interviews that visitors and residents can watch and listen to while visiting the different POIs in the cities where these are enabled.

TreSight is another relevant project that uses IoT for Smart Tourism. It has been set up in the cosmopolitan city of Trento (Italy), and it uses IoT technologies to contribute to the immersion of visitors in the experience. The purpose of this solution is to alternate the innovation in the field of tourism with the cultural charm typical of the area, taking care of their cultural heritage. It consists of a tool that alternates the offer of services in the area,

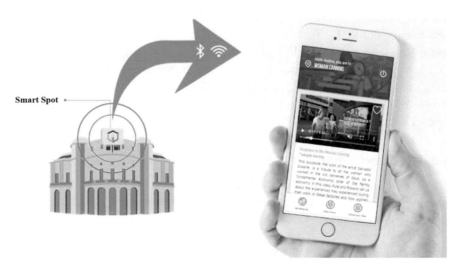

Smart POI: Smart Point of Interaction (area)

Smart Spot

Figure 8.7 Smart POI of Be Memories project.

Figure 8.8   Web-App of Be Memories.

shops, etc. making use of Open Data from Trento (Open Data Trentino) where you can find information about points of interest, temperature, typical restaurants, etc. [21]. To use *Tresight*, visitors receives a Wearable bracelet and a link to download an app that offers them information and suggestions. The bracelet sends data about the user's location and environment, therefore enhancing the Open Data. The role of the Wearable is to be a tool that analyses the zones in real-time and provides a database with more real updates than the Open Data [12].

8.3.4 IoT and Smart Nature Destinations

The Urban Eco Islands project [22] will turn the Vasikkasaari island in Helsinki and on Aegna island in Tallinn into new types of smart nature tourism destinations.

The islands will be developed based on the principles of sustainable tourism, taking into account the sensitive archipelago nature. At the same time, the project will examine how experience-based and sustainable tourism can be promoted with new, innovative digital solutions.

The aim is to develop the areas together with the visitors by increasing environmental and nature awareness and attractiveness for both locals and tourists. The accumulated sensor data is visualised and processed further to show more extensively how the area interests' visitors. IoT applications supporting smart nature destinations are listed below:

- Smart IoT solutions can create new value for nature tourists as well as help to conserve the fragile environment on a local and global level. Based on measuring weather condition data, e.g., wind, temperature, humidity, and UV-intensity, recommendations about clothing and personal protection can be given.
- Acoustic and video sensors on sensitive nature areas could provide detailed up-to-date information to tourists as well as the scientific community about nature values and the biodiversity of the area. Virtual view of an interesting spot, e.g., a sensitive remote bird area with nesting birds is a way to reduce tourism-related pressure to the environment since in this way tourists can enjoy the experience without physically visiting the spot.
- Drone and UAV based imaging can be used to monitor the impact of visitors in the area. Time series of images can reveal new paths created as a consequence of erosion caused by visitors.

8.4 Conclusion

IoT still has a lot of evolving to do, and it will revolutionise the travel industry and the tourism sector in general. With IoT, things will be a whole lot easier, we will be able to check-in and out of hotels automatically, find our travel destination a lot easier, monitor the performance of airline engines, etc. IoT will help to improve customer services and increase revenues as well as customer loyalty.

With IoT, vast amounts of data will be at our disposal, but this data needs to be analysed and understood; so, for appropriate solutions to be implemented, investments need to be made in suitable technology.

For companies in the travel industry to benefit from future innovations, they need to start incorporating IoT into their systems now. IoT technology is very relevant in the tourism industry and both tourists and managers are adapting to this new era because sooner than we think, IoT will overtake the whole cycle of operations in the travel industry and this is what tourists expect more and more. Very soon, this will not only be what they expect, but it will become a real need and requirement.

Nonetheless, the deployment of IoT in tourist destinations needs to have adequate communications and data management infrastructure. Vast amounts of data will be at the disposal of tourist operators, but this data needs to be analysed and understood. In this sense, it will also be necessary to have information quality control systems and a good knowledge of data management. Likewise, it will be a priority developing a framework for collaboration between public and private initiatives to promote the development and exploitation of IoT platforms.

For companies in the travel industry to benefit from future innovations, they need to start incorporating IoT into their systems as soon as possible. IoT technology is essential in the tourism industry, and both tourists and managers are adapting to this new era sooner than what was expected.

Acknowledgements

The fact of it being possible for us to include an article about IoT and Tourism is in itself a milestone for the acknowledgment of the importance of this sector in IoT as well as in other disruptive technologies. This would not have been possible if projects like CREATE-IoT had not identified its enormous potential and the importance of this sector for the future of IoT. For this reason, we would like to give a very special thank you to all the CREATE-IoT partners for supporting us and making it possible for us to write this article and lay the foundations for the development of a sector that is as strategic as this one in the European Union. HOP Ubiquitous thanks to Ceutí City Council as a one of our City Labs, Synchronicity Large-Scale Pilot Project (732240) and Walk a Story ERASMUS + project (2018-1-DK01-KA202-047095).

List of Notations and Abbreviations

UNWTO	United Nations World Tourism Organisation
SEGITTUR	State Mercantile Society for Tourism Technologies and Innovation Management
IoT	Internet of Things
STD	Smart Tourism Destination
UNE	Spanish Association for Standardisation
ICT	Information and Communication Technology
VR	Virtual reality
AR	Augmented Reality

References

[1] "International Tourist Arrivals Reach 1.4 billion Two Years Ahead of Forecasts," UNWTO (United Nations World Tourism Organization), 21st January 2019, online at: https://www2.unwto.org/press-release/2019 -01-21/international-tourist-arrivals-reach-14-billion-two-years-ahead-forecasts.

[2] C. R. a. E. S. M. Spanish Government Presidency, "IET/2481/2012," Spanish Official State Bulletin (BOE – Boletín Oficial del Estado), 20 November 2012, online at: https://www.boe.es/diario_boe/txt.php?id= BOE-A-2012-14286.

[3] E. C. o. S. Tourism, "Home," European Capital of Smart Tourism, 2019, online at: https://smarttourismcapital.eu/.

[4] A. Lopez de Ávila and S. García, "Destinos turísticos Inteligentes," *Economía Industrial*, vol. 295, no. ISSN 0210-900X, pp. 61–69, 2015.

[5] D. Buhalis and A. Aditya, "Smart Tourism Destinations," Cham, 2014.

[6] U. Gretzel, M. Sigala, Z. Xiang and C. Koo, "Smart tourism: foundations and developments," *Electronic Markets*, vol. 25, no. 3, pp. 179–188, 2015.

[7] C. Chen, C. Murhpy and S. Knecht, "An Importance Performance Analysis of smartphone applications," *Journal of Tourism Management*, vol. 29, pp. 69–79, 2016.

[8] F. Gonzalez-Reverte, P. Diaz-Luque, J. Gomis-López, and S. Morales-Pérez, "Tourists' risk perception and the use of mobile devices in beach tourism destinations," *Sustainability*, vol. 10, p. 413, 2018.

[9] B. Brown and M. Chalmers, "Tourism and mobile technology," in *Proceedings of the European Conference on Computer-Supported Cooperative Work*, Helsinki, 2003.

[10] A. Gomez-Oliva, J. Alvarado-Uribe, M. C. Parra-Meroño, and A. J. Jara, "Transforming Communication Channels to the Co-Creation and Diffusion of Intangible Heritage in Smart Tourism Destination: Creation and Testing in Ceutí (Spain)," *Sustainability*, vol. 11, no. 14, p. 3848, 2019.

[11] T. Nam and T. A. Pardo. "Conceptualizing smart city with dimensions of technology, people, and institutions," in *Proceedings of the 12th Annual International Digital Government Research Conference: Digital Government Innovation in Challenging Times*, 2011.

[12] A. Gomez-Oliva, M. Parra-Meroño and A. Jara, "Diseño de un canal de difusión del patrimonio inmaterial para Destinos Turísticos Inteligentes, basado en Internet de las Cosas," Murcia, 2019.

[13] D. Boswarthick, O. Elloumi, and O. Hersent, "M2M communications: a systems approach," *ETSI World Class Standards*, 2012.

[14] P. Pathak, "Internet of Things: A look at paradigm shifting applications and challenges," *International Journal of Advanced Research in Computer Science*, vol. 7, no. 2, 2016.

[15] N. Dave, "8 ways in which IoT is Shaping the Future of Travel Industry," Digital Doughnut, 29th January 2018, online at: https://www.digitaldoughnut.com/articles/2018/january/ways-in-which-iot-is-shaping-the-future-of-travel.

[16] A. Aluri, HITEC 2016 Special Report HITEC 2016 Special Report, 2016.

[17] A. Verma and V. Shukla, "Analyzing the Influence of IoT in Tourism Industry," in *International Conference on Sustainable Computing in Science, Technology & Management*, 2019.

[18] D. Buhalis, "Marketing the competitive destination of the future," *Tourism management*, vol. 21, no. 1, pp. 97–116, 2000.

[19] *Estrategias de marketing para destinos*, 2016.

[20] E. Binkhorst, "The co-creation tourism experience" Sitges.

[21] Y. Sun, H. Song, A. J. Jara and R. Bie, "Internet of Things and Big Data Analytics for Smart and Connected Communities," *IEEE Access*, vol. 4, pp. 766–773, 2016.

[22] Forum Virium Helsinki, "Home," Forum Virium Helsinki, online at: https://forumvirium.fi/en/urban-eco-islands-develops-archipelago-tourism-with-digital-solutions-2/.

[23] D. Buhalis and A. Amaranggana, "Smart tourism destinations enhancing tourism experience through," in *Information and Communication Technologies in Tourism*, Cham, 2015.

Index